Designing with Geosynthetics

6th Edition

Volume I

Designing with Geosynthetics

6th Edition

Volume I

Robert M. Koerner

Emeritus Director, Geosynthetic Institute

Emeritus Professor, Drexel University

Library of Congress Control Number: 2011909102

ISBN:	Hardcover	978-1-4628-8289-2
	Softcover	978-1-4628-8288-5
	eBook	978-1-4628-8290-8

Print information available on the last page.

Rev. date: 08/11/2016

To order additional copies of this book, contact:
Xlibris
1-888-795-4274
www.Xlibris.com
Orders@Xlibris.com
591434

Dedication

To the member organizations of the Geosynthetic Institute for their confidence, interaction, friendship, and financial support over the past twenty-five years and particularly to my wife, Paula, for her tender love and care over the past fifty-three years.

Contents

* The Index is at the end of Volume 2.

** Volume 2 has the remaining Chapters 5 through 8 in it.

Preface

The first book on what we know today as *geosynthetics* was coauthored by this book's author in 1980. While it was very qualitative in the presentation of the subject matter, it was instantly successful indicating a clear need for information on the subject. Its impact spurred the author into writing a much more quantitative book on the subject, which resulted in the first edition of *Designing With Geosynthetics* published in 1986. It was specifically targeted as a student textbook, yet many professionals used it as well. As time progressed, the robust growth of the technology and associated industry required regular updates in the form of subsequent editions as indicated below.

Publication History of the *Designing with Geosynthetics* Book

Edition	Published	Pages	Sales
First	1986	424	3197
Second	1990	652	2645
Third	1994	783	4686
Fourth	1998	761	5460
Fifth	2005	796	3500

Within the twenty-five years to date, the various editions have been used around the world and translated, in whole or part, into numerous languages. While each edition changed over time, so did universities insofar as teaching advanced level courses (such as geosynthetics) is concerned. Three university-related changes come to mind: fewer credits are required for student graduation, proportionately more credits are required in areas of nontechnical courses, and less allocation in the curriculum is reserved for specialized courses.

These occurrences led to the direction for this sixth edition in that the emphasis now somewhat shifts from an academic textbook to also being a professional users book. In so doing, a number of items have been removed from the previous edition, e.g., many descriptive photographs and the chapter on geopipe. Conversely, there is also information that has been added—e.g., sustainability is included in every chapter, information is given on geosynthetic field failures, current literature references are included, etc. The net result of these changes in this sixth edition is hoped to be a user-friendly professional book focused on the credible use of geosynthetics in its major applications. That said, the book can still be used as a university textbook and the Geosynthetic Institute has a companion solutions manual keyed to each of problems in the eight chapters of the combined two volumes of this 6th Edition. Contact <www.geosynthetic-institute.org> for information in this regard.

As with previous editions, it should be noted that the first chapter entitled "Overview of Geosynthetics" encompasses polymeric principles of all the specific types of geosynthetic materials described in subsequent chapters. Thus, if one desires detailed information on either geotextiles, geogrids, geonets, geomembranes, geosynthetic clay liners, geofoam or geocomposites, the first chapter should be read first and then one can skip to the specific geosynthetic's chapter for the detailed engineering information. As always, the author wishes all an informed and successful venture into, and use of, geosynthetics.

Robert M. Koerner, PhD, PE, NAE, DGE, Dist. M. ASCE
Director—Geosynthetic Institute
Emeritus Professor—Drexel University

Acknowledgments

The first edition of this book, and all subsequent ones including this sixth edition, has been indelibly tied to the formation and growth of the Geosynthetic Research Institute (GRI) and then the superseding Geosynthetic Institute (GSI). Both enterprises have been outstanding successes, and as such, the author expresses sincere appreciation to the current (and past) members, affiliated members and associate members. The current list is available on the GSI website at <www.geosynthetic-institute.org> along with the current board of directors and other contact individuals.

The colleagues of the author who have been invaluable in these undertakings are Drs. George R. Koerner (director designate) and Y. Grace Hsuan (professor of Civil Engineering) who regularly share technical information and ongoing discussions, along with Mrs. Paula Koerner (treasurer), Ms. Marilyn Ashley (executive secretary) and Mrs. Jamie Koerner (special projects). They are to be congratulated for making both of the above mentioned enterprises great successes. My heartfelt gratitude is particularly extended to this specific group of talented individuals.

Chapter 1

Overview of Geosynthetics

1.0 INTRODUCTION

According to ASTM D4439, a geosynthetic is defined as follows:

> **geosynthetic**, *n*—a planar product manufactured from polymeric material used with soil, rock, earth, or other geotechnical engineering related material as an integral part of a human-made project, structure, or system.

Since 1977, the time of the first geosynthetics conference in Paris, geosynthetics have emerged as exciting engineering materials in a wide array of civil engineering applications, e.g., transportation, geotechnical, geoenvironmental, hydraulics, and private development. The rapidity at which the related products have been and continue to be developed is nothing short of amazing. At no time in the author's experience has a new engineering material come on so strong. The reasons for this explosion of geosynthetic materials onto the civil engineering market are numerous and include the following:

- They are quality-control manufactured in a factory environment.
- They can be installed rapidly.
- They generally replace raw material resources.
- They generally replace difficult designs using soil or other construction materials.
- Their use is required by regulations in many cases.
- They have made heretofore impossible designs and applications possible.
- They are being actively marketed and are widely available.
- Their technical database (both design and testing) is nicely established.
- They are being integrated into the profession via generic specifications.
- They are invariably cost competitive against soils or other construction materials.
- Their carbon footprint is very much lower than traditional solutions.

The professional groups most strongly influenced are transportation, geotechnical, environmental, hydraulics and private development engineering communities, although all soil-, rock-, and groundwater-related activities fall within the general scope of the various applications. This being the case, the term *geosynthetics* seems appropriate. *Geo*, of course, refers to the earth. The realization that the materials are almost exclusively from human-made products gives the second part to the name—*synthetics*. The materials used in the manufacture of geosynthetics are almost entirely from the plastics industry—that is, they are polymers made from hydrocarbons, although fiberglass, rubber, and natural materials are occassionally used. Interestingly, the case could easily be made that the entire technology could better be called *geopolymers*, but the term *geosynthetics* took hold before this realization. The sections that follow in this opening chapter present an overview of the various type of geosynthetic materials and serve as an introduction to the remainder of the text—namely chapters 2 through 9—where each particular type of geosynthetic will be focused on in depth.

1.1 BASIC DESCRIPTION OF GEOSYNTHETICS

Lost to history are the initial attempts to reinforce soils; the adding of materials that would enhance the behavior of the soil itself was no doubt done long before our first historical records of this process. It seems reasonable to assume that first attempts were made to stabilize swamps and marshy soils using tree trunks, small bushes, and the like. These soft soils would accept the fibrous material until a somewhat stable mass was formed that had adequate properties for the intended purpose. It also seems reasonable to accept that either the continued use of such a facility was possible because of the properly stabilized nature of the now-reinforced soil (probably by trial and error), or was impossible due to a number of factors, among which were

- insufficient reinforcement material for the loads to be carried;
- the pumping of the soft soil up through the reinforcement material; and
- the degradation of the fibrous material with time, leading back to the original unsuitable conditions.

Such stabilization attempts were undoubtedly continued with the development of a more systematic approach in which timbers of nearly uniform size and length were lashed together to make a mattressed surface. Such split-log corduroy roads over peat bogs date back to 3000 BC [1]. This art progressed to the point where the ridged surface was filled in smooth. Some of these systems were surfaced with a stabilized soil mixture or even paved with stone blocks. Here again, however, deterioration of the timber and its lashings over time was an obvious problem.

The concept of reinforcing soft soils has continued until the present day. The first use of fabrics in reinforcing roads was attempted by the South Carolina Highway Department in 1926 [2]. A heavy cotton fabric was placed on a primed soil subgrade, hot asphalt was applied to the fabric, and a thin layer of sand was placed on the asphalt. In 1935, results were published of this work, describing eight separate field experiments. Until the fabric deteriorated, the results showed that the roads were in good condition and that the fabric reduced cracking, raveling, and localized road failures. This project was certainly the forerunner of the separation and reinforcement functions of geosynthetic materials as we know them today. The separation and reinforcement of unsuitable soils is a major topic area of this book.

A second major topic area is that of providing an intermediate barrier between two dissimilar materials for the purpose of liquid (usually water) drainage and soil filtration. When requiring liquid flow across such a barrier, it must obviously be porous, yet the voids must not be open so much as to lose the retained soil—thus the necessity of using some sort of intermediate filter. Again, the historical development of filtration provides an important background for understanding the work that followed. Naturally occurring sands and gravels, which were found to be well graded, had been used as filter material since ancient times. The idea of systematizing filtration criteria was originated by Karl Terzaghi and Arthur Casagrande in the 1930s and brought to use by Bertram [3] shortly thereafter. This idea of soil filters, even multiple-graded soil filters, is a target area for the geosynthetic materials described in this book—now for reasons of construction quality control and cost effectiveness.

A last major topic area is that of providing a waterproof barrier for preventing liquid or gas movement from a given containment area. Such liners have historically been made using low-permeability clay soils. The Roman aqueducts were lined in such a manner, and the technology undoubtedly preceded them by many years [4]. Liners made from bitumen, and various cements have been used since the 1900s, but it was the synthetic rubber material "butyl" in the 1930s that ushered in polymeric liners [5]. Today, such liners (there are many different types) are regulatory-mandated for use in certain environment-related applications. Interestingly, the newest barrier materials are combinations of both geosynthetics and bentonite soil used as a composite material.

Thus, geosynthetic materials perform five major functions: (1) separation, (2) reinforcement, (3) filtration, (4) drainage, and (5) containment (of liquid and/or gas). The use of geosynthetics has basically two aims: to perform better (e.g., with no deterioration of material or excessive leakage) and to be more economical than using traditional materials and solutions (either through lower initial cost or through greater durability and longer life, thus reducing maintenance and replacement costs).

1.1.1 Types of Geosynthetics

There are seven specific types of geosynthetics: (1) geotextiles, (2) geogrids, (3) geonets, (4) geomembranes, (5) geosynthetic clay liners, (6) geofoam, and (7) geocomposites. They are shown in figure 1.1 and are discussed next.

Geotextiles. Geotextiles (see section 1.3 and chapter 2) form one of the two largest groups of geosynthetics described in this book. Their rise in growth during the past thirty-five years has been nothing short of extraordinary. They are indeed textiles in the traditional sense, but consist of synthetic fibers rather than natural ones, such as cotton, wool, or silk. Thus, biodegradation and subsequent short lifetime is not a problem. These synthetic fibers are made into flexible, porous fabrics by standard weaving machinery or are matted together in a random nonwoven manner. Some are also knitted. The major point is that geotextiles are porous to liquid flow across their manufactured plane and also within their thickness, but to a widely varying degree. There are at least one hundred specific application areas for geotextiles that have been developed; however, the fabric always performs at least one of four discrete functions: separation, reinforcement, filtration, and/or drainage.

Geogrids. Geogrids (see section 1.4 and chapter 3) represent a rapidly growing segment within the geosynthetics. Rather than being a woven, nonwoven, or knitted textile fabric, geogrids are plastics formed into a very open gridlike configuration—i.e., they have large apertures between individual ribs in the machine and cross machine directions. Geogrids are formed in three ways: (1) stretched in one or two directions for improved physical properties, (2) made on weaving or knitting machinery by standard and well-established methods and then coated, or (3) made by bonding rods or straps together. There are many application areas; however, they function almost exclusively as reinforcement materials.

Geonets. Geonets, called *geospacers* by some (see section 1.5 and chapter 4), constitute another specialized segment within the geosynthetics area. They are formed by a continuous extrusion of parallel sets of polymeric ribs at acute angles to one another. When the ribs are opened, relatively large apertures are formed into a netlike configuration. Their design function is completely within the drainage area where they are used to convey liquids of all types.

Geomembranes. Geomembranes (see section 1.6 and chapter 5) represent the other largest group of geosynthetics described in this book. Their growth in the United States and Germany was stimulated by governmental regulations originally enacted in the early 1980s. The materials themselves are relatively thin, impervious sheets of polymeric material used primarily for linings and covers of liquid—or solid-storage facilities. This includes all types of landfills, reservoirs, canals, and other containment facilities. Thus, the primary function is always containment as a liquid or vapor barrier or both. The range of applications, however, is very great; and in addition to the environmental area, applications are rapidly growing in geotechnical, transportation, hydraulic, and private development engineering.

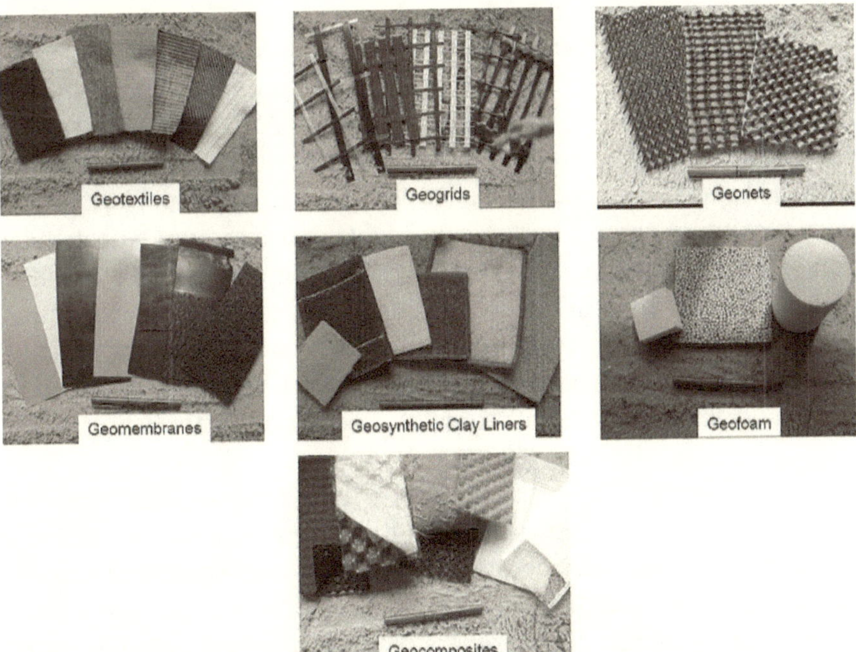

Figure 1.1 Collage of the basic types of geosynthetic products.

Geosynthetic Clay Liners. Geosynthetic clay liners (GCLs) (see section 1.7 and chapter 6) are an interesting juxtaposition of polymer and natural soil materials. They are rolls of factory-fabricated thin layers of bentonite clay sandwiched between two geotextiles or bonded to a geomembrane. Structural integrity of the composite is obtained by needle-punching, stitching, or adhesive bonding. GCLs are used as a composite component beneath a geomembrane or by themselves in environmental and containment applications as well as in transportation, geotechnical, hydraulic, and various private development applications.

Geofoam. Geofoam (see section 1.8 and chapter 7) is a product created by polymeric expansion processes resulting in a "foam" that consists of many closed but gas-filled cells. The skeletal nature of the cell walls is the unexpanded polymeric material. The resulting product is generally in the form of large, but extremely light, blocks that are stacked side by side, providing lightweight fill in numerous applications. Although the primary function is dictated by the application, separation is always a consideration, and geofoam will be included in this category rather than creating a separate one.

Geocomposites. Geocomposites (see section 1.9 and chapter 8) consist of a combination of geotextiles, geogrids, geonets, and/or geomembranes in a factory-fabricated unit. Also, any one of these four materials can be combined with another synthetic material (e.g., deformed plastic sheets or steel cables) or with soil. For example, a geonet with geotextiles on both surfaces and a GCL consisting of a geotextile/bentonite/geotextile sandwich are both geocomposites. This exciting area brings out the best creative efforts of the engineer, manufacturer, and contractor. The application areas are numerous and growing steadily. They encompass the entire range of functions listed for geosynthetics: separation, reinforcement, filtration, drainage, and containment.

Geo-Others. The general area of geosynthetics has exhibited such innovation that many systems defy categorization. For want of a better phrase, *geo-others* describes items such as threaded soil masses, polymeric anchors, and encapsulated soil cells. As with geocomposites, the primary function of geo-others is product-dependent and can be any of the five major functions of geosynthetics. These materials will be discussed in the chapters that they are related the closest to, or in chapter 8 on the basis of their primary function.

1.1.2 Functions Provided by Geosynthetics

The juxtaposition of the various types of geosynthetics just described with the primary function that the material is called on to serve allows for the creation of a matrix that will be used throughout the book. In essence, this matrix is the "scorecard" for understanding the entire geosynthetic field and its design-related methodology. Table 1.1 illustrates the primary function that each of the geosynthetics can be called on to serve. Note that these are primary functions and in many (if not most) cases there are secondary functions, and perhaps tertiary ones as well. For example, a geotextile placed on soft soil will usually be designed on the basis of its reinforcement capability, but separation and filtration might certainly be secondary and tertiary considerations. A geomembrane is obviously used for its containment capability, but separation will always be a secondary function.

The greatest variability from a manufacturing and materials viewpoint is the category of geocomposites. The primary function will depend on what is actually created, manufactured, and installed.

Note that table 1.1 will be constantly referred to throughout this book. It will clearly identify each geosynthetic material vis-à-vis the primary function (usually by application) that is being served.

1.1.3 Market Activity

To say that the geosynthetic market activity, as indicated by sales volume, is strong is decidedly an understatement. All existing application areas are seeing constant growth, albeit at different rates.

TABLE 1.1 IDENTIFICATION OF USUAL PRIMARY FUNCTION VERSUS TYPE OF GEOSYNTHETIC

Type of Geosynthetic (GS)	Primary Function					Book Chapter
	Separation	Reinforcement	Filtration	Drainage	Containment	
Geotextile (GT)	√	√	√	√		2
Geogrid (GG)		√				3
Geonet (GN)				√		4
Geomembrane (GM)					√	5
Geosynthetic Clay Liner (GCL)					√	6
Geofoam (GF)	√					7
Geocomposite (GC)	√	√	√	√	√	8

Note: This table will be referred to in every chapter of this book. Note that Chapters 5 through 8 are in Volume 2 of this book.

The general motivators for such growth are increased benefit/cost ratio over conventional materials and solutions (for geotextiles and geogrids), and requirements by federal or state regulations (for geomembrane and geonets). Current geosynthetic sales are difficult to assess, but the author's estimate for the year 2010 on a worldwide basis is as follows (note that the values are in millions of square meters and millions of US dollars):

Geotextiles	1400 M m^2 @ \$0.75/m^2	=	\$1,050 M
Geogrids	250 M m^2 @ \$2.50/m^2	=	625
Geonets	75 M m^2 @ \$2.00/m^2	=	150
Geomembranes	300 M m^2 @ \$6.00/m^2	=	1,800
Geosynthetic Clay Liners	100 M m^2 @ \$6.50/m^2	=	650
Geofoam	5 M m^3 @ \$75.00/m^3	=	375
Geocomposites	100 M m^2 @ \$4.00/m^2	=	400
	Total =		\$5,050 M

While the total expenditure is impressive and indicates that geosynthetics are well-entrenched construction materials, the situation could, perhaps even should, be much larger than indicated. If one factors in the concept of sustainability, it will be seen in the chapters to follow that the carbon footprint of each type of geosynthetic in comparison to use of traditional materials is distinctively lower. The future will see this as an additional factor in furthering the use of geosynthetics.

It also should be mentioned that relatively few colleges and universities teach geosynthetics as a specialized course. It appears as though the only way that the subject is being introduced to students (often only graduate students) is as part of existing courses, which is a logical way to contrast geosynthetics with traditional materials. Geosynthetics can also, of course, be studied via professional courses taken after graduation. These are offered on intermittent schedules by numerous associations, institutes, and continuing education organizations. It appears as though the Internet will eventually be the educational outreach vehicle of the future for geosynthetics, particularly in the form of *webinars*. Whatever vehicle form it takes, education in geosynthetics is still a major objective and the one in which this book is focused on.

1.2 POLYMERIC MATERIALS (AKA, PLASTICS)

The vast majority (well over 95%) of the geosynthetics discussed in this book are made from synthetic polymers, broadly characterized as plastics. Thus, a brief discussion on the topics of polymer composition, structure, and identification is in order. This section is not meant to make a polymer engineer or a polymer scientist out of the reader, but only to afford an appreciation of (1) the wealth of information that is available, (2) the sophistication of the topic area, and (3) the need for at least a rudimentary understanding of geosynthetics at the molecular level, which will prove beneficial for the topics discussed throughout the book.

To begin with, recognize that the plastics industry is enormous. Worldwide sales are over $1 trillion per year and the distribution reflects both the strength and diversity of consumption. Fortunately, of the thousands of commercialized polymers in existence the area of geosynthetics utilizes very few. The following are the most commonly used in the manufacturing of geosynthetics:

- High-density polyethylene (HDPE)—developed in 1941
- Linear low-density polyethylene (LLDPE)—developed in 1956
- Polypropylene (PP)—developed in 1957
- Polyvinyl chloride (PVC)—developed in 1927
- Polyester (PET)—developed in 1950
- Expanded polystyrene (EPS)—developed in 1950
- Chlorosulphonated polyethylene (CSPE)—developed around 1965
- Thermoset polymers such as ethylene propylene diene terpolymer (EPDM)—developed in 1960

1.2.1 Brief Overview

The basic "feedstock" for almost all the polymers used to make geosynthetics is ethylene gas. As can be seen in figure 1.2, almost all the polymers mentioned previously are included in the various branches. Ethylene is reacted by a catalyst to form discrete particles called "flake" in a huge refinery. To say that the chemistry involved in the reaction is complex is a vast understatement, as evidenced by Ziegler and Natta who shared the Noble Prize for their respective discoveries of the catalysts for polyethylene and for polypropylene in 1963. Subsequently, Flory received the Noble Prize in 1974 for understanding the physical chemistry of polymers. Polymer manufacturing and properties are a significant component of chemistry and materials engineering departments at every college.

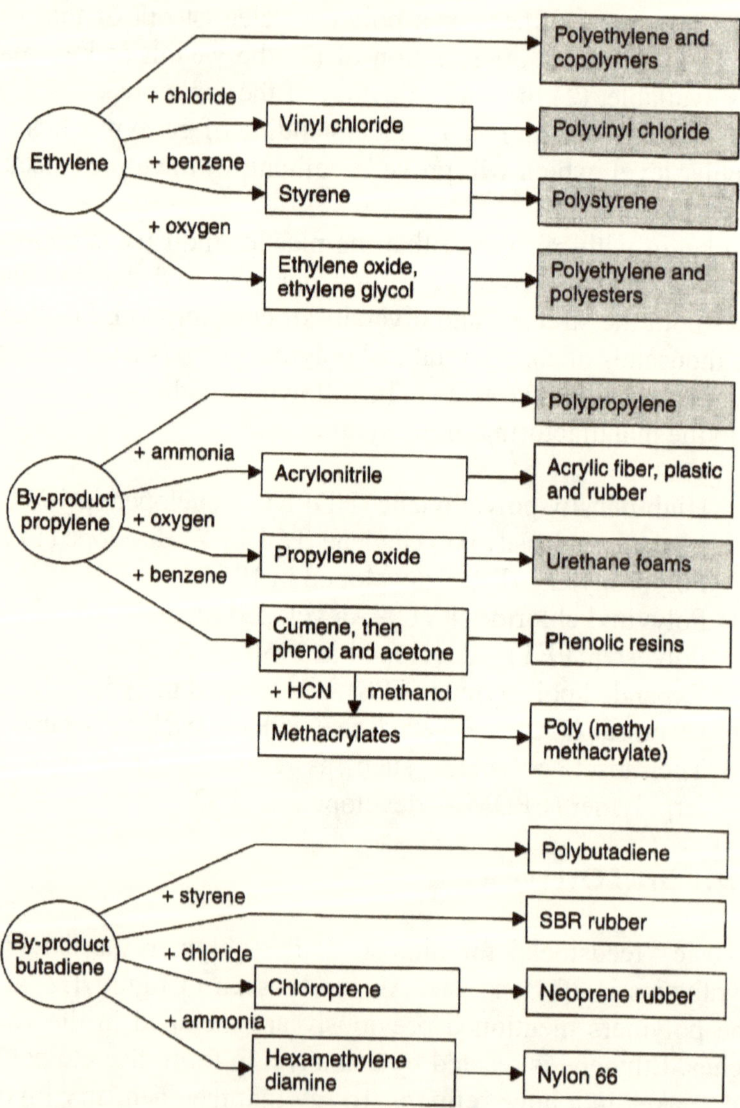

Figure 1.2 Major polymeric products derived from ethylene and its by-products, after Rodriguez [6]. The major polymers used in geosynthetics manufacturing are shaded.

The word *polymer* comes from the Greek *poly*, meaning "many" and *meros* meaning "parts." Thus, a polymeric material consists of many parts joined together to make the whole. Each part, or unit, is called a *monomer*, the molecular compound used to produce the polymer. It should be recognized that the monomers and the repeating molecular units are different. This is due to the polymerization process. The functionality (i.e., the number of sites at which a monomeric molecule can link with other monomer molecules) determines the type and length of the chain.

The molecular weight of a polymer is the degree of polymerization (i.e., the number of times a repeating unit occurs) multiplied by the molecular weight of the repeating unit. The average molecular weight and its statistical distribution are very important in the resulting behavior of the polymer, since increasing *average molecular weight* has several results: increased textile strength, increased elongation, increased impact strength, increased stress crack resistance, increased heat resistance, decreased flow behavior, and decreased processability. Narrowing the *molecular weight distribution* also has several results: increased impact strength, decreased stress crack resistance, decreased flow behavior, and decreased processability.

While most of the polymers used in the manufacture of geosynthetics are from one type of monomer, thus called *homopolymers*, there are other possibilities. A polymer made from two repeating units in its chain is called a *copolymer*. It is important to note here the manner of linking or joining the repeating units. This can be random, alternating, block, or branch (graft). Such copolymerization greatly expands the structural properties of the resulting polymer. Furthermore, it is possible to have three repeating units in the chain in what is called a *terpolymer*. It is easy to see that the options are essentially limitless, which explains why there are approximately fifty thousand commercialized polymers in existence.

As previously mentioned, there are only a few polymers that make up the majority of geosynthetic materials. Table 1.2 presents the repeating molecular units of polymers used in manufacture of geosynthetics. Among the groups shown, polyethylene and polypropylene are the most common and are collectively called *polyolefins*.

Bonding between polymer molecules and their chains is critically important in understanding their behavior and performance. A number of excellent references are available for in-depth study [6-9]; however; Moore and Kline [10] is particularly well suited for an introduction to the subject. The bonds between polymer molecules are van der Waals forces, permanent dipoles, or hydrogen bonds. Between molecular chains, however, the bonds are usually much weaker and often must be supplemented by some form of cross-linking by means of covalent bonds or covalent bonding systems. Cross-links can be formed by the use of

- monomers having a functionality greater than two,
- chemical agents (sometimes called *curing*), and
- nuclear radiation methods.

Cross-linking is an important conceptual consideration because it separates the two major types of polymeric materials—i.e., thermoplastic and thermoset. A *thermoplastic polymer* can be repeatedly heated to its softening point, shaped, or worked as desired, and then cooled to preserve that remolded shape; the polymer structure remains essentially unaltered. In a *thermoset polymer*, the process cannot be repeated. Any additional heat after first forming will lead only to charring and degradation of the material. The key to this behavior in thermoset materials is, of course, cross-linking, which does not exist to such a degree in thermoplastic materials. Examples of thermoplastic materials are polyethylene (PE), polypropylene (PP), polyvinyl chloride (PVC), and polyester (PET); examples of thermoset materials are nytrile, butyl, and ethylene polypylene diene terpolymer (EPDM). As mentioned previously, however, geosynthetics consist almost entirely of thermoplastic materials. With the exception of EPDM geomembranes, there are essentially no thermoset materials currently used in geosynthetic applications.

Crystallinity can indeed exist in polymeric materials but does so to widely varying degrees. In a rather difficult-to-visualize manner, the aligned portions of the polymer chain(s) in small regions are called "crystallites." The nonaligned regions are called *amorphous*. The crystallinity patterns are very complex and are still being researched. For example, the aligned molecular chain can loop back on itself in a series of folds called *spherulites* and can form exotic configurations such as "snowflakes" and "shish kebabs." The amount of crystallinity gives rise to a further polymer classification of semicrystalline or noncrystalline, hence amorphous. (No polymer is completely crystalline.) Thus, the three major classifications of polymers that can be used for geosynthetic materials are (1) semicrystalline thermoplastic, (2) amorphous thermoplastic, and (3) thermoset; the latter is rarely used for geosynthetics, and essentially all the polymers used in the manufacture of geosynthetics are of the first two varieties.

TABLE 1.2 REPEATING UNITS OF POLYMERS USED IN THE MANUFACTURE OF GEOSYNTHETICS

Polymer	Repeating Unit	Types of Geosynthetics
Polyethylene (PE)	$\left[-CH_2-CH_2-\right]_n$	Geotextiles, geomembranes, geogrids, geopipe, geonets, geocomposites
Polypropylene (PP)	$\left[-CH_2-CH(CH_3)-\right]_n$	Geotextiles, geomembranes, geogrids, geocomposites
Polyvinyl chloride (PVC)	$\left[-CH_2-CHCl-\right]_n$	Geomembranes, geocomposites, geopipe
Polyester (polyethylene terephthalate) (PET)	$\left[-O-R-O-C(=O)-R'-C(=O)-\right]_n$	Geotextiles, geogrids
Polyamide (PA) (nylon 6/6)	$\left[-NH-(CH_2)_6-NH-C(=O)-(CH_2)_4-C(=O)-\right]_n$	Geotextiles, geocomposites, geogrids
Polystyrene (PS)	$\left[-CH_2-CH(C_6H_5)-\right]_n$	Geocomposites, geofoam

The amount of crystallinity varies from nil to 30% in some polyvinyl chlorides (PVCs) to as high as 65% in high-density polyethylene (HDPE). Crystallinity is significant and, in some instances, critical in the behavior of polymeric geosynthetics. It can be shown that increasing crystallinity results in the following: increased stiffness or hardness, increased heat resistance, increased tensile strength, increased modulus, increased chemical resistance, decreased diffusive permeability (or vapor transmission), decreased elongation or strain at failure, decreased flexibility, decreased impact strength, and decreased stress crack resistance.

Finally, in this brief section on polymer chemistry, there are two fundamental temperatures that are important to keep in mind: (1) the glass transition temperature, T_g, and (2) the melting temperature, T_m.

These values are given in table 1.3 for the common polymers that are made into geosynthetics. Although the melting temperature is intuitive, the glass transition temperature is not. In a physical sense, T_g is the temperature below that the polymer is glassy, hence rigid and essentially brittle. Above the T_g, the polymer is rubbery, hence flexible and essentially fluidlike. The implications with respect to mechanical properties such as creep and stress relaxation are important in this regard.

1.2.2 Polymer Identification

There are a number of possible ways to identify the specific polymer from which a material (in our case, a geosynthetic) has been made. Table 1.4 gives an indication of polymer type on the basis of its burning characteristics.

It must be recognized, however, that the burning tests described in table 1.4 are very subjective. For a significantly more accurate identification of the particular type of synthetic polymer, there are a number of *chemical analysis tests*. Such tests are finding a place in geosynthetic materials analysis for the following reasons:

- They are used in quality assurance and product certification.
- They are used to evaluate the estimated lifetime of field-retrieval samples.
- They are used in laboratory investigations into material degradation and subsequent lifetime prediction.
- They are valuable in the forensic analysis of field failures.
- They are used for research and development into new additive packages (stabilizers, antioxidants, plasticizers, and additives).
- They are used for new geosynthetic product development and application investigations.

TABLE 1.3 TRANSITION TEMPERATURES FOR SELECTED POLYMERS

Monomer Unit	T_g, °C	T_m, °C
Entylene (linear)	-125	141
Propylene (isotactic)	-7	187
Styrene (isotactic)	100	240
Vinyl chloride (syndiotactic)	81 to 98	273
Vinyl alcohol	85	265
Ethylene terephthalate	60 to 85	280
Nylon 66	50	280

TABLE 1.4 BURNING CHARACTERISTICS OF POLYMERIC MATERIALS USED IN GEOSYNTHETICS (compiled by Dr. Y. Grace Hsuan)

Polymer Type*	Behavior Beginning, During, and After Burning
POLYETHYLENE (HDPE and LLDPE)	Before touching the flame, the material melts, shrinks, and curls. Burns readily and is not self-extinguishing. Burns rapidly when moved away from the flame. Becomes clear when molten and tends to drip. When the flame is extinguished the smell is of molten wax. Ash is soft and same color as material.
POLYPROPYLENE (PP)	Before touching the flame, the material shrinks, melts, and curls. Burns in a manner similar to polyethylene and is not self-extinguishing. Burns slowly when moved away from the flame. Burns with no clear blue color at the base of the flame, except with carbon black. Faint odor of burning-asphalt is given off. Ash is hard and light tan.
POLYVINYL CHLORIDE (PVC) unplasticized	Burns with difficulty and is self-extinguishing. Flame is yellow, green at the bottom edges, with spurts of green and yellow. White smoke is given off. The material softens on ignition and has an unpleasant acidic smell.
POLYVINYL CHLORIDE (PVC) plasticized	Flammability behavior depends on amount of plasticizer present; most plasticizers burn readily with a yellow, smoky flame. Black smoke is given off. Odors are mostly floral (ester-like) but with an unpleasant acidic smell.
POLYESTER (PET)	Burns slowly with a yellow smoky flame. Floral (ester) odor. Flame jitters and dances. Molten material drips.

TABLE 1.4 - CONTINUED

Polymer Type*	Behavior Beginning, During, and After Burning
NYLON (PA)	Difficult to ignite and is self-extinguishing. Slight green flame occurs due to chlorine. Flame is blue with a yellow top. The material froths on ignition. Odor is of burning wool or hair.
POLYSTYRENE (PS)	Burns readily and is not self-extinguishing. Flame is orange-yellow, and black dense smoke containing soot is given off. Flame jitters and dances somewhat. On ignition, the material softens and the odor is characteristic of benzene.
CHLOROSUPHONATED POLYETHYLENE (CSPE) (Hypalon®)	Difficult to ignite. White smoke while in flame. Black smoke while burning. Slight green flame. Selt extinguishing No drops. Wax odor.
MOST THERMOSETS, e.g., EPDM	Rapid burning, very intense. No drops. Chars at edges. Orange flame with black smoke. Char is tacky.

*To conduct final check on a chlorine-containing polymer: Heat copper wire in a flame and press the heated wire into the sample; then slowly put the wire back into the flame. A green flame indicates a chlorine-containing polymer.

A brief description of the most frequently used chemical analysis tests as applied to the polymers used in the manufacture of geosynthetics follows (see Halse et al. [11] for additional insight into these methods).

Thermogravimetric Analysis (TGA). Thermogravimetric analysis (TGA) is one of a series of thermal methods in which a property of the polymer is tracked as a function of a controlled temperature program (see Thomas and Verschoor [12] for a review). TGA follows mass change as a function of temperature. Continuous weighing of the decreasing mass of a specimen that is being subjected to a constantly increasing temperature produces curves such as shown in figure 1.3. The pronounced decreases in weight at specific temperatures signify vaporization of specific components within the formulation. The plasticizer and hydrogen chloride in the PVC is removed at about 300°C, while the resin is removed between 450°C and 500°C. What remains beyond 500°C is carbon black and ash, since the tests were performed in a nitrogen atmosphere. The weight percentage of each component is computer obtained, since the device automatically normalizes the vertical axis.

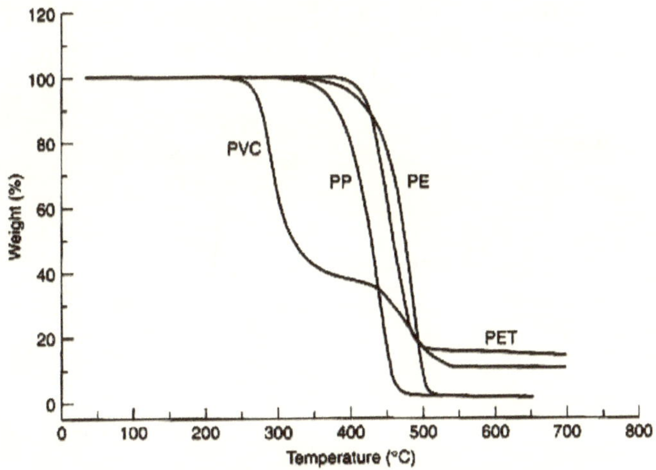

Figure 1.3 Thermogravimetric analysis curves of some common geosynthetic polymers. (After Thomas and Veschoor [12])

The technique can also be used to determine kinetic information concerning the stability of the polymer and the energy of activation for thermal decomposition. This latter piece of information can be used in an Arrhenius plot to predict in-service lifetime at a specific temperature.

Differential Scanning Calorimetry (DSC). Using a differential scanning calorimeter (DSC), a temperature balance between a reference cell and a test specimen cell can be maintained and the heat flow into and out of a specimen can be monitored and plotted as a function of temperature. Figure 1.4 shows such the response for polyester [12]. The glass transition temperature is at 80.36°C, the exothermic crystallization of the polymer backbone is at 164.20°C, and the endothermic melting of the crystallites is at 251.43°C. The area under the curve of the crystallization melt—for example, the value of 31.56 J/g—is proportional to the percent crystallinity of the polymer. Reference standards are used as a calibration for obtaining the actual percent crystallinity value. Lastly, the beginning of melting of the crystalline portion of the polymer at 236.43°C is important as it relates to proper field seaming. This can be seen in the DSC curves of figure 1.5 for different types of polyethylene [13]. Here, the crystalline melting zone is clearly defined, and the higher the density of the PE, the higher the melting temperature, and the narrower the temperature window over which melting takes place. In this regard, it is inferred that HDPE can be a challenging material to properly seam.

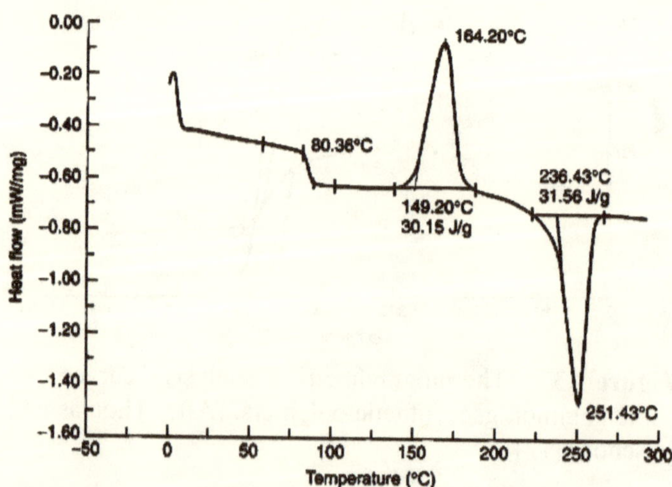

Figure 1.4 Differential scanning calorimeter curves for quenched polyester. (After Thomas and Vescoor [12])

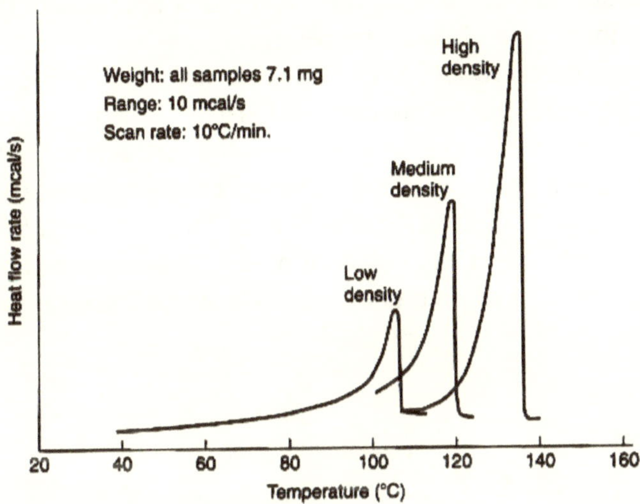

Figure 1.5 Differential scanning calorimeter curves for various densities of polyethylenes. (Compliments of Perkin-Elmer Instrument Co.)

The curves of figure 1.5 are very instructive with respect to field welding of different polyethylenes. For example, LLDPE geomembranes are essentially impossible to weld to HDPE pipe since the melting "windows" for the two materials do not overlap. The net result of such an attempt would be to either completely melt the LLDPE, or not sufficiently melt the HDPE. It is even difficult to weld LLDPE geomembranes to HDPE geomembranes (which are made from MDPE resin) since the welding windows overlap so little, e.g., from 90°C to 130°C.

Oxidative Induction Time (OIT). The oxidative induction time (OIT) test uses a differential scanning calorimeter with a special testing cell capable of sustaining pressure. In the standard OIT (Std-OIT) test, per ASTM D3895, a 5 mg specimen is heated from room temperature to 200°C at a rate of 20°C/mm under a nitrogen atmosphere. Oxygen is then introduced, and the test is terminated when an exothermal peak is reached (see figure 1.6a), and the OIT time is readily obtained. This time is related to the quantity and type of antioxidants used in the polymer formulation. As seen in figure 1.6b, the OIT time is also related to laboratory incubation time of HDPE geomembrane samples in forced-air ovens at elevated temperatures,

[14]. Data such as this can be used to predict antioxidant depletion lifetime at in situ (and lower) temperatures.

The high-pressure OIT (HP-OIT) test uses higher pressure and lower temperature than the standard test just described. It is designated ASTM D5885. Unless otherwise stated by the parties involved, the test is conducted at a pressure of 3.4 MPa and a temperature of 150°C. A response similar to figure 1.6a is obtained. This test is felt to be more representative of the in situ behavior of low-temperature-functioning antioxidants insofar as prediction methods are concerned.

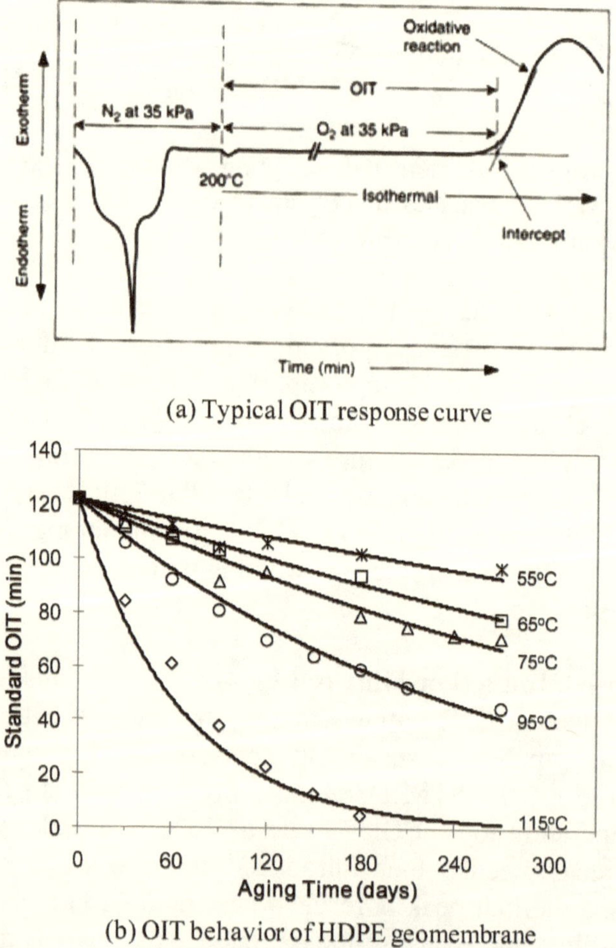

(a) Typical OIT response curve

(b) OIT behavior of HDPE geomembrane

Figure 1.6 Oxidative induction time response and test behavior. (After Hsuan and Guan [14]).

Still further, Li and Hsuan [15] have followed the oxidation of polyolefin geosynthetics by OIT procedures after incubation at extremely high pressures. The pressures, up to 6.3 MPa, provide for reduced laboratory incubation times.

Thermomechanical Analysis (TMA). Thermomechanical analysis (TMA) measures a particular dimension of the polymer specimen under a controlled increase in temperature. A quartz probe rests on the test specimen, and its displacement is precisely measured as the temperature increases (or decreases). The modes of operation are expansion, penetration, or shear flow of the polymer. The most straightforward property to obtain using TMA is the coefficient of thermal expansion. It is simply the slope of the temperature-deformation curve. Figure 1.7 shows the results of such a test for PET, where the linear coefficient of thermal expansion is 76.4 μm/m°C for the initial (called the "glassy" state) stage and 132 μm/m°C for the final (or "rubbery" state) stage. The transition value between the two stages clearly defines the glass transition temperature, which is 80.55°C.

Dynamic Mechanical Analysis (DMA). Dynamic mechanical analysis (DMA) is a thermal technique measures the mechanical response of a polymer as it is deformed under a periodic stress in a controlled temperature environment. Thus, the viscoelastic properties can be evaluated. The test measures the dynamic storage modulus, E' (a measure of stiffness), the dynamic loss modulus, E" (to measure glass transition and softening points) and the ratio of storage-to-loss moduli, which is called the damping, or tan δ value. Figure 1.8 gives the response of a HDPE geomembrane sample under the following conditions: fixed frequency of 1.0 Hz, heated at 4°C per minute, under 5.7 J of energy. The amplitude of the frequency is 0.20 mm (peak to peak).

DMA devices are of additional interest in evaluating the viscoelastic engineering properties via either creep or stress-relaxation test modes. Since DMA units are computer controlled, one can either preset the load and measure changing deformation (i.e., the creep mode) or preset the deformation and measure the sustaining force (i.e., the stress-relaxation mode). By repeating the particular test at a series of increasing (or decreasing) temperature increments, a family of curves results. Additionally, a master curve can be generated using the time-temperature superposition principle for use in long-term studies and lifetime prediction.

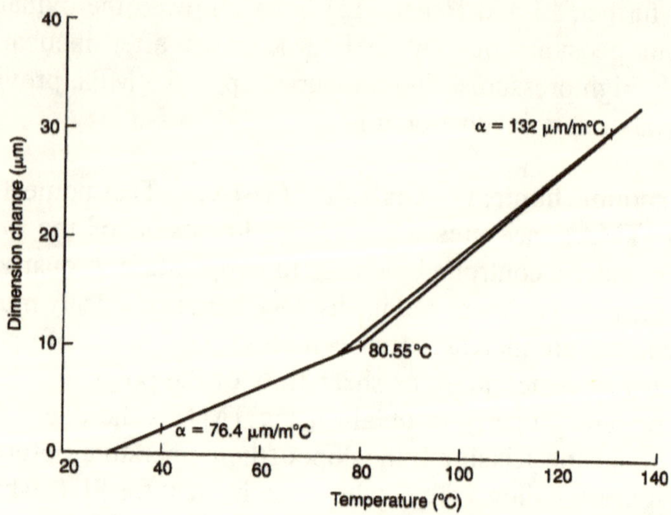

Figure 1.7 Thermomechanical analysis curves for PET under a temperature increase of 10°C/min. (After Thomas and Verschoor [12])

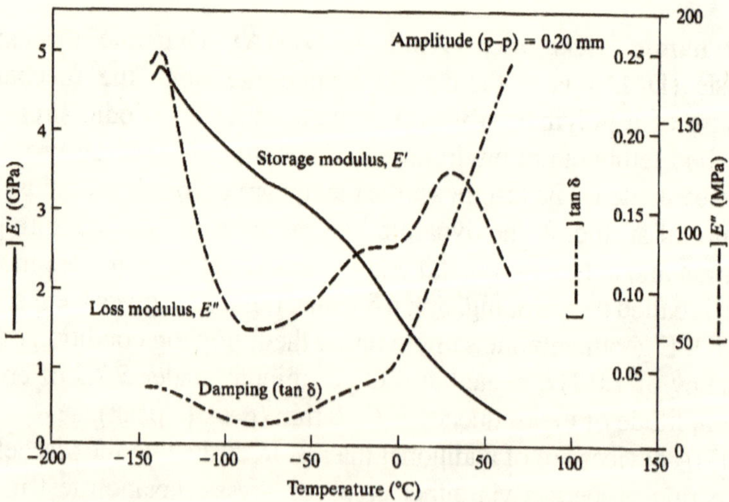

Figure 1.8 Dynamic mechanical analysis curves at fixed frequency for HDPE geomembrane test specimen.

Infrared Spectroscopy (IR). The concept of infrared spectroscopy (IR), generally used as Fourier transform infrared spectroscopy (FTIR), is based on the realization that the functional groups in molecules (such as the -CH- group in polyethylene) are always in motion. During an FTIR analysis, the polymer test specimen is subjected to radiation. The frequency of the incident radiation is in the infrared region. If the frequency matches a natural motion of the functional group, the polymer will absorb this energy and an absorption band will appear on the FTIR frequency sweep.

Figure 1.9 shows the spectrum of a polyethylene specimen without compounding agents. Each peak in the spectrum represents the motion of a functional group either in bending or in stretching. For example, the strong peak at a frequency of 2850 cm^{-1} is the absorption peak due to C-H stretching motion. The area under the curve in this region is represented as I_{2850}.

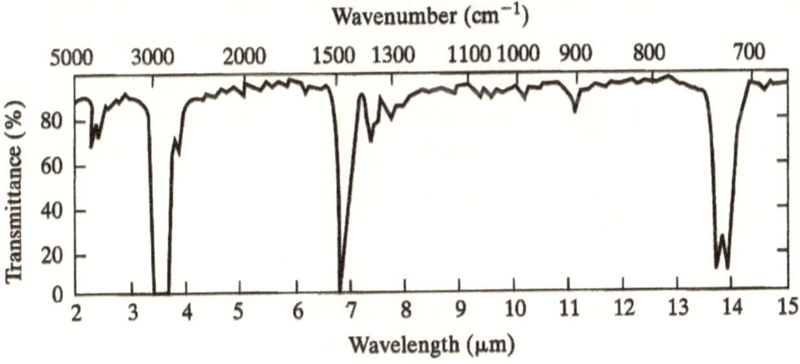

Figure 1.9 Infrared spectrum of a HDPE geomembrane test specimen. (After Halse, et al. [11])

For a majority of polymers, carbonyl groups (-C=0-) would be produced after a certain amount of oxidation reaction (degradation) has occurred. The frequency corresponding to the motion of this molecular group is approximately 1715 cm^{-1}. (This frequency range does not appear in figure 1.9 since it was a virgin nondegraded material). The area of the peak (I_{1715}) is proportional to the amount of carbonyl groups formed in the polymer. Hence, it can be used to monitor the progress of oxidation. Often the results are normalized with another peak area taken from the polymer spectrum. For example, using polyethylene, the I_{1715} value would be normalized with the peak area at 2850 cm^{-1}—i.e., a ratio of I_{1715}/I_{2850} would be obtained. It is then defined as the *carbonyl index*.

Chromatography (GC, LC and HPLC). Chromatography is an analysis method that allows for the separation, isolation, and identification of complex mixtures. After the polymer specimen is liquefied in a solvent carrier, the components of the mixture are carried through a stationary column, and the migration rates indicate fundamental differences. The soluble mobile phase either dissolves, absorbs, or reacts with the stationary phase within the column.

In gas chromatography (GC) the mobile phase is a gas, which is passed through the column, and a detector produces a plot of concentration versus time. The position of the peaks serves to identify the components, and the area under the peaks represents the concentration. Plasticizers in PVC geomembranes have been identified by this method, which is standardized as ASTM D7083. That said, the more common method is D2257, which is a "soxlet" extraction method for such identification.

In liquid chromatography (LC) the mobile phase is liquid, and the stationary phase is either liquid or solid. The separation process is very time-consuming, which has led to a technique known as high-pressure liquid chromatography (HPLC), which results in much-improved flow rates. Additives in various polymers have been identified and quantified by HPLC.

Molecular Weight Determination (GPC and MI). There are four molecular weight averages in common use [9]: (1) the number-average molecular weight, M_n; (2) the weight-average molecular weight, M_w; (3) the z-average molecular weight, M_z; and (4) the viscosity-average molecular weight, M_v. These are defined below in terms of the number of molecules, N_i, having molecular weights M_i; or the weight of species, w_i, with molecular weights M_i.

$$M_n = \frac{\Sigma_i N_i M_i}{\Sigma_i N_i} = \frac{\Sigma_i w_i}{\Sigma_i \left(w_i / M_i\right)} \tag{1.1}$$

$$M_w = \frac{\Sigma_i N_i M_i^2}{\Sigma_i N_i M_i} = \frac{\Sigma_i w_i M_i}{\Sigma_i w_i} \tag{1.2}$$

$$M_z = \frac{\Sigma_i N_i M_i^3}{\Sigma_i N_i M_i^2} = \frac{\Sigma_i w_i M_i^2}{\Sigma_i w_i M_i} \tag{1.3}$$

$$M_v = \left[\frac{\Sigma_i N_i M_i^{1+a}}{\Sigma_i N_i M_i}\right]^{1/a} \tag{1.4}$$

These values are seen on the molecular weight distribution curves of figure 1.10, which is for a HDPE geomembrane. M_n is seen to be close to the mean value (approximately 50,000 for this material) while M_w is near the upper inflection point (approximately 170,000). M_z is an

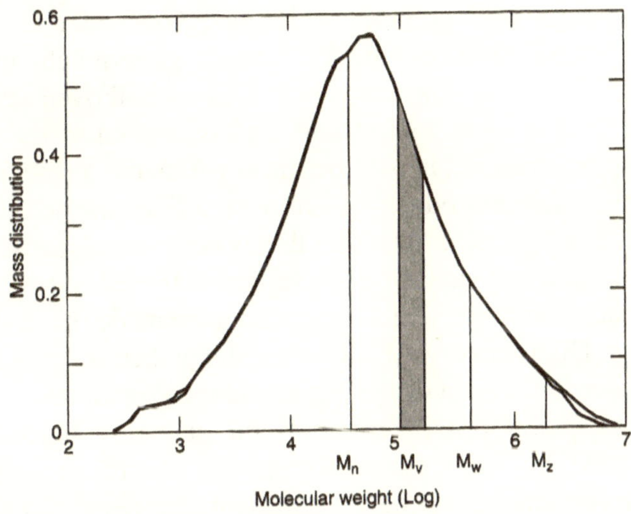

Figure 1.10 Gel permeation chromotograph results of normalized molecular weight on HDPE geomembrane test specimens. There are two curves shown: one with one melt cycle, the other with two melt cycles. (After Struve [16])

indicator of the upper end of the curve (above M_w), and M_v is a zone varying between M_n and M_w. The curve (along with its various descriptors) is probably the most powerful indicator of any available method regarding the molecular structure of polymers and its subsequent degradation behavior. There are actually two curves lying almost on top of one another illustrating the noneffect of an additional heat cycle in degrading the molecular weight of the geomembrane [16].

To determine the entire distribution of molecular weight, gel permeation chromatography (GPC) is sometimes performed. This is essentially a process for the fractionation of polymers according to their molecular size and, therefore, according to their molecular weight. The molecular weight is determined indirectly by calibration of the system in terms of the elution time expected for a particular polymer molecular-weight fraction with a particular piece of equipment. The column packages are made with microporous glass beads and powdered, swelled, cross-linked polystyrene. It is a tedious test, requiring care and precision, but it is one that has had a dramatic effect on the procedures for polymer characterization and molecular weight determination [17].

It should be noted that a qualitative test to indirectly measure the molecular weight of a polymer is the *melt flow index* test. ASTM D1238 or ISO 01133 are commonly used for geosynthetic materials. In this type of test, the polymer is heated in a small oven attached to the device until it melts. A constant load is applied to the polymer melt and pushes the molten polymer from within the oven through an orifice. The weight extruded in 10 min. is defined as the melt index (MI) value. The lower the MI, other things being equal, the higher the molecular weight and vice versa. By repeating the test at two different constant loads, the respective MI values can be made into a flow-rate ratio (FRR). High values of FRR, other things being equal, indicate broader molecular weight distributions and vice versa.

These melt flow tests, particularly the melt flow index test, are routinely used by the industry for quality control, conformance, and quality assurance testing of the incoming resin and the final geosynthetic product. Melt flow index testing, along with density testing (which is simply the specimen's weight in air divided by its weight in water [ASTM D792 or ISO1183], are considered by many to be the "workhorses" of all types of polymer manufacturing, including geosynthetic materials).

Intrinsic Viscosity Determination (IV). The molecular weight of the polymer from which geosynthetics are made can also be indirectly determined using solution viscosity methods. The results are empirically related to the molecular weight in that higher viscosity values come about from higher molecular-weight resins, all other things being equal. The intrinsic viscosity value is particularly applicable to polyester (PET) resins, fibers, yarns and straps, and rods used in the manufacture of reinforcement geotextiles and geogrids.

According to ASTM D4603, the inherent viscosity is obtained using a 0.50% concentration of PET resin in a 60/40 phenol/1, 2, 3, 3-tetrachloroethane solution. The flow time of the solution is measured in a capillary viscometer at 30°C. The pure solvent is also measured under the identical test conditions and the ratio of the two values is the relative viscosity, i.e.,

$$\eta_r = t / t_o, \tag{1.5}$$

where

η_r = relative viscosity,
t = average solution flow time (sec),
t_o = average solvent flow time (sec).

This value can then be used to calculate a inherent viscosity, if desired.

$$\eta_{inh} = \frac{\ln \eta_r}{C} \tag{1.6}$$

where

η_{inh} = inherent viscosity at 0.5% and 30°C,
C = polymer solution concentration, g/dL.

Finally, the intrinsic viscosity is obtained, i.e.,

$$[\eta] = \frac{0.25(\eta_r - 1 + 3 \ln \eta_r)}{C} \tag{1.7}$$

where

$[\eta]$ = intrinsic viscosity.

In its specification for PET geosynthetic reinforcement of walls and slopes, the Association of State Highway and Transportation Officials (AASHTO) requires that the minimum number average molecular weight (M_n) to be 25,000 or higher. To obtain this value from intrinsic viscosity, the Mark-Honwock-Sakmada equation is used.

$$[\eta] = KM_n^a, or \tag{1.8a}$$

$$M_n = exp\left[\frac{[\eta]}{K}\right]^{1/a} \tag{1.8b}$$

where K and a are constants for a particular solvent and temperature. For example, for a test performed at 35°C, K = 125 ml/g and a = 0.65.

Carboxyl End Group Analysis (CEG). An analysis of carboxyl end groups for polyester (PET) resins, fibers, yarns, straps, and rods is an important indicator of the polymer's long-term durability. According to Pohl [18], a procedure for the rapid determination of the number of carboxyl end groups of a polyester specimen entails dissolving the polymer in benzyl alcohol rapidly at a high temperature (e.g., at 200°C) then quickly mixing the solution with chloroform and titrating with sodium hydroxide and the aid of a phenol red indicator. The result of the procedure is expressed in equivalents per million grams. A maximum value of 30 is sometimes required for PET used in reinforcement geosynthetics (e.g., in the AASHTO specifications).

Commentary on Chemical Fingerprinting. The previously described series of chemical analysis tests are sometimes referred to as "chemical fingerprinting." However, *fingerprinting* is perhaps too descriptive of a word since no two products will give identical response curves to each other. Nevertheless, such response curves should be close enough to one another to substantiate their equivalency, or fundamental differences. In this sense perhaps *signature* would be a better descriptive term, but the term generally used is *fingerprinting*.

Taken collectively, the tests are very strong in their identification capability. Table 1.5 gives a summary of advantages and disadvantages of each method. While this collection of tests may initially be felt by the engineering designer to be excessive and unwarranted, there are numerous instances where the typical physical and mechanical engineering tests (weight, strength, elongation, puncture, etc.) are simply not sensitive enough to evaluate the situation under study. The fallback position, which is invariably the use of these chemical analysis tests, will be used frequently. They have an ongoing importance in many facets of geosynthetic engineering (see Halse et al. [11] for specifics on most of the tests).

1.2.3 Polymer Formulations

No geosynthetic product is 100% of the polymer resin associated with its name. In all cases, the primary resin (from which the name is derived) is mixed, or formulated, with antioxidants, screening agents, fillers, and/or other materials for a variety of purposes. The total amount of each additive in a given formulation varies widely—from a minimum of 1% to as much as 50%. The additives, either in particulate or liquid form, are used as ultraviolet (UV) light absorbers, antioxidants, thermal stabilizers, plasticizers, biocides, flame retardants, lubricants, colorants, foaming agents, or antistatic agents. The resulting mixture can be homogeneous or heterogeneous, depending on the solubility parameters of the additives versus the primary resin polymer. Heterogeneous mixtures can also be particulate or fibrous [10].

Common particulate additives include carbon blacks; various antioxidants; calcium carbonate; metallic powders and flakes; silicate minerals such as clay, talc, and mica; silica minerals, such as quartz, diatomaceous earth, and novaculite; metallic oxides, such as alumina, biocides, and other synthetic polymers. Carbon blacks of different particle sizes are very common additives. Carbon black content is determined according to ASTM D1603 or ISO-6964. In addition to the amount of carbon black, its uniform dispersion in the final product is important. The test method for dispersion is ASTM D5596 or ISO-11420.

Common liquid additives include plasticizers, fillers, and colorants. Common fibrous additives (although rarely used in geosynthetic materials) include glass, carbon and graphite, cellulosics such as alpha cellulose, synthetic polymers such as nylon, metals such as steel fibers and strands, and boron.

The resulting formulation varies from product to product, but can be generalized for the most common polymers used to manufacture geosynthetics, as shown in table 1.6. Manufacturing of the particular polymer product is addressed in many polymer and materials engineering books.

Thus, an understanding of a polymeric material insofar as its formulation is concerned is a complex and formidable task, but a "doable" one. Unfortunately, it is rarely given a high priority in engineering curricula, the obvious exception being in a polymer (or materials) engineering program. Rarely (if at all) does a civil, mechanical, or industrial engineer have any formal training in polymers, and even many chemical engineering programs are quite lean in this area. Future college curricula must be more attuned to the necessities of modern material systems, in which polymers play a key and ever-expanding role.

TABLE 1.5 CHEMICAL IDENTIFICATION METHODS AND RELATED COMMENTS

Method	Information Obtained	Advantages*	Disadvantages
TGA	• Polymer, additives and ash content • Carbon black amount • Decomposition temperatures	• Straightforward measurement • High accuracy • All polymers	• Qualitative results • High cost
DSC	• Melting point • Crystallinity • Oxidative induction time • Glass transition	• Straightforward measurement • High accuracy • All polymers	• Qualitative results • Limited to chlorinated polymers • High cost
TMA	• Coefficient of linear thermal expansion • Softening point • Glass transition	• Straightforward measurement • High accuracy • All polymers	• High cost
DMA	• Elastic constants • Loss modulus • Creep behavior • Stress relaxation behavior	• High accuracy • Versatile • All polymers • Temp. controlled	• High cost • High maintenance • Complex unit
IR	• Identifies additives • Identifies fillers • Identifies plasticizers • Rate of oxidation reaction	• All polymers	• Difficult specimen preparation • No resin information • High cost
GC, LC and HPLC	• Identifies additives • Identifies plasticizers	• Straightforward measurement	• Difficult specimen preparation • No resin information • High cost
GPC	• Molecular weight distribution • Average molecular weight	• Accurate values • Only valid technique for molecular weight • All polymers	• Tedious test • Difficult specimen preparation • Uses strong solvents • Very high cost
MF and FRR	• Indication of molecular weight • Indication of molecular weight distribution	• Straightforward • Low cost • Used throughout industry	• None
Density	• Density	• Straightforward • Low cost • Used throughout the industry	• None
IV	• Intrinsic viscosity • Average molecular weight	• Straightforward measurement • Common test • Assesses susceptibility to hydrolysis	• Indirect measurements • Need correlations • Uses strong solvents
CEG	• Titration method • Carboxyl end group	• Assesses susceptibility to hydrolysis	• Tedious test • Uses strong solvents

*An advantage common to all methods is the extremely small specimen size required for testing in comparison to traditional physical and mechanical test specimen sizes.

TABLE 1.6 COMMONLY USED GEOSYNTHETIC POLYMERS AND THEIR APPROXIMATE WEIGHT PERCENTAGE FORMULATIONS

Polymer Type	Resin	Filler	Carbon Black or Pigment	Additives	Plasticizer
Polyethylene (PE)	95-98	0	2-3	0.5 – 2.0	0
Polypropylene (PP)	85-96	0-13	2-3	1-2	0
Polyvinyl chloride (PVC) (unplasticized)	70-85	5-15	5-10	2-3	0
Polyvinyl chloride (PVC) (plasticized)	30-40	20-30	5-10	2-3	25-30
Polyester (PET)	96-98	0	2-3	0.5-1.0	0
Polyamide (PA) (nylon)	96-98	0	2-3	0.5-1.0	0
Polystyrene (PS)	96-98	0	2-3	0.5-1.0	0
Chlorosuphonated Polyethylene (CSPE)	40-60	40-50	5-10	5-15	0
Ethylene propylene diene terpolymer (EPDM)	25-30	20-40	20-40	1-5	0

(Note: All values are percent on the basis of weight measurement)

1.3 OVERVIEW OF GEOTEXTILES

1.3.1 History

Geotextiles, as they are known and used today, were originally intended to be an alternative to granular soil filters. Thus, the original, and still sometimes used, term for geotextiles is *filter fabrics*. Barrett [19], in his now classic 1966 paper, tells of work originating in the late 1950s using geotextiles behind precast concrete seawalls, under precast concrete erosion control blocks, beneath large stone riprap, and in other erosion control situations. He used different styles of woven monofilament fabrics, all characterized by a relatively high percentage open area (varying from 6 to 30%). He discussed the need for both adequate permeability and soil retention, along with adequate fabric strength and proper elongation and set the tone for geotextile use in filtration situations. Note should be made that an earlier paper by Agerschou [20] discussed applications along the same general lines.

In the late 1960s Rhone-Poulenc Textiles in France began working with nonwoven needle-punched fabrics for quite different applications. Here emphasis was on unpaved roads, beneath railroad ballast, within embankments and earth dams, and the like. The primary function in many of these applications was that of separation and/or reinforcement. Additionally, a distinctly different use of this particular style of fabric was also recognized—that is, that thick feltlike fabrics can also transmit water within the plane of their structure, acting as drains. Such uses as dissipation of pore-water pressures, and horizontal and vertical flow interceptors, grew out of this particular drainage function. Today's use of the word *geotextiles* recognizes these many possible functions of fabrics when used within a soil mass.

Credit for early investigations into the use of geotextiles should also be given to the Dutch and the English. ICI Fibres was a major influence in the use of nonwoven, heatbonded fabrics in a wide variety of uses. The first nonwovens used in the United States were imported in the late 1970s from ICI Fibres by Mirafi Inc. Rankilor [21] describes this worldwide movement of geotextiles in the formative years. ICI Fibres provided early design-related literature that was very significant in proper use of geotextiles in a variety of applications. Chemie Linz (now TenCate) in Austria, NAUE Geosynthetics in Germany, du Pont (now Fiberweb) in the United States and TenCate in Europe were also early leaders in the technology. These firms and many others have continued to introduce geotextiles on a worldwide basis. Today hundreds of manufacturers are involved in the production, sales, and distribution of geotextiles.

A number of early conferences were held exclusively on the subject of geotextiles. More recently, the conferences have branched into the entire breadth of geosynthetics. They began in Paris in 1977 and have continued to be held every four years under the auspices of the International Geosynthetics Society (IGS) in locations around the world. The original books on the subject—those of Koerner and Welsh [22] in 1980, Rankilor [21] in 1981, and van Zanten [23] in 1986—gave credibility to the emerging technology. Today, additional books dealing with geotextiles along with thousands of separate papers and reports are available. In addition, dedicated journals have been launched dealing with all types of geosynthetics [24, 25]. This massive generation and dissemination of information was led initially by geotextile manufacturers. Their influence in this market continues to be active and, indeed, is very positive and welcome. It has been followed by an entire community of governmental, industrial, consulting, research, testing and academic institutions interested in this technology. The culmination of this activity is evidenced by two long-standing organizations; the International Geosynthetics Society (IGS) founded in 1983 and the Geosynthetic Institute (GSI) founded in 1986.

1.3.2 Manufacture

As noted, the role of the fabric manufacturer in the stimulation and growth of the geotextile market has been both large and positive. Many fiber types and fabric styles have been developed both for general use and for specific applications. In fact, it seems that these two approaches to the marketing of geotextiles typify all geotextile manufacturers: manufacturers tend to target products either for the larger customary (or commodity) market or for the smaller specialized (or engineered) market. Whatever the case, three points are relevant insofar as manufacturing is concerned: (1) type of polymer, (2) type of fiber, and (3) fabric style. Each will be discussed separately.

Type of Polymer. The polymers used in the manufacture of geotextile fibers are made from the following polymeric materials and, as seen below, the large majority is made from polypropylene:

- Polypropylene ($\simeq$ 95%)
- Polyester ($\simeq$ 2%)
- Polyethylene ($\simeq$ 2%)
- Polyamide (nylon) ($\simeq$ 1%)

Their respective repeating units were given in table 1.2; table 1.7 presents some of their relevant properties. Note that moisture plays a relatively minor role in strength that only polyolefins (polypropylene and polyethylene) are lighter than water, that polyester absorbs the least amount of water, and all polymeric materials have quite high melting points. While an extremely large database [27, 28] is available on these and other polymers, it is the final manufactured product that is of primary interest to the engineering designer and end user.

TABLE 1.7 SOME PHYSICAL PROPERTIES OF SYNTHETIC FIBERS
(STANDARD LABORATORY CONDITIONS FOR FIBER TESTS: 20°C AND 65% RELATIVE HUMIDITY)

Fiber	Breaking Tenacity (g/denier)*		Specific Gravity	Standard Moisture Regain (%)	Coefficient of Thermal Expansion ($\times 10^{-5}$ per 1°C)	Effect of Heat
	Standard	Wet				
Polyethylene (high-density)	-	-	0.96	2.0	13	Melts at 110- 140°C
Polypropylene (filament and staple)	4.8-7.0	4.8-7.0	0.91	3.0	6	Melts at 160- 170°C
Polyester						
Regular-tenacity filament	4.0-5.0	4.0-5.0	1.22 or 1.38	0.4 or 0.8	4 to 5	Melts at 250- 290°C
High-tenacity filament	6.3-9.5	6.2-9.4	1.22 or 1.38	0.4 or 0.8	4 to 5	Melts at 250- 290°C
Regular-tenacity staple	2.5-5.0	2.5-5.0	1.22 or 1.38	0.4 or 0.8	4 to 5	Melts at 250- 290°C
High-tenacity staple	5.0-6.5	5.0-6.4	1.22 or 1.38	0.4 or 0.8	4 to 5	Melts at 250- 290°C
Nylon						
Nylon 66 (regular-tenacity filament)	3.0-6.0	2.6-5.4	1.14	4.0-4.5	5.5	Sticks at 230°C Melts at about 260°C
Nylon 66 (high-tenacity filament)	6.0-9.5	5.0-8.0	1.14	4.0-4.5	5.5	Same as above
Nylon 66 (staple)	3.5-7.2	3.2-6.5	1.14	4.0-4.5	5.5	Same as above
Nylon 6 (filament)	6.0-9.5	5.0-8.0	1.14	4.5	5.0	Melts at 210°C-220°C
Nylon 6 (staple)	2.5	2.0	1.14	4.5	5.0	Melts at 160- 220°C

Source: Modified from Shreve and Brink [26].
*Denier is equivalent to the grams per 9000 meters of the thread used to make synthetic fabrics. The higher the denier, the heavier the fabric.

Type of Fiber. There are five principal types of fibers used in the construction of geotextiles: monofilament, multifilament, staple fiber yarn, slit-film monofilament, and slit-film multifilament (see figure 1.11). The properly formulated polymers are made into fibers (or yarns where a yarn consists of one or more fibers) by melting them and forcing them through a spinneret, similar in principle to a bathroom showerhead. The resulting fiber filaments are then hardened or solidified by one of three methods: wet, dry, or melt. Most geotextile fibers are made by the melt process; these include polyolefins, polyester, and nylon. Here hardening is by cooling, and simultaneously or subsequently the fibers are stretched. Stretching reduces the fiber diameter and causes the molecules in the fibers to arrange themselves in an orderly fashion. When this happens, the fiber's strength increases, its elongation at failure decreases, and its modulus increases. A wide range of stress versus strain responses can be achieved. These monofilaments can also be twisted together to form a multifilament yarn. Note that the diameter of the fiber is characterized by its *denier*—the weight in grams of 9000 m of fiber or yarn. The related textile term, *tex*, is the weight in grams of 1000 m of yarn.

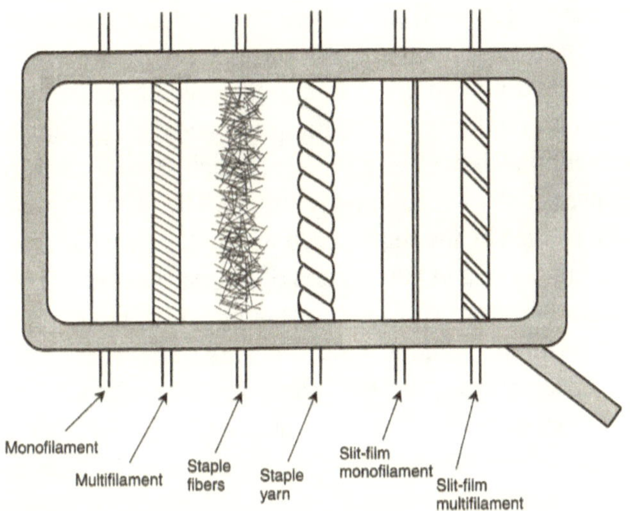

Figure 1.11 Types of polymeric fibers (or yarns) used in the manufacturing of geotextiles.

Staple fibers are very different than that described above. They are produced by continuous filaments of specific denier gathered in a large ropelike bundle called a *tow*. A tow can contain thousands of continuous filaments, and it can be converted directly into yarn. More often, however, these filament bundles are then crimped and cut into short lengths of 25 to 100 mm. The short fibers, or *staple*, are then opened and subsequently twisted or spun into long yarns for eventual fabric manufacturing.

The last type of fiber to be mentioned is made completely differently from those discussed above. These fibers, called slit (or split) film or tapes, are made from a continuous sheet of polymer that is cut into fibers by knives or lanced by air jets. The resulting ribbonlike fibers are referred to as *slit-film monofilament fibers*. Obviously, these fibers can be twisted together to make a slit-film multifilament.

Fabric Style. Once the *yarns*, as they would be referred to in the textile industry, are made, they must be manufactured into fabrics. The basic manufacturing choices are woven, nonwoven, or knit, although knit fabrics are seldom used as geotextiles. Various woven and nonwoven types are shown in figure 1.12. The woven fabrics are made on conventional textile weaving machinery into a wide variety of fabric weaves. Kaswell [27] gives an excellent review of weaving technology in which each of the various fabric weaves is clearly illustrated. Variations are many and most have direct influence on the physical, mechanical, and hydraulic properties of the fabric.

For conventional industrial fabrics (of which geotextiles form a subset) the weaves are usually kept relatively simple. The particular pattern of the weave is determined by the sequence in which the warp yarns are threaded into the weaving loom and the position of the warp harness for each filling pick (see figure 1.13a). As shown in figure 1.13b, reeds shed the warp yarns up, allowing a shuttle to insert the weft yarn. The reeds then shed downward, encapsulating the weft yarn and allowing the return of the shuttle in the opposite direction with another weft yarn. The reeds then shed back upward, and the process continues as a cycle. This action gives rise to nomenclature in woven fabrics of warp direction (the direction the fabric is being made, or machine direction), weft or fill direction (the cross direction, or cross machine direction), and selvedge (edges of the fabric where the weft yarns reverse direction and gather the outer warp yarns on each side of the fabric). This action gives rise to the various types of weaves common to the formation of fabrics for use as geotextiles.

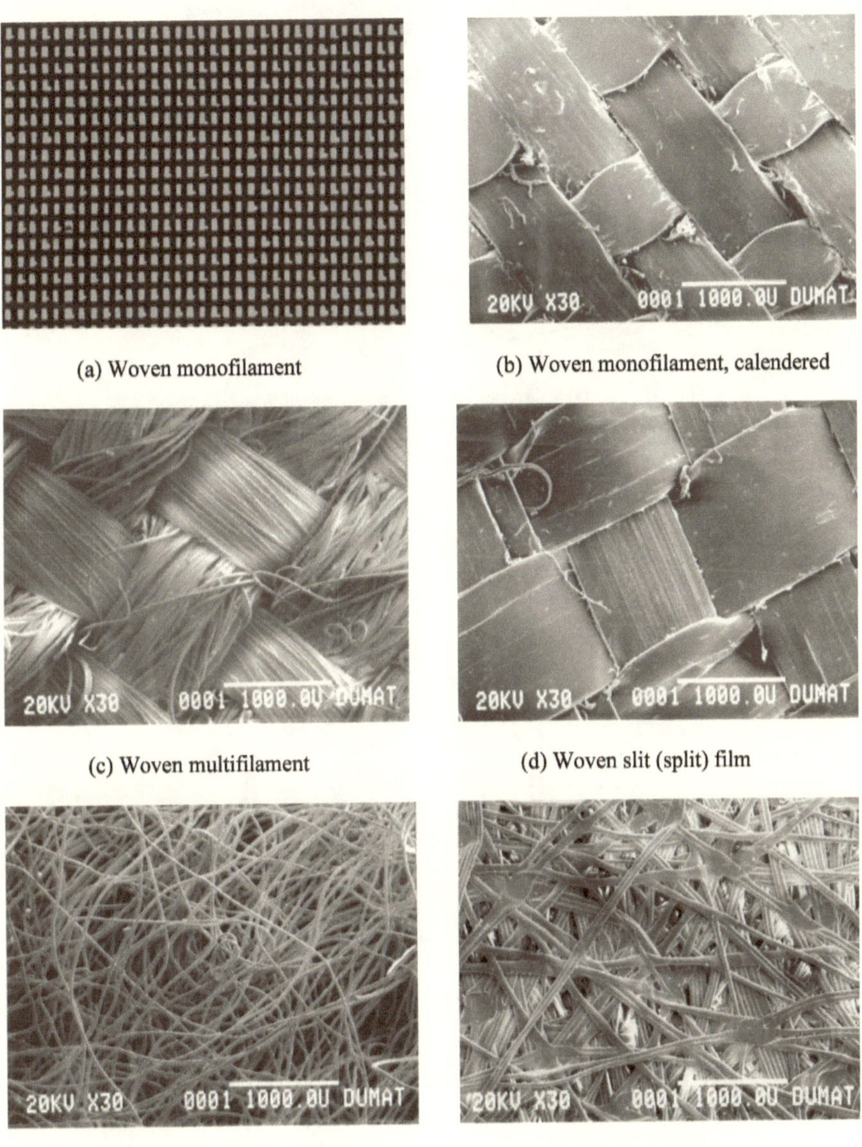

(a) Woven monofilament

(b) Woven monofilament, calendered

(c) Woven multifilament

(d) Woven slit (split) film

(e) Nonwoven needle punched

(f) Nonwoven heat bonded

Figure 1.12 Photomicrographs of various fabrics used as geotextiles. Magnification of (a) is × 5; all others are × 30.

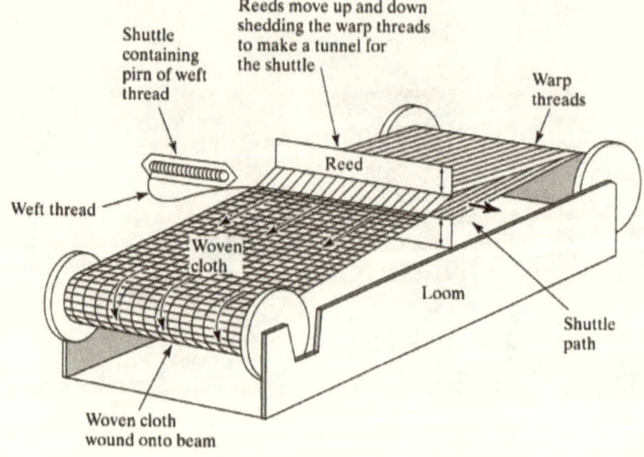

(a) Main components and identification of terms

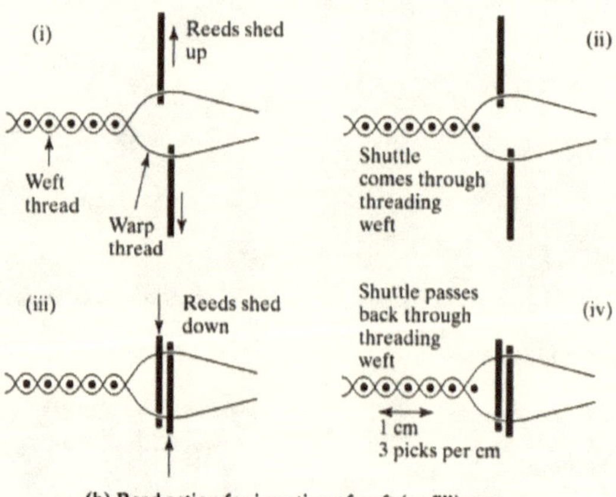

(b) Reed action for insertion of weft (or fill) yarn

Figure 1.13 Basic functioning of a weaving loom. (after Rankilor(21))

- *Plain weave*: The simplest and most common weave; also known as "one up and one down." Most woven geotextiles are of this type.
- *Basket weave*: A weave using two or more warp and/or filling yarns as one. For example, a "two-by-two basket weave" takes two warp and two weft yarns acting as individual units.
- *Twill weave*: A weave in which a diagonal or "twill" line moves across the fabric by moving yarn intersections one pick higher

on successive warp yarns. Related patterns—e.g., steep twills, broken twills—can also be formed.

- *Satin weave*: A weave in which the warp (or weft) yarn is carried over many weft (or warp) yarns, resulting in a smooth and shiny fabric surface.

Additional details on the weaving process using both natural and synthetic fibers are found in Kaswell [27].

The manufacture of nonwoven fabrics is very different from that of woven fabrics. Each nonwoven manufacturing system generally includes four basic steps: fiber preparation, web formation, web bonding, and post-treatment. Within each category are many possibilities, so only those most common to current geotextiles will be described.

Of the four basic steps, fiber preparation has already been discussed. The process of *spun bonding* (used to manufacture heat-bonded fabrics) encompasses the remaining three steps in one operation. Spun bonding is a continuous process used to produce a finished fabric from a polymer. The polymer formulation is fed into an extruder. As the polymer melt flows from the extruder, it is forced through a spinneret or a series of spinnerets. The fibers are then stretched, usually by air, and, after cooling, are laid on a moving conveyor belt to form a continuous web. In the lay-down process, the desired orientation of the fibers is achieved by various means, such as rotation of the spinneret, electrical charges, introduction of controlled airstreams, or by varying the speed of the conveyor belt. Of course, a random orientation is possible and is very common. The mat of fabric is then bonded (i.e., the filaments are made to adhere to one another) by thermal, chemical, or mechanical treatment before being wound up into finished roll form (see figure 1.14).

Alternatively, the web can be formed by starting the process with short crimped fibers—i.e., with staple fibers—of 25 to 100 mm in length. The fibers are directly made or purchased by the geotextile manufacturer in the form of bales, which are opened by forced air in what is referred to as a "carding" process. The discrete fibers are then moved by conveyer in a lay-down process to form a web of desired width, orientation, and mass per unit area. The process has enormous flexibility, including the choice of initial fiber selection. Once the loose web is formed, one of three processes are used to bond the filaments of the web together: needle punching, resin bonding, and melt bonding.

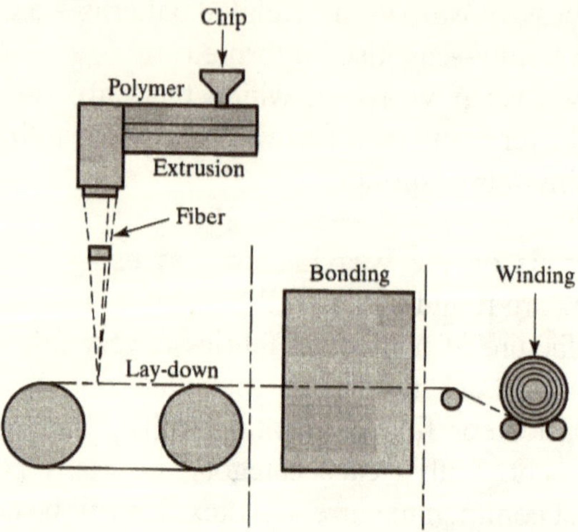

Figure 1.14 Diagram of the spun bonding process to manufacture geotextiles. Note that bonding can be by needle punching, heat bonding, or resin bonding. (Compliments of INDA)

In needle punching, which is the most common nonwoven geotextile bonding method, a fibrous web is introduced into a machine equipped with hundreds of specially designed needles. The needles are about 75 mm long and each have three or four downward-oriented barbs (see figure 1.15). While the web is trapped between a bedplate and a stripper plate, the barbed needles punch through it and reorient the fibers so that mechanical bonding is achieved throughout the length and width of the fabric. It is generally on the downstroke where the entanglement process occurs. Often, the web or *batt* of laid-down fibers is carried into the needle punching section of the machine on a lightweight support material (e.g., an open woven fabric) or substrate. This is done to improve finished fabric strength and integrity. The needle-punching process is generally used to produce fabrics that have high mass per unit area yet retain considerable bulk. Fabric weights up to 750 g/m^2 can be made in a single pass, and if desired, such fabrics can be stacked and needled together, forming weights in excess of 2000 g/m^2. Of course, the thickness and related properties increase proportionately.

In the resin bonding process, a fibrous web is either sprayed or impregnated with an acrylic resin. After curing and/or calendering, bonds are formed between filaments. Often a forced air-drying operation is used to reestablish the fabric's open-pore structure before the resin has hardened or cured. The resulting surface texture of the fabric is quite rough and abrasive.

In the melt-bonding process (also called *heat bonding* or *heat setting*), the web, which is composed of continuous filaments or staple fibers, is melted together at filament or fiber crossover points. The resultant fabrics are rather stiff in texture and feel. Somewhat higher fabric strength can be achieved with this type of manufacture at lower fabric weights than for other fabric styles, owing to the fiber bonding

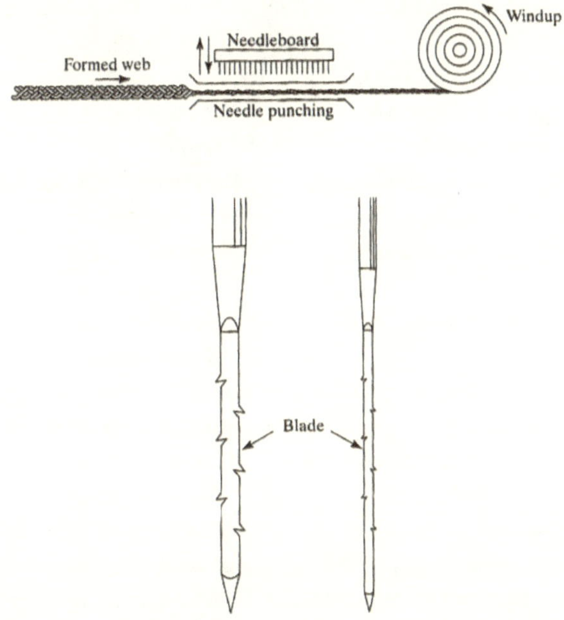

Figure 1.15 Diagram of needle punched process and details of typical needles. (Compliments of INDA)

utilized in the process. The bonding operations differ between the commercially available fabrics, depending on the basic fiber characteristics.

A major point to be emphasized is that the textile industry is a very mature and sophisticated one that can produce a tremendous variety of fabrics. Indeed, tailoring a fabric for a specific purpose or property is well within the state of the practice.

1.3.3 Current Uses

Although chapter 2 will deal with the applications of geotextiles more thoroughly, a few remarks can be made here. As already mentioned, the functions of geotextiles are separation, reinforcement, filtration, and drainage. Within these functions, however, there are a huge number of specific use areas or applications, as shown in table 1.8. Obviously this is not an all-inclusive list and it is constantly growing, but it does give an idea of the scope of the geotextile market.

TABLE 1.8 MAJOR APPLICATIONS OF GEOTEXTILES

Separation of Dissimilar Materials

- Between subgrade and stone base in unpaved roads and airfields
- Between subgrade and stone base in paved roads and airfields
- Between subgrade and ballast in railroads
- Between landfills and stone base courses
- Between geomembranes and sand drainage layers
- Between foundation and embankment soils for surcharge loads
- Between foundation and embankment soils for roadway fills
- Between foundation and embankment soils for earth and rock dams
- Between foundation and encapsulated soil layers
- Between foundation soils and rigid retaining walls
- Between foundation soils and flexible retaining walls
- Between foundations soils and storage piles
- Between slopes and downstreams stability berms
- Beneath sidewalk slabs
- Beneath curb areas
- Beneath parking lots
- Beneath sport and athletic fields
- Beneath precast blocks and panels for aesthetic paving
- Between drainage layers in poorly graded filter blankets
- Between various zones in earth dams
- Between old and new asphalt layers

Reinforcement of Weak Soils and Other Materials

- Over soft soils for unpaved roads
- Over soft soils for airfields
- Over soft soils for railroads
- Over soft soils for landfills
- Over soft soils in sport and athletic fields
- Over thermokarst areas
- Over unstable landfills as closure systems
- For lateral containment of railroad ballast
- To wrap soils in encapsulated fabric systems
- To construct fabric-reinforced walls

- To reinforcement embankments
- To aid in construction of steep slopes
- To reinforce earth and rock dams
- To reinforce stacked gabions
- To reinforce stacked geofoam
- To stabilize slopes temporarily
- To halt or diminish creep in soil slopes
- To reinforce jointed flexible pavements
- As basal reinforcement over soft soils
- As basal reinforcement over karst areas
- As basal reinforcement over thermokarst areas
- As basal reinforcement between pile foundation caps
- As lateral confinement for sand supported columns
- To bridge over cracked or jointed rock
- To hold graded-stone filter mattresses
- As a substrate for articulated concrete blocks
- To stabilize unpaved storage yards and staging areas
- To anchor facing panels in reinforced earth walls
- To anchor concrete blocks in small retaining walls
- To prevent puncture of geomembranes by subsoils
- To prevent puncture of geomembranes by landfill materials or stone base
- To create more stable side slopes due to high frictional resistance
- To contain soft soils in earth dam construction
- For use in membrane-encapsulated soils
- For use in in-situ compaction and consolidation of marginal soils
- To bridge over uneven landfills during closure of the site
- To aid in bearing capacity of shallow foundations

Filtration (Cross-Plane Flow)

- In place of granular soil filters
- Beneath stone base for unpaved roads and airfields
- Beneath stone base for paved roads and airfields
- Beneath ballast under railroads
- Around crushed stone surrounding underdrains
- Around crushed stone without underdrains (i.e., French drains)
- Around perforated underdrain pipe
- Around stone and perforated pipe in tile fields
- Beneath landfills that generate leachate
- To filter hydraulic fills
- As a silt fence
- As a silt curtain
- As a snow fence
- As a flexible form for containing sand, grout, or concrete in erosion control systems
- As a flexible form for reconstructing deteriorated piles
- As a flexible form for restoring underground mine integrity
- As a flexible form for restoring scoured bridge pier bearing capacity
- To protect chimney drain material
- To protect drainage gallery material
- Between backfill soil and voids in retaining walls
- Between backfill soil and gabions
- Around molded cores in fin drains

- Around molded cores in strip drains
- Against geonets to prevent soil intrusion
- Against geocomposites to prevent soil intrusion
- Around sand columns in sand drains
- Around porous tips for wells
- Around porous tips for piezometers
- As a filter beneath stone riprap
- As a filter beneath precast blocks

Drainage (In-Plane Flow)

- As a chimney drain in an earth dam
- As a drainage gallery in an earth dam
- As a drainage interceptor for horizontal flow
- As a drainage blanket beneath a surcharge fill
- As a drain behind a retaining wall
- As a drain at the base of a retaining wall
- As a drain beneath railroad ballast
- As a water drain beneath geomembranes
- As an air drain beneath geomembranes
- As a drain beneath sport and athletic fields
- As a drain for roof gardens
- As a pore water dissipator in earth fills
- As a replacement for sand drains
- As a capillary break in frost-sensitive areas
- As a capillary break for salt migration in arid areas
- To dissipate seepage water from exposed soil or rock surfaces

1.3.4 Sales

Between 1977 (the date of the first conference on geosynthetics) and 2010, geotextiles have experienced enormous growth. Actual sales are roughly estimated at 1.4 billion m^2 representing approximately $1 billion worldwide. Unit prices vary considerably from $0.50/$m^2$ for lightweight separation and filtration fabrics to $10.00/$m^2$ for high-strength reinforcement fabrics. In total use and sales, North America and Europe are well established and approximately equal to each other. The greatest increase in growth rates, however, is in Asia, Latin America, and Africa.

Within the above total, the distribution on the basis of end use is approximately as follows. Distribution of geotextiles from the manufacturer to the ultimate user is handled from the mill directly, by means of commissioned agents, and through individual distributors. Generally, but certainly not always, direct-mill sales efforts are focused on unusually large jobs, where competition is very intense. Commissioned agents, who are often very well versed in geotextile applications, functions, properties, and designs, work with design consultants and specifiers, and generally service the engineered job applications. Individual distributors service the standard applications and are often wired into certain segments of the industry (e.g., specific roadwork or erosion-control projects.)

Sales of geotextiles, and to a similar extent the other geosynthetics, are strongly related to government spending and private sector development. Both are tied to the general economy and reflect slow or accelerating cycles. Growth also reflects educational outreach programs; and the more activity in professional courses, seminars, conferences, tutorials, webinars, the more vibrant will be sales and use of geotextiles and related geosynthetics.

1.4 OVERVIEW OF GEOGRIDS

1.4.1 History

The development of methods of preparing high-modulus polymer materials by tensile drawing [29], in a sense "cold working," raised the possibility that such materials could be used in the reinforcement of a soils for walls, steep slopes, roadway bases, and foundation soils. Thus, the major function of such geogrids is in the area of reinforcement. This area, as in many others, is very active, with a number of different products, materials, and connections making up today's geogrid market. The key feature of all geogrids is that the *apertures*—the openings between the adjacent longitudinal and transverse ribs—are large enough to allow for soil communication, or strike-through, from one side of the geogrid to the other. The ribs of geogrids are often quite stiff compared to the fibers of geotextiles. As will be discussed later, not only is rib strength important but also is junction strength. The reason for this is that in certain situations the soil strike-through within the apertures bears against the transverse ribs, which transmits the load to the longitudinal ribs via the junctions. The junctions are, of course, where the longitudinal and transverse ribs meet and are connected. They are sometimes called *nodes*.

The original geogrids (which are categorized as unitized or homogeneous) were made in the United Kingdom by Netlon Ltd. (now Tensar) and were brought in 1982 to North America by the Tensar Corporation. A similar type of drawn geogrid, which originated in Italy by Tenax (now Syntec Inc. in the USA), is also available as are products by new manufacturers in Asia. More flexible, textilelike geogrids using bundles of polypropylene-coated polyester fibers as the reinforcing component were developed by ICI in the United Kingdom around 1980. This led to the development of polyester yarn geogrids made on textile weaving machinery. In this process hundreds of continuous fibers are gathered together to form yarns that are woven into longitudinal and transverse ribs with large open spaces between. The crossovers are joined by knitting or intertwining before the entire unit is protected by a subsequent coating. Geogrids within this group are manufactured by many companies having various trademarked products. There are possibly as many as twenty-five companies manufacturing coated yarn-type polyester (or other resin, like PVA) geogrids on a worldwide basis. The third group of geogrids are made by laser or ultrasonically bonding together polyester or polypropylene rods or straps in a gridlike pattern. Both Colbond and NAUE manufacture this type of geogrid.

1.4.2 Manufacture

Each of the above mentioned three different types of geogrids will be described. The polymers used to manufacture unitized or homogenous geogrids are HDPE for the unidirectional types and PP for the bidirectional and tridirectional types. The former is used in walls and slopes where the principal stress direction is known; the latter two are used in base and foundation reinforcement where load orientation can be in all directions. The process begins with heavy-gauge sheet of the appropriate polymer. Typical thicknesses are 4 to 6 mm. Holes are then punched into the sheeting on a regular pattern, and the sheet is then drawn uniaxially, biaxially, or triaxially (see figure 1.16). Drawing is done under controlled temperatures and strain rates so as to avoid fracture while allowing ductile flow of the molecular chains into an elongated condition. The key variable in the process is the draw ratio, but other variables—such as molecular weight, molecular weight distribution, and degree of branching or cross-linking—are also important [30, 31]. Aside from significant increases in modulus and strength, the creep sensitivity of the elongated ribs is greatly reduced by the drawing process. The resulting geogrids are referred to as unitized or homogeneous geogrids and are relatively stiff with respect to the coated yarn types.

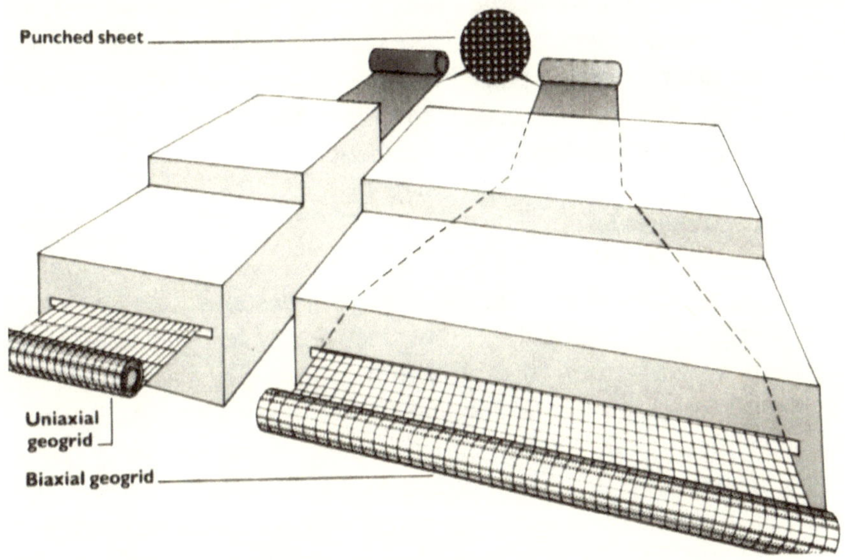

Figure 1.16 Method of manufacturing homogeneous, unitized geogrids. (After Netlon/Tensar [31])

Coated yarn-type geogrids are made from high-tenacity polyester yarns, woven into an open structure with the junctions being knitted together or physically intertwined to link the transverse and longitudinal ribs. The entire geogrid is then coated with PVC, latex, or bitumen for dimensional stability and to provide protection for the ribs soil backfilling [32]. Although PET yarns are by far the most widely used, fiberglass, nylon, and PVA yarns and also possible. The resulting geogrids are referred to as coated yarn types and are relatively flexible with respect to the others. They are available in the widest widths of the various classes of geogrids.

Polyester and/or polypropylene rod or strap geogrids are made from the same material used to package, bind, and ship various articles and materials. Parallel sets of straps approximately 10 mm wide are joined to perpendicular sets by laser or ultrasonic welding [33, 34]. Depending on the number and spacing of the rods or straps, these are the stiffest of all types of geogrids and, depending on the rib spacing, can be the highest strength.

There are other types of prototype geogrids currently under development. Many are composite materials with intriguing junction assemblies, while some are continuous sheets of "super tuff" polymeric materials with large holes punched in them. This activity in the geogrid area stems from the excellent anchorage and pullout resistance afforded when placed in a soil system. As will be seen in the design portion of chapter 3, the reinforcement function can profit handsomely from this type of geosynthetic material.

1.4.3 Current Uses

The geogrids that result from the processes described are relatively high-strength, high-modulus, low-creep-sensitive polymers with apertures varying from 10 to 100 mm in size. These apertures are either elongated ellipses, near squares with rounded corners, squares, or rectangles. Under some circumstances, separation may be a function, but only with very coarse gravels and large particle size materials. Invariably, geogrids are involved in the primary function of reinforcement. The following uses have been reported in the literature.

- Beneath or within aggregate in unpaved roads
- Beneath or within ballast in railroad construction
- Beneath or within surcharge fills or temporary construction sites
- As mechanically stabilized earth for a variety of wall facings
- As wraparound walls providing reinforcement and facing
- Reinforcement of embankment fills and earth dams
- Repairing slope failures and landslides
- As gabions for wall and bridge abutment construction
- As gabions for erosion control structures
- As basal reinforcement over soft soils
- As basal reinforcement over karst areas
- As basal reinforcement over thermokarst curves
- As basal reinforcement between pile and other deep foundation caps
- As lateral confinement to stone for constructing stone columns
- As a bridge over cracked or jointed rock
- To construct mattresses for fills over soft soils
- To construct mattresses over peat, tundra, and muskeg
- As sheet anchors for retaining wall-facing panels
- As sheet anchors and facing panels to form an entire retaining wall
- As asphalt reinforcement in pavements
- As cement or concrete reinforcement in a wide variety of applications
- To reinforce disjointed rock sections

- To reinforcement disjointed concrete sections
- As composite forms with nonwoven geotextiles
- As inclusions between geotextiles
- As inclusions between geomembranes
- As inclusions between a geotextile and a geomembrane
- To reinforce landfills to allow for vertical expansion
- To reinforce landfills to allow for lateral expansion
- To stabilize leachate collection stone as veneer reinforcement
- To stabilize landfill cover soil as veneer reinforcement
- As three-dimensional mattresses for landfill bearing capacity
- As three-dimensional mattresses for embankments over soft soils

1.4.4 Sales

In serving as a reinforcement material, geogrids compete directly against geotextiles in many of the above uses. Some of the manufacturers of geogrids also manufacture high-strength geotextiles. As such, sales are difficult to separate out for the different products. We estimate the global geogrid market to be approximately 250 million square meters representing $625 million in sales. For example, a recent survey of walls and steep soil slopes reinforced by geosynthetics shows that over 90% of such mechanically stabilized earth systems walls built worldwide have been reinforced by geogrids. This particular area, particularly among private owners and developers, is seeing tremendous growth.

1.5 OVERVIEW OF GEONETS

1.5.1 History

Geonets were originally developed by Bryan Mercer, of Netlon Ltd., in the United Kingdom. Mercer patented the machinery and processing methods for the lightweight plastic nets commonly seen in supermarkets for carrying produce, fruits, and vegetables. Experimentation with gradually thicker ribs in various configurations led to drainage nets of the type used in geosynthetic engineering. The first known use was in 1984 for the environmental application of leak detection in a double-lined hazardous liquid waste impoundment. Although geonets are indeed gridlike materials and were included in the geogrid chapter of the first edition of this book, current use dictates a separate identity. The reason for this separate treatment lies not in the material or its configuration, but in its function. Geonets are used for their *in-plane drainage* capability, while geogrids (as just discussed) are used for *reinforcement*. It should be stated at the outset, however, that geonets are not weak, flimsy materials. They have reasonable tensile strength, but they are used exclusively in drainage applications. Note that geonets are generally used with a geotextile, geomembrane, or other material on their upper and lower surfaces to prevent soil intrusion into the apertures that would block the in-plane drainage function of the material. Hence, they are used as a composite and could equally as well as be included in the chapter on geocomposites, but they deserve mention in their own right. They can also be used by themselves—for example, when placed between two geomembranes.

1.5.2 Manufacture

Essentially all geonets are made of polyethylene. The specific gravity of most geonets is in the range of 0.937 to 0.947; thus they are in the upper range of medium density or lower range of high density, depending on the classification system used. The separation between medium and high density polyethylene established by the American Society for Testing and Materials (ASTM) is 0.940/0.941. The only additives in geonets are carbon black (1 to 2%) and a processing/antioxidant package (0.5 to 1.0%); thus the material is almost pure resin.

In the manufacture of *biplanar geonets*, the ingredients are mixed and forced through an extruder that ejects the melt into a die with slotted counter-rotating segments. This is called a "stenter" (see figure 1.17). Here the polymer melt flows at angles forming discrete parallel ribs in two planes. As continuous pressure on the ejected material forces the semisolid mass forward, it is forced over a gradually increasing diameter core (or mandrel), which separates the ribs and opens the net. Thus, diamond-shaped apertures are formed that are approximately 12 mm long by 8 mm wide. The resulting angles between sets of ribs are on the order of 70 to 110 degrees, resulting in a diamond-shaped pattern. By the time the net has cooled completely, its full dimensions are realized. The geonet is then quenched in a water bath, cut along its manufactured axis, opened up into a flat sheet, and formed into rolls for shipment. Final widths have been increasing with the development of newer production facilities so as to produce geonets up to 4.5 m wide. Because of this formation process, the intersecting ribs generally are not perpendicular to one another but are at slight angles. This is an important consideration when it comes to normal load-carrying capability. Newer round ribs have significantly higher capacity in this regard.

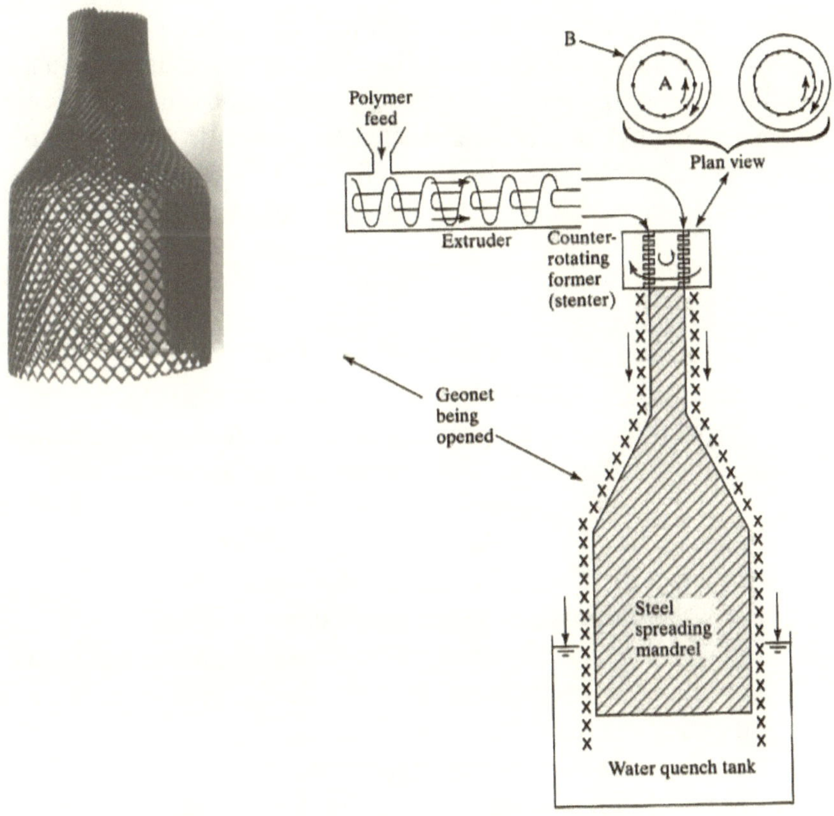

Figure 1.17 Diagram of geonet manufacturing process along with prototype shape as expanded over a steel spreading mandrel.

An alternative geonet is known as a *triplanar geonet* [35]. The die used to manufacture triplanar nets has three segments. The largest and central set of ribs are oriented in the principal flow or machine direction. Thus, field placement must be oriented hydraulic gradientwise in this direction. The upper and lower sets of ribs are smaller and closer together to maintain the structure of the central ribs and to minimize geotextile intrusion into the flow channels. The resulting triplanar geonets have flow rates considerably higher in their manufactured direction than the conventional diamond-shaped biplanar geonets. The cross-machine direction, however, has lower flow rates; hence these geonets are used on slopes where flow is unidirectional and known in its orientation.

A variety of newer geonets has also been formed by casting, injection molding, and other methods and can serve equally well, provided their mechanical and hydraulic properties are adequate. These include boxlike channel types as well as protruding column types. We can anticipate future geonets being quite different than the biplanar and triplanar types just described. Their cost/benefit ratio, however, will have to be comparable or better than those currently available [36].

1.5.3 Current Uses

As just described, geonets are used almost exclusively for their drainage capability. As such, they are single-function geosynthetics. The following uses have been documented in the literature.

- For water drainage behind retaining walls
- For water drainage of seeping rock slopes
- For water drainage of seeping soil slopes
- For water drainage behind geomembranes in dams and canals
- For water drainage beneath sport fields, golf courses, and other athletic facilities
- For water drainage of frost-susceptible soils
- For water drainage beneath building foundations
- For water drainage of plaza decks
- For water drainage beneath highways and airfields
- For leachate collection in landfills and waste piles
- For leachate collection in heap leach pads
- To detect leaks between double liners in landfills and surface impoundments
- As underdrain systems beneath landfills
- As surface water drains in landfills caps and closures

- To detect leaks between two geomembranes in vertical containment walls
- As drainage blankets beneath a surcharge fill

1.5.4 Sales

Geonets as drainage materials fall at an intermediate point in their flow capability between needle-punched nonwoven geotextiles and thick-molded types of drainage geocomposites. These other types of drainage geocomposites will be described later. As such, geonets compete with each of these materials at each end of their use spectrum. Yet their use has increased dramatically, from virtually nil in 1984 to an estimated 75 million square meters values at $150 million today, of which approximately 75% of the usage is in North America. Other types of drainage materials are more common elsewhere in the world, and these materials will be covered in section 1.9 and chapter 8. They are, indeed, viable geosynthetic materials in their own right.

1.6 OVERVIEW OF GEOMEMBRANES

1.6.1 History

In 1839, Charles Goodyear used vulcanization to cure natural rubber with sulfur, resulting in a synthetic rubber that is the current classification of *thermoset polymers*. The impetus was the inherent instability of natural (gum) rubber—i.e., it was brittle in cold weather and sticky in hot weather. Today, the production of various synthetic rubber materials is a major industry. The original geomembrane was a rubber product and was used as a water reservoir pond liner. It was butyl rubber, which is a copolymer of isobutylene with approximately 2% isoprene. Butyl rubber is quite impermeable and has its major use as inner tubes and as the liners of tubeless tires. Many other combinations and variants of rubber materials are possible—for example, nitrile and EPDM. Seaming, however, cannot be done thermally, and an adhesive tape must be used. In part from the seaming, as well as the complexity of manufacturing, the geosynthetics industry has shifted from thermoset polymers to *thermoplastic polymers* (the exception being EPDM geomembranes). Thus, almost all the geomembrane materials we will discuss fall into the category of polymers classified as thermoplastic materials. By definition, these are materials that become soft and pliable when heated without any substantial change in inherent properties and when cooled revert to their original properties. Thus, they are readily seamed by thermal methods, such as hot wedges or extrusion methods.

Polyethylene is formed by the polymerization of compounds containing an unsaturated bond between two carbon atoms. Production in quantity began in 1943. Its main original uses were (and continue to be) in the packaging and molding industries. In its various densities

- high-density polyethylene (HDPE) ≥ 0.941 g/cc,
- linear medium-density polyethylene (MDPE) = 0.940 to 0.926 g/cc, and
- linear low-density polyethylene (LLDPE) = 0.925 to 0.919 g/cc,

polyethylene is the most widely used polymer in the manufacturing of geomembranes. The development of crystallizing polypropylene is an outgrowth of low-pressure polymerization of ethylene and is the basic material from which many geosynthetics are made (recall figure 1.2). Polyvinyl chloride is another member of this group commonly used to manufacture plastic pipe and, when plasticized, geomembranes. This resin was developed in 1939 and has extensive uses. It ranks second in use to the various density polyethylenes. It is interesting to note that polyethylene geomembranes were first used in Europe and South Africa and moved to North America, while polyvinyl chloride used for geomembranes had its roots in the United States and moved to Europe and elsewhere. Other types of geomembranes were being developed in the 1960s and were used by the US Bureau of Reclamation. These geomembranes served primarily as canal liners, and their use spread to Canada, Russia, Taiwan, and Europe. Another early geomembrane, chlorosulfonated polyethylene (CSPE), resulting from the reaction of chlorine and sulfur chloride on polyethylene, was introduced for reservoir and landfill liners in the late 1960s. This geomembrane type was used in Europe shortly thereafter.

Flexible polypropylene (fPP) was introduced to the geomembrane market in about 1990 by LyondellBasell, and most recently, a hybrid polymer called thermoplastic polyolefin (TPO). This polymer is a blend (reactorwise or other) of flexible polypropylene and a thermoset polymer like polyolefinic rubber. We will only refer to fPP geomembranes in this book since the resin is tightly controlled, e.g., see the GRI-GM18 specification. Today's polymeric geomembranes are made from various thermoplastic resins and are manufactured and distributed throughout the world, making all types of products readily available. However, what matters most to the owner/designer/specifier, and what is the focus of this book, is to use the proper material for the particular project. *That* is the essence of the design-by-function concept.

1.6.2 Manufacture

The manufacturing of geomembranes begins with the production of the raw materials, which include the polymer resin itself: various additives such as antioxidants, plasticizers, fillers, carbon black, and lubricants (as a processing aid). Recall table 1.6, which gave the approximate amounts of different materials used to make geosynthetic materials. These raw materials (i.e., the "formulation") are then processed into geomembrane sheets of various widths and thicknesses by one of three ways shown in figure 1.18: extrusion, calendering, and spread coating.

High-density polyethylene (HDPE), linear low-density polyethylene (LLDPE), and flexible polypropylene (fPP) geomembranes are manufactured by an *extrusion* method. The polymer resin in pelletized form is mixed with a pelletized master batch that contains carbon black, stabilizers, and antioxidants in a carrier resin. The two pelletized material systems are carefully metered to result in the proper weight percentages and pneumatically loaded into the feed hopper of an extruder (see figure 1.19). The extruder contains a heated rotating continuous flight screw. The formulation passes successively through a feed section, compression section, and metering section where it finally emerges as a mixed and molten material, which is passed through a breaker plate and filter screen and then fed directly into a die. Two variations of extrusion processing are then used to make geomembranes. One process uses a flat die (also called *cast sheet*), which forces the polymer formulation between two horizontal die lips, in a coat hanger-like manner, resulting in polymer sheet of closely controlled thickness from 0.75 to 3.0 mm. The sheet widths vary from 1.8 to 4.6 m. When two parallel extruders are used, the width can be increased to 9.5 m (see figure 1.20). The other process uses a circular die (called *blown sheet*), which forces the polymer formulation between two

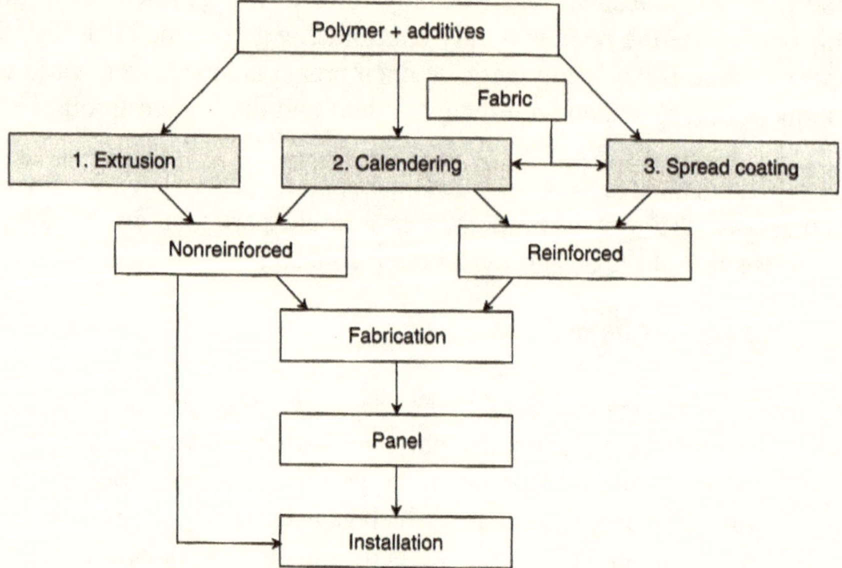

Figure 1.18 Three methods used to manufacture geomembranes. (After Haxo [37])

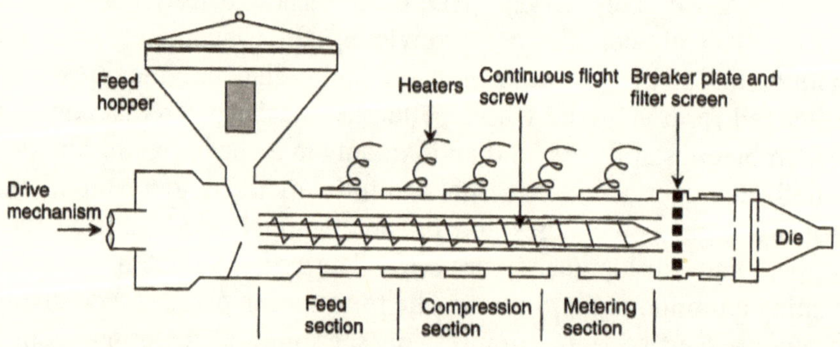

Figure 1.19 Cross-section diagram of a horizontal single-screw extruder for polymer processing.

concentric die lips oriented vertically upward. As seen in figure 1.21a, the polymer formulation exits the die and is supported by a large circular internal mandrel as it eventually extends upward in an enormous cylinder (see figure 1.21b). Typical cylinders of geomembrane are up to 10 m in circumference and 30 to 40 m in height. At the top of the system, two counter rotating rollers draw the cylinder upward and maintain stability. After passing over the rollers, the sheet is longitudinally cut, unfolded to its full width, and rolled onto a take-up core. Geomembranes produced by both types of extrusion are rolled onto a stable core and stored appropriately. They are transported to the job site and installed in the field where they are field-seamed together into a complete liner system.

By creating a roughened surface on a smooth HDPE, LLDPE, or fPP sheet through a process called *texturing* in this book, a high-friction surface can be created. Figure 1.22 illustrates the four methods that have been used to texturize geomembranes: coextrusion, impingement, lamination, and structuring.

The most common method for texturing polyethylene geomembranes is *coextrusion*. It utilizes a blowing agent (typically nitrogen gas)

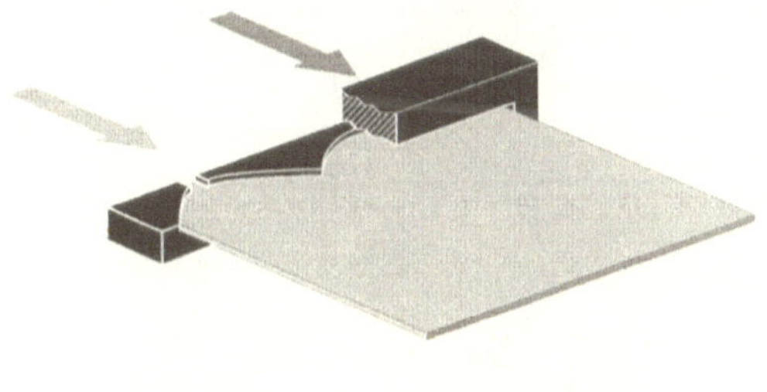

Figure 1.20 Processing of geomembranes by flat die extrusion using two extruders in parallel. (Compliments of NAUE Geosynthetics, Inc.)

in the molten extrudate and delivers it from a small extruder immediately adjacent to the main core extruder. When both sides of the sheet are to be textured, two small extruders are used. (For blown film, one internal and one external to the main extruder are needed; for cast sheet, one above and below the main extruder.) As the extrudate from these smaller extruders meets the cooler air, the blowing agent expands, opens to the atmosphere and creates the textured surface(s). A small width (approximately 300 mm) of the cylinder's circumference or flat sheet can be left smooth, which after central cutting or trimming, becomes the two lengthwise edges of the roll for ease of seaming.

The second method of texturing is *impingement*, a process in which hot polyethylene particles are actually projected onto the previously

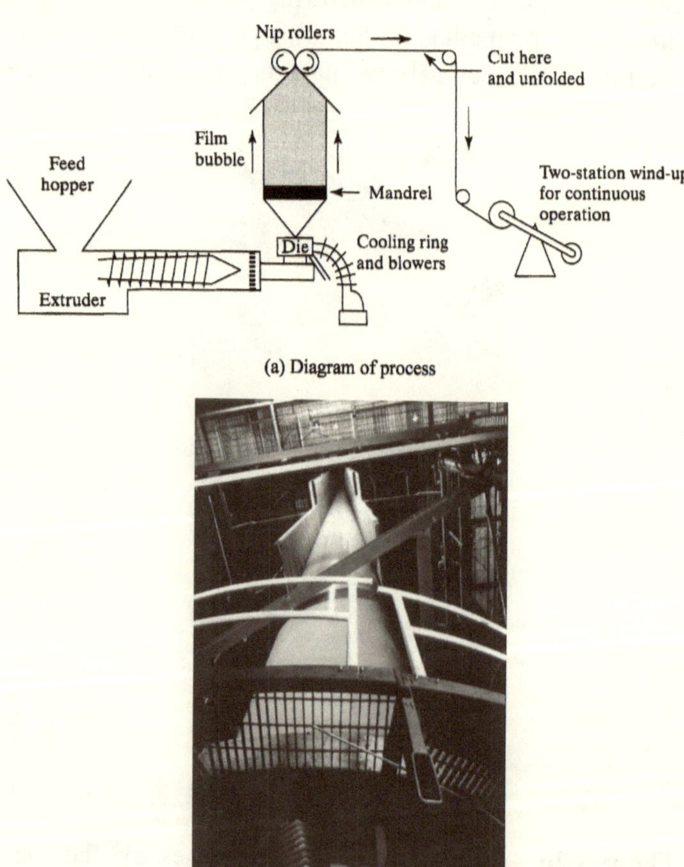

(a) Diagram of process

(b) Geomembrane bubble

Figure 1.21 Processing of geomembranes by blown sheet extrusion. (Compliments of GSE Lining Technology, Inc.)

manufactured smooth sheet on one or both of its surfaces in a secondary operation. The adhesion of the hot particles to the cold surface(s) should be as great, or greater, than the shear strength of the adjacent soil or other abutting material. The lengthwise edges of the sheets are left nontextured for approximately 150 mm for ease of seaming. The method is common in Europe but not often used in North America due to its relatively high cost during manufacturing.

The third method of texturing is *lamination*, a process that involves a foam on the previously manufactured smooth sheet in a secondary operation. In this method a foaming agent contained within molten polyethylene provides a froth that is adhered to the previously manufactured smooth sheet providing a rough textured surface. The degree of adhesion is important with respect to the shear strength of the adjacent soil or other abutting material. If texturing on both sides of the geomembrane is necessary, the roll must go through another cycle but now on its opposite side. The lengthwise edges of the sheets are left nontextured for approximately 150 mm so that field seaming can be readily accomplished. This method is relatively expensive and is rarely used.

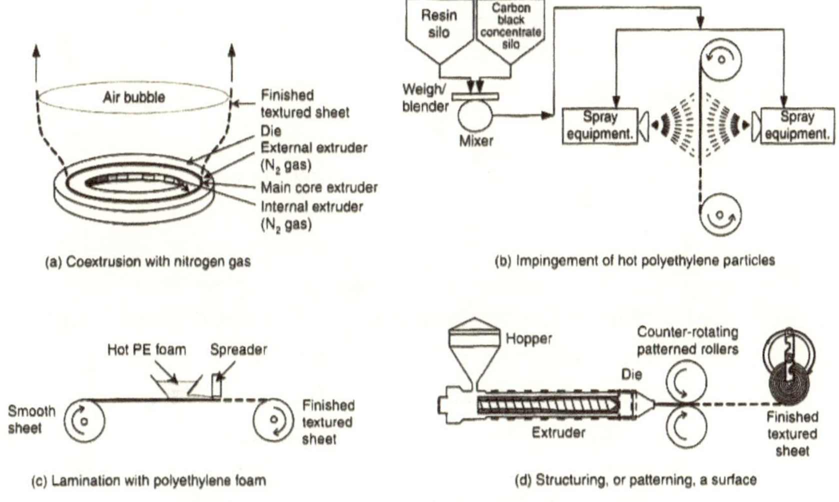

Figure 1.22 The four methods used to produce textured surfaces on HDPE, LLDPE, and fPP geomembranes.

The fourth method of texturing is *structuring*, or *patterning*. In this method a smooth sheet is made by the flat die method and immediately upon leaving the die lips passes between two counter-rotating patterned rollers. These rollers have patterned surface(s) allowing the still-hot sheet (approximately at 120°C) to pass between and deform into the inverse pattern of the rollers. This gives a single or multiple raised surface patterns on the sheet as it exits the rollers; one type being a box and point pattern, but the variations are endless. For example, the patterned rollers can have a very knurled and rough pattern, and the subsequent sheet will reflect the inverse of this pattern. The lengthwise edges of the sheets are left nontextured for approximately 150 mm so that field seaming can be readily accomplished. This texturing method is a common one, second only to coextrusion and they are shown together in figure 1.23.

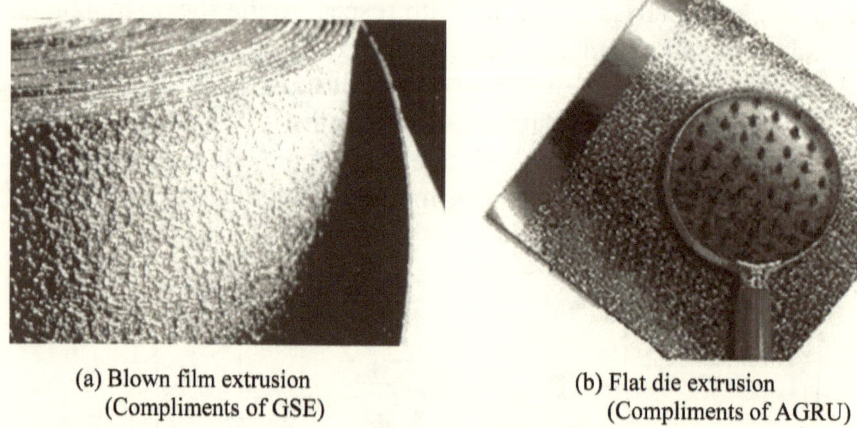

(a) Blown film extrusion (Compliments of GSE)

(b) Flat die extrusion (Compliments of AGRU)

Figure 1.23 Examples of texturing geomembrane sheets for increased friction to adjacent materials.

PVC, CSPE, and scrim-reinforced geomembranes including CSPE-R and fPP-R are not produced as described above and are manufactured by *calendaring*, a method in which the polymer resin, carbon black, filler, plasticizer (if any), and additive package are weighed (recall table 1.6) and mixed in a batch (Banbury type) or continuous mixer (Farrel type). During mixing, heat is added, which initiates a reaction between the components. The material exits the mixer and moves by conveyor to a roll mill where it is further blended and homogenized or "masticated." Now in the form of a continuous mass, it is conveyed through a set of counter-rotating rollers (called a "calender") to form the final sheet. The versatility of calenders is seen in figure 1.24. This type of manufacturing gives rise to the concept of multiple plies of laminated geomembranes, sometimes with an open-weave fabric (called *scrim*) inserted between the individual plies.

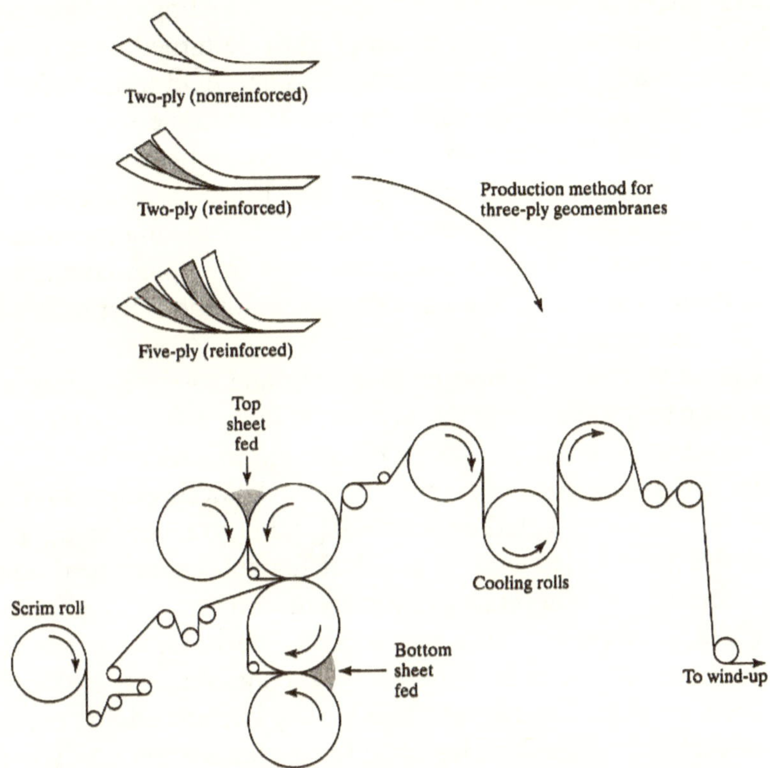

Figure 1.24 Processing of geomembranes by calendering, utilizing multiple plies of material. (Compliments of Dow Chemical Co.)

The openings in the scrim must be large enough to allow the plies to adhere to one another (called "*strike-through*") with adequate ply adhesion to prevent delamination. When the fabric scrim is included, such geomembranes are called "reinforced" and carry the designation accordingly—for example, CSPE-R and fPP-R. Geomembranes produced by calendering are available in widths up to 2 to 3 m. To produce much wider widths (called *panels*), the rolls are sent to a fabricator who factory-seams the roll edges together and packages them in a double accordion-folded manner for shipment to the field.

Reinforced geomembranes can also be made by a manufacturing method called *spread coating*. In this method the molten polymer (whatever its formulation) is spread in a relatively thin coating over a tightly woven fabric or even a nonwoven fabric. Generally, the open pore spaces of the fabric are insufficient to allow for penetration to the opposite side; hence, if coating on both sides of the fabric is required, the material must be turned over, and the process repeated. Ongoing research and some field trials have been using a spray coating with various types of elastomers, one of them being polyurea. The elastomer may also be applied directly to the subgrade soil as an in situ geomembrane. There are a large number of possibilities for spread coating and/or spray coating manufacture of geomembranes, but they are rarely seen in critical containment applications, extrusion, and calendering production methods being much more common.

The only type of thermoset geomembrane currently being used is ethylene propylene diene terpolymer, both nonreinforced (EPDM) and scrim reinforced (EPDM-R). The initial processing is very similar to that described in a conventional calendering operation. The sheets are then fabricated on a splicing table to form large panels, which are then rolled onto a mandrel for subsequent heat curing or vulcanization. The curing equipment consists of the standard autoclave, preferably with a heated jacket to reduce condensation and a closed chamber in which a rack containing several curing mandrels are placed and the steam is slowly introduced. In operating a closed-chamber steam vulcanizer, the curing system consists of a rise to the predetermined pressure, a definite period at the required curing pressure, and a reduction to atmospheric pressure. The rack of curing mandrels are allowed to set approximately 30 to 45 minutes before preparing each mandrel for observation (inspection, cut plan, etc.) on the finishing/inspection floor. Finally, the cured geomembranes are packaged (rolled onto a core) for transportation and distribution.

From the previous description of extrusion, calendered, spread (or spray) coated, and autoclaved manufacturing, it is important to appreciate that a complete geomembrane production process includes several separate companies: the resin producer, the additive producers, the formulators, the manufacturers, the fabricators, and (eventually) the installers. Communication between each party and proper liaison is critical in arriving at an acceptable and properly functioning installation. Problems and misunderstandings can arise because of the relatively large number of parties involved. It is critical that proper manufacturing quality control (MQC) measures be taken by the manufacturer and fabricator in bringing to the job site the geomembrane that was designed, specified, and purchased. The quality procedures embodied in ISO 9000 and ISO 14,000 indicate to the designer, specifier, and purchaser that a manufacturing quality (MQC) system has been developed and is being practiced by the manufacturer. In this same light, manufacturing quality assurance (MQA)—seeing that the proper geomembrane has been manufactured per the project plans and specifications—is important and routinely practiced in the geomembrane industry.

1.6.3 Current Uses

A wide range of uses of geomembranes have been developed, all of which relate to the primary function of a material being "impermeable." Note at the outset that nothing is *strictly* impermeable in an absolute sense. Here we are speaking of relative impermeability compared to that of competing materials. In the case of solid—or liquid-waste containment geomembranes, the competing material is often natural or amended clay, which usually has a targeted hydraulic conductivity (permeability) of approximately 1×10^{-9} m/s. By contrast, the equivalent diffusion permeability of a typical geomembrane will be 1×10^{-11} m/s to 1×10^{-14} cm/s. Thus, we speak of geomembranes as being relatively impermeable.

Geomembranes have been used in the following environmental, geotechnical, hydraulic, transportation, and private development applications:

- As liners for potable water
- As liners for reserve water (e.g., safe shutdown of nuclear facilities)
- As liners for waste liquids (e.g., sewage sludge)
- Liners for radioactive or hazardous waste liquid
- As liners for secondary containment of underground storage tanks
- As liners for solar ponds
- As liners for brine solutions
- As liners for the agriculture industry
- As liners for the aquiculture industry
- As liners for golf course water holes and sand bunkers
- As liners for all types of decorative and architectural ponds
- As liners for water conveyance canals

- As liners for various waste conveyance canals
- As liners for primary, secondary, and/or tertiary solid-waste landfills and waste piles
- As liners for heap leach pads
- As covers (caps) for solid-waste landfills
- As covers for aerobic and anaerobic manure digesters in the agriculture industry
- As covers for power plant coal ash
- As liners for vertical walls: single or double with leak detection
- As cutoffs within zoned earth dams for seepage control
- As linings for emergency spillways
- As waterproofing liners within tunnels and pipelines
- As waterproof facing of earth and rockfill dams
- As waterproof facing for roller-compacted concrete dams
- As waterproof facing for masonry and concrete dams
- Within cofferdams for seepage control
- As floating reservoirs for seepage control
- As floating reservoir covers for preventing pollution
- To contain and transport liquids in trucks
- To contain and transport potable water and other liquids in the ocean
- As a barrier to odors from landfills
- As a barrier to vapors (radon, hydrocarbons, etc.) beneath buildings
- To control expansive soils
- To control frost-susceptible soils
- To shield sinkhole-susceptible areas from flowing water
- To prevent infiltration of water in sensitive areas
- To form barrier tubes as dams
- To face structural supports as temporary cofferdams
- To conduct water flow into preferred paths
- Beneath highways to prevent pollution from deicing salts
- Beneath and adjacent to highways to capture hazardous liquid spills
- As containment structures for temporary surcharges
- To aid in establishing uniformity of subsurface compressibility and subsidence
- Beneath asphalt overlays as a waterproofing layer
- To contain seepage losses in existing aboveground tanks
- As flexible forms where loss of material cannot be allowed

1.6.4 Sales

Although there are always new resins being developed in the geomembrane market, the US market currently is divided between HDPE, LLDPE, fPP, PVC, CSPE-R, EPDM-R, and others such as ethylene interpolymer alloy (EIA-R), and can be summarized as follows:

(Note that $M\ m^2$ refers to millions of square meters.)

- high-density polyethylene (HDPE) $\simeq$ 35% or 105 M m^2
- linear low-density polyethylene (LLDPE) $\simeq$ 25% or 75 M m^2
- polyvinyl chloride (PVC) $\simeq$ 25% or 75 M m^2
- flexible polypropylene (fPP) $\simeq$ 10% or 30 M m^2
- chlorosulphonated polyethylene (CSPE) $\simeq$ 2% or 6 M m^2
- Ethylene propylene diene terpolymer (EPDM) $\simeq$ 3% or 9 M m^2

The above represents approximately $1.8 billion in worldwide sales. Projections for future geomembrane usage and sales are strongly dependent on the application and geographic location. Landfill liners and covers in North America and Europe will probably see modest growth ($\simeq$ 5%), while in other parts of the world growth could be dramatic (10-15%). Perhaps the greatest increases will be seen in the containment of coal ash and heap leach mining for precious metal capture. Other applications depend on education of the potential users and marketing, but in the aggregate should be very strong in the future.

1.7 OVERVIEW OF GEOSYNTHETIC CLAY LINERS

1.7.1 History

Geosynthetic clay liners (or GCLs) are factory-manufactured hydraulic barriers consisting of a thin layer of bentonite supported by geotextiles and/or geomembranes, being mechanically held together by needling, stitching, or chemical adhesives.

The use of GCLs as a separate category of geosynthetics first occurred in the United States in 1986 in the field of solid-waste containment as a backup to a geomembrane [38]. The product was Claymax, which is bentonite mixed with an adhesive so as to bond the clay between two geotextiles—one below (the *carrier*) and the other above (the *cover*).

About the same time a different product in Germany, named Bentofix, was manufactured by placing bentonite powder between two geotextiles and then needle-punching the three-component system together.

Other names for GCLs are clay blankets, bentonite blankets, bentonite mats, prefabricated bentonite clay blankets, and clay geosynthetic barriers, the latter currently favored by the International Standards Organization (ISO). The engineering function of a GCL is containment as a hydraulic barrier to water, leachate, or other liquids and sometimes gases. As such, GCLs are used as replacements to either compacted clay liners (CCLs) or geomembranes, or they are used in a composite manner to augment the more traditional liner materials.

1.7.2 Manufacture

There are many GCLs available today. In addition to the above-mentioned products, these would include Bentomat, which consists of two geotextiles needle-punched together containing bentonite powder; Gundseal, which uses an adhesive to bond bentonite powder onto a HDPE or LLDPE geomembrane; NaBento, which consists of two geotextiles containing bentonite powder and stitch-bonded together. Figure 1.25 illustrates the related products available in North America with a small hydrated sample placed above the as-received products.

Figure 1.26 presents the production concept for GCLs, with 1.26a depicting the two adhesive bonded products and 1.26b depicting the needle-punched and stitch-bonded products. All the GCL products manufactured in North America use a sodium bentonite in the mass per unit area range of 3.2 to 6.0 kg/m^2. The bentonite thickness varies in the range of 4.0 to 6.0 mm. The hydraulic conductivity (permeability) is typically in the range of (1 to 5) $\times 10^{-11}$ m/s, thus 100 to 20 times lower than the typical compacted clay liner. The various products come to the job site at a humidity-equilibrated moisture state that varies from 10 to 35%. This is due to the extremely high-hydrophilic nature of bentonite and to purposely added moisture during manufacturing. The types of geotextiles used with the different products vary widely in their manufacturing (e.g., needle-punched nonwoven, woven silt film, spunlaced, composite, etc.) and in their mass per unit area (e.g., varying from 100 g/m^2 to 600 g/m^2). The particular product with a geomembrane backing can also vary in its type, thickness, and surface texture. Some recent

Figure 1.25 Various types of geosynthetic clay liners currently available, showing the corresponding hydrated product directly on top of the as-received product.

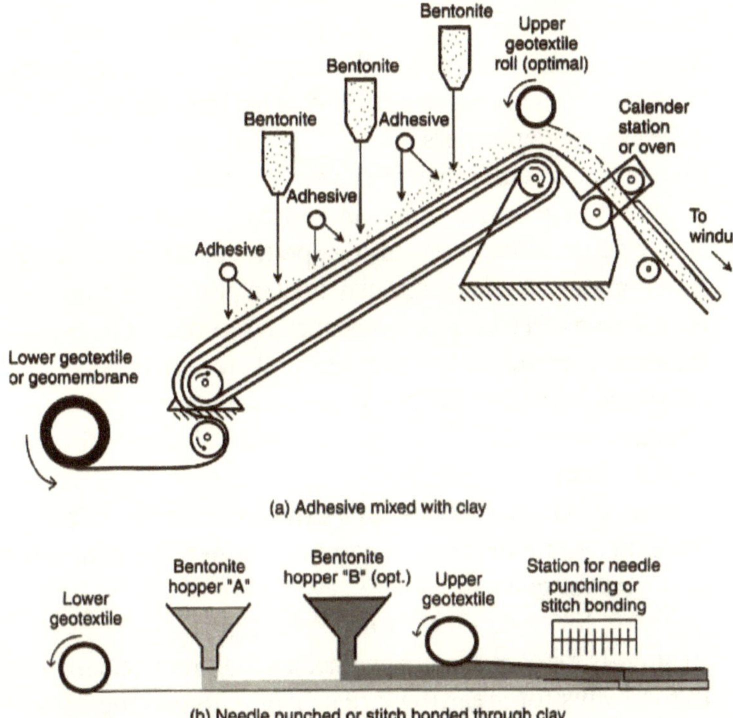

Figure 1.26 Methods of manufacturing different types of geosynthetic clay liners.

GCL variations include a thin polymeric film on or beneath the cover geotextile and a polymer impregnated cover geotextile. In addition, research on polymer modified bentonite and nanotechnology with GCL is ongoing [39,40].

GCLs are factory-made in widths of 4.0 to 5.2 m and lengths of 30 to 60 m. Upon manufacturing, they are wrapped around a core and are covered with a plastic film to prevent additional moisture absorption or wetting (i.e., premature hydration) during storage, transportation, and placement prior to their eventual covering with an overlying layer.

1.7.3 Current Uses

GCLs are indeed hydraulic barriers to liquid movement and, as such, are competitive or complementary wherever geomembranes and compacted clay liners are used. GCL applications reported in the literature are as follows:

- Beneath geomembranes in the primary liners of landfills
- Beneath geomembranes in the secondary liners of landfills
- Beneath geomembranes and above clay liners of landfills—i.e., three-component liners
- Beneath geomembranes in the covers of landfills
- Adjacent to geomembranes in vertical cutoff walls
- Above geomembranes as puncture protection against coarse gravel
- As a portion of a compacted clay liner in primary composite liners
- As a portion of a compacted clay liner in secondary composite liners
- As secondary liners for underground storage tanks
- As single liners for surface impoundments
- Beneath geomembranes as composite liners for surface impoundments
- Beneath geomembranes as composite liners for heap leach ponds
- Beneath geomembranes as composite liners for coal ash piles
- As liners for canals
- As liners for agricultural waste treatment
- At airports for containment and neutralization of deicing solutions
- In roadways for areas that are sensitive to deicing salts

1.7.4 Sales

The use of geosynthetic clay liners has moved from conception to application quite rapidly. It is estimated that approximately 100 M m^2 of GCL was installed worldwide in 2002. Within this total, the application breakout is estimated as follows:

- Landfill and heap leach liners (usually beneath a geomembrane) $\simeq$ 40% or 40 M m^2
- Landfill waste pile covers (often beneath a geomembrane) $\simeq$ 30% or 30 M m^2
- Other environmental applications $\simeq$ 15% or 15 M m^2
- Other geotechnical, transportation, hydraulic and private development applications $\simeq$ 15% or 15 M m^2

This total represents approximately $650 million in total worldwide sales.

1.8 OVERVIEW OF GEOFOAM

1.8.1 History

Geofoam is a foamed polymeric geosynthetic material (generally expanded polystyrene) manufactured in slab or block form and used primarily for its lightweight and sometimes for its insulating properties. The original applications as a lightweight fill were in Norway in 1972 and, subsequently, throughout Scandinavia. Early use has also been in Japan and Southeast Asia. These uses invariably take advantage of geofoam fills being only 1 to 3% as heavy as conventional soil fills. Thus, carrying loads, such as highways over compressible soils, frost-sensitive soils, and other settlement-prone situations favor the use of geofoam.

The first use in the United States was to relieve lateral earth pressures behind a retaining wall [41]. An international conference in 1995 [42] and subsequent publications by Horvath [43] and Negussay [44] have been influential in the proper positioning of geofoam for the civil engineering designer and specifier.

1.8.2 Manufacture

The resin styrene was developed in 1937. Expanded polystyrene (EPS), the general material for the manufacture of geofoam, was developed by the BASF Corporation in 1950. Their manufacturing process is described.

The molecular structure of polystyrene (PS) is given in table 1.2 and its genesis from ethylene is shown in figure 1.2. The public knows the material best as packaging material and insulating container material. Geofoam production follows three processing stages, shown in figure 1.27.

In the first stage, styrene with various additives is introduced with pentane as a blowing agent via water into a polymerization unit. Using steam as the heat transfer medium, the resin softens and the increased vapor pressure of the blowing agent expands the resin beads to about fifty times their original size. During this increase in volume, the close-cell foam structure of the beads is formed. The apparent density is an extremely low value ranging between 10 and 30 kg/m^3.

During the second stage, which is called intermediate bead processing, steam is again used along with rotary blowers or blower injectors for the following purposes:

- Air diffuses through the cell walls making the particles mechanically stable.
- Moisture is dissipated to the atmosphere, aiding in the flow properties of the beads.
- Any residual blowing agent diffuses out through the walls of the beads.

The now-stabilized beads are then stored in a silo for 5 to 28 hours.

The third stage is final processing in which slabs or blocks of geofoam are made. Again, steam is used on the beads placed in the appropriate forms, causing the polymer beads to soften. The free space between beads is closed, causing a polyhedral structure to form and the touching surfaces bond together. When cooled, the finished slabs or blocks are removed, sent to a sizing operation, and eventually shipped to the customer. Modern production also allows for continuous production (rather than the batch process described above).

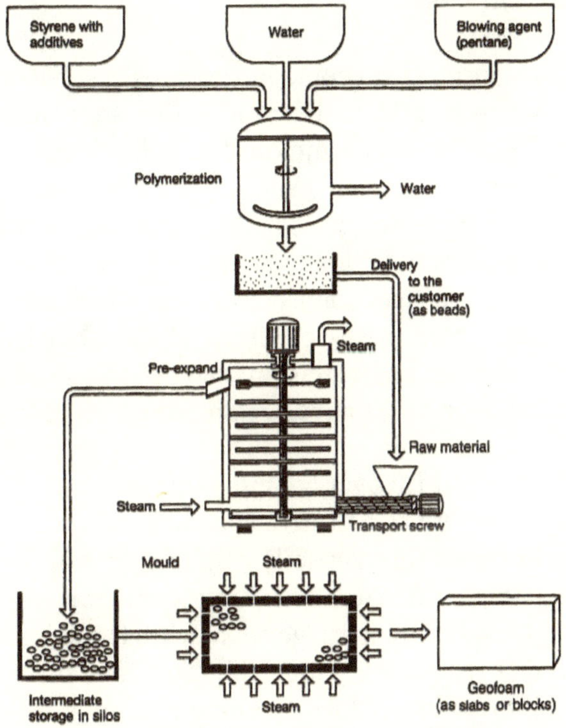

Figure 1.27 BASF Inc. production method of Styropor expanded polystyrene (EPS) slabs and blocks.

A somewhat different geofoam manufacturing process produces extruded polystyrene (XPS). Note that EPS and XPS are referred to in ASTM standards as rigid cellular polystyrene (RCPS), but we will use the term *geofoam* throughout. XPS is manufactured by expanding the polystyrene and shaping the final product in a continuous process. The blowing agents are in the family of fluorocarbons, which are currently under scrutiny from the perspective of depletion of the ozone layer. XPS products are relatively thin in comparison to EPS products—for example, 100 mm versus 1 m, or more. Thus, XPS is usually limited to insulation rather than lightweight fill where large volumes of material are an asset. Additionally, the unit price for XPS is generally higher than EPS. However, in discussing geofoam in chapter 7, both EPS and XPS will be included. Detailed commercial literature on manufacturing and subsequent properties is available from the respective manufacturers.

1.8.3 Current Uses

EPS blocks used as geoform come in various sizes, but cross sections of 1.25 m in width and 0.5 to 1.0 m in height are common. Lengths also vary, but 2.0 to 3.0 m is common. The bulk, rather than weight, is the limiting factor as far as field handling and placement is concerned. A 1.25 × 0.5 × 2.5 m geofoam block at 15 kg/m^3 density weighs only 23 kg. Such blocks are placed in the field by hand and attached to one another by galvanized steel connector plates, which are cleated such that the sharp metal barbs easily penetrate adjacent blocks of geofoam. Any size and shape can be field-assembled depending on the intended purpose. The following uses have been reported in the literature:

- As lightweight fill over compressible soils
- As lightweight fill over frost sensitive soils
- As lightweight fill over soils with uneven bearing capacity
- As lightweight fill over both karst and thermokarst areas
- As lightweight fill for bridge approaches to minimize settlement
- For use as temporary bridge abutments and piers
- For compressible inclusion behind retaining walls
- For compressible inclusion beneath (or above) pipelines
- For compressible inclusion beneath hydraulic structures
- For compression inclusion beneath building foundations
- For compressible inclusion in seismic-prone situations
- For compressible inclusion for vibration damping
- As thermal insulation for below-grade construction
- As thermal insulation to avoid frost pressures
- For use as a fluid transmission medium when properly detailed

1.8.4 Sales

It is difficult to estimate sales of geofoam since it is regularly used for "nongeo" uses. The estimate of 5 M m^3 represent worldwide sales of $375 million given earlier is largely not in North America. The major users are in Scandinavian countries and Japan. The growth, however, certainly justifies the individual treatment of geofoam as a separate geosynthetic material. Early editions of this book included only brief commentary on the subject, but this edition will hopefully correct the oversight.

1.9 OVERVIEW OF GEOCOMPOSITES

The basic philosophy behind geocomposite materials is to combine the best features of different materials in such a way that specific applications are addressed in an optimal manner. In conforming with the theme of the book, these geocomposites will generally be geosynthetic materials, but not always. In some cases it may be more advantageous to use a nonsynthetic material with a geosynthetic one for optimum performance or lowest cost. As you will see, the number of possibilities is huge—the only limits being one's ingenuity and imagination.

In considering the following geocomposites, keep in mind that the five basic functions presented in table 1.1 are separation, reinforcement, filtration, drainage, and containment.

1.9.1 Geotextile-Geonet Composites

When a geotextile is used on one or both sides of a geonet, the separation and filtration functions are automatically satisfied and the drainage function is vastly improved over geotextiles by themselves. Such geocomposites are regularly used to intercept and convey leachate in landfill liner and cover systems and to conduct vapor or water beneath pond liners of various types. Such drainage geocomposites also make excellent drains to intercepted water in a capillary zone where frost heave or salt migration is a problem. In all cases the liquid enters through the geotextile and then travels horizontally within the geonet to a suitable exit.

1.9.2 Geotextile-Polymer Core Composites

A core in the form of a quasi-rigid plastic sheet can be extruded or deformed in such a way as to allow very large quantities of liquid to flow within its structure; it thus acts as a drainage core in the same manner as a geonet. As with a geonet composite, the core must be protected by a geotextile, acting as a filter and separator, on one or both sides. Three main systems are available. The first type is called a wick drain in the United States and a prefabricated vertical drains (PVD) elsewhere. The 100 mm wide by 5 mm thick polymer core is often fluted for ease of conducting water. A geotextile acting as a filter and separator is socked around the core. The emergence of such geotextile polymer-core composites has all but eliminated traditional sand drains as a rapid means of consolidating fine-grained saturated soils.

The second type of system, called a sheet drain, is in the form of panels, the rigid polymer core being nubbed, columned, dimpled, or formed as a three-dimensional net. With a geotextile on one side, it makes an excellent drain on the backfilled side of retaining walls, basement walls, and plaza decks. The cores are often vacuum-formed dimples or stiff 3-D meshes. As with wick drains, the geotextile is the filter/separator, and the thick polymer core is the drain. Many systems of this type are available, the latest addition having a thin pliable geomembrane on the side facing the wall and functioning as a vapor barrier.

The third system within this area of drainage geocomposites is the category of prefabricated edge drains. These materials, typically 500 mm high by 20 to 30 mm wide, are placed adjacent to highway pavements, airfield pavements, or railroad rights-of-way, for lateral drainage out of and away from the pavement section. The systems are incredibly rapid in their installation and extremely cost effective.

1.9.3 Geotextile-Geomembrane Composites

Geotextiles are laminated on one or both sides of a geomembrane for a number of purposes. The geotextiles provide increased resistance to puncture, tear propagation, and friction related to sliding, as well as providing tensile strength in and of themselves. Quite often, however, the geotextiles are of the nonwoven, needle-punched variety and are of relatively heavy weight. In such cases the geotextile component acts as a drainage media, since its in-plane transmissivity feature can conduct water, leachate or gases away from direct contact with the geomembrane.

1.9.4 Geomembrane-Geogrid Composites

Since some types of geomembranes and geogrids can be made from the same material (e.g., high-density polyethylene), they can be bonded together to form an impervious barrier with enhanced strength and friction capabilities.

1.9.5 Geotextile-Geogrid Composites

A needle punched, nonwoven geotextile bonded to a geogrid provides in-plane drainage while the geogrid provides tensile reinforcement. Such geocomposites are used for internal drainage of low-permeability backfill soils for reinforced walls and slopes. The synergistic properties of each component enhances the behavior of the final product.

1.9.6 Geosynthetic-Soil Composites

As typified by the geosynthetic clay liners described in section 1.7, many other variations of geosynthetic products and soil can be developed. For example, geocells are geomembrane or geotextile strips that have been cleverly arranged vertically in a boxlike fashion, placed horizontally (standing upright), and filled with soil. Thus, the material forms a cellular structure and, acting with the contained sand or gravel, makes an impressively strong and stable mattress for vehicular trafficking. Sizable earth embankments have been built on such systems with the possibility of supporting structures over weak soils in the near future (i.e., an inexpensive mat foundation). Three-dimensional geotextile cells ($\simeq$ 1.0 m high) can be stacked on one another forming walls and steep soil slopes. Used by themselves or with geogrid reinforcement major structures can be realized.

Another variation of using geosynthetics and soil together is to use polymer fibers and sand to form steep soil slopes with excellent strength properties. The fibers give the composite material a very pronounced apparent cohesion. The area called *geofibers* has resulted in impressive shear strength gains for deep-seated soil stability as well as near-surface soil stability in an area called *turf reinforcement*. The latter applications include horse racetracks, sport fields, golf courses, and parking and picnicking areas.

1.9.7 Other Geocomposites

Weaving steel strands within a geotextile matrix can result in incredible composite material strengths. Used as a substrate, extremely large loads can be sustained. Large composite mattresses have been constructed in this manner. A measurable increase in bearing capacity for the support of light structures is also possible.

Geotextiles with prefabricated holes for the insertion of steel rod anchors (called anchor spider netting) have been used to stabilize slopes and as in-situ compaction and consolidation systems. The rods act as anchors, stressing the geotextile against the soil, which is put into compression. The geotextile thus acts dually as a tensile-stressing mechanism and as a filter allowing the pore water to escape while retaining the subsurface soil particles.

Another example of a new geocomposite trend called an *articulated concrete mattress*, interconnected steel strand concrete paving blocks that are bonded to a geotextile acting as a separator and filter. It competes against traditional hard armor alternatives like rock riprap. The possibilities of such geocomposite systems are essentially endless.

1.9.8 Sales

Considering the plethora of products within the geocomposite area, an estimate of total sales is very difficult. Furthermore, some products are fully entrenched (such as wick drains) while others are still emerging. To give an approximation, however, it is felt that the worldwide activity is perhaps 100 M m^2 representing about $400 million. This relatively large amount readily reflects the great variety of products that are available.

1.10 OUTLINE OF BOOK

To the author, and hopefully to many others, the area of geosynthetics is a vibrant, exciting, and rapidly growing field within geotechnical, transportation, environmental, hydraulics, and private development engineering. The sales information presented in section 1.1.3 (which is based on the author's best estimates) reflects this dynamic growth in each geosynthetic area. The data are approximate, but are indicative of the general worldwide pattern.

The field is in a constant state of activity, and the designs and test methods to be presented here might be superseded in the future. This is to be expected. Nevertheless, the area demands a specific *design-by-function* methodology—an end to which this book is committed. It is felt that time will validate the effort.

The remaining seven chapters of this two volume book will each focus on one of the following topics: geotextiles, geogrids, geonets, geomembranes, geosynthetic clay liners, geofoam, and geocomposites.

Each chapter will refer to the relevant part of this opening chapter for details on polymer properties, manufacturing and the like. Each chapter is also relatively self-contained and independent, although there is a logic and plan to the sequence in which the chapters appear. For those who are interested in only one type of geosynthetic, that chapter can be studied in-isolation, but the background given in this first chapter should be referenced continuously, particularly those parts related to materials manufacturing. Design-by-function is emphasized throughout the book, with illustrative examples and problems in each chapter. Reference lists unique to each chapter are included. The author extends best wishes as you go forward into the details and idiosyncrasies of the various types of geosynthetics.

REFERENCES

1. Dewar, S., "The Oldest Roads in Britain," *The Countryman*, vol. 59, no. 3, 1962, pp. 547-555.
2. Beckham, W. K., and Mills, W. H., "Cotton-Fabric-Reinforced Roads," *Engineering News-Record*, October 3, 1935, pp. 453-455.
3. Bertram, G. E., "An Experimental Investigation of Protective Filters," Graduate School of Engineering, Harvard University, Publ. 267, January 1940, 140 pgs.
4. Kays, W. B., *Construction of Linings for Reservoirs, Tanks and Pollution Control Facilities*, 2nd ed., John Wiley and Sons, 1982.
5. Lauritzen, C. W., "Plastic Films for Water Storage," Journal of American Water Works Assoc., vol. 53, no. 2, February, 1963, pp. 135-140.
6. Rodriguez, F., *Principles of Polymer Systems*. 4th ed. New York: McGraw Hill, 1996.
7. Mascia, L., *Thermoplastics: Engineering*. New York: Applied Science Publishers, 1982.
8. Rosen, S. L., *Fundamental Principles of Polymeric Materials*. New York: Wiley 1982.
9. Sperling, L. H., *Introduction to Physical Polymer Science*. New York: Wiley, 1986.

10. Moore, G. R. and Kline, D. E., *Properties and Processing of Polymers for Engineers*. Englewood Cliffs, NJ: Prentice Hall, 1984.
11. Halse, Y., Wiertz, J., Rigo, J.-M. and Cazzuffi, D. A., "Chemical Identification Methods Used to Characterize Polymeric Geomembranes," In *Geomembranes: Identification and Performance Testing*, edited by A. Rollin and J-M Rigo RILEM Report 4, London: Chapman and Hall 1991, pp. 316-336.
12. Thomas, R. W. and Verschoor, K. L., *Thermal Analysis of Geosynthetics*, Geotechnical Fabrics Report, IFAI, vol. 6, no. 3, May/June, 1988, pp. 24-30.
13. Brennan, W. P., *Characterization of Polyethylene Films by Differential Scanning Calorimetry*, Norwalk, CT: Perkin-Elmer Instrument Division, March, 1978, 25 pgs.
14. Hsuan, Y. G. and Guan, Z., "Evauation of Oxidation Behavior of Polyethylene Geomembranes Using Pressure Different Scanning Calorimetry," In *Oxidative Behavior of Materials by Thermal Analytical Techniques*, ASTM STP1326, ed. A. T. Riga and G. H. Patterson, eds., American Society for Testing and Materials, 1997, pp. 76-90.
15. Li, M. and Hsuan, Y. G., "Pressure Effects on the Oxidation of High Density Polyethylene Geogrids," *Proceedings from Geosynthetics '04*, Winnipeg, Canada, 2004, 8 pgs.
16. Struve, F., "Extrusion Fillet Welding of Geomembranes," *Journal Geotextile and Geomembranes*, vol. 9, nos. 4-6, 1990, pp. 1-14.
17. Allcock, H., *Contemporary Polymer Chemistry*, Englewood Cliffs, NJ: Prentice Hall, 1981.
18. Pohl, H. A., "Determination of Carboxyl End Groups in a Polyester (Polyethylene Terephthalate)," *Journal of Analytical Chemistry*, vol. 26, no. 10, October 1954, pp. 1614-1616.
19. Barrett, R. J., "Use of Plastic Filters in Coastal Structures," *Proceedings from the 16th International Conference Coastal Engineers*, Tokyo, September 1966, pp. 1048-1067.
20. Agerschou, H. A., "Synthetic Material Filters in Coastal Protection," *Journal of Waterways Harbors Division*, ASCE, vol. 87, no. WW1, Februay 1961, pp. 111-124.
21. Rankilor, P. R., *Membranes in Ground Engineering*. New York: Wiley 1981.

22. Koerner, R. M., and Welsh, J. P., *Construction and Geotechnical Engineering Using Synthetic Fabrics*. New York: Wiley, 1980.
23. van Zantan, R. V., ed., *Geotextiles and Geomembranes in Civil Engineering*. A. A. Balkema, 1986.
24. Rowe, R. K., ed., *Journal of Geotextiles and Geomembranes*. Elsevier Applied Science Publishers.
25. Ingold, T., ed., *Geosynthetics International*. T. Telford Publishers Ltd.
26. Shreve, R. N., and Brink, J. A., Jr., *Chemical Process Industries*, 4th ed., McGraw-Hill, 1977.
27. Kaswell, E. R., *Handbook of Industrial Textiles*. New York: West Point Pepperall, 1963.
28. INDA, Association of Nonwoven Fabrics Industry, 10 East 40th Street, New York, NY 10016.
29. Capaccio, G. and Ward, I. M., "Properties of Ultra-high Modulus Linear Polypropylene," *Nature Physical Sciences*, vol. 243, 1974, pp. 130-143.
30. Ward, I. M., "The Orientation of Polymers to Produce High Performance Materials," *Proceedings of the Symposium on Polymer Grid Reinforcement in Civil Engineering*, Institute of Civil Engineers, UK,1984.
31. Netlon Ltd. Blackburn, United Kingdom (product literature) and Tensar Inc., Morrow, GA (product literature).
32. Paulson, J. N., "Flexible Geogrids," *Proceedings GRI-8, Geosynthetic Resins, Formulations and Manufacturing*, IFAI Publishers, 1995, pp. 199-204.
33. Heerten, G., Floss, R. and Brau, G., "Safe and Economical Soil Reinforcement Using a New Style Geogrid," *Landmarks in Earth Reinforcement*, edited by H. Ochiai, J. Otani, N. Yasufuku and K. Omine, Rotterdam: A. A. Balkema, IS Kyushu 2001, pp. 43-48.
34. Voskamp, W., "Index and Performance Testing of a New Geogrid Made of Highly Oriented Straps," *Proceedings of GeoDenver 2000, Advances in Transportation and Geoenvironmental Systems Using Geosynthetics*, edited by J. G. Zornberg and B. R. Christopher, GSP No. 103, GeoInstitute of ASCE, 2000, pp. 360-372.
35. Austin, R. A., "The Manufacture of Geonets and Composite Products," *Proceedings of Geosynthetic Resins, Formulations*

and Manufacturing, edited by Y. G. Hsuan and R. M. Koerner, IFAI, 1995, pp. 127-138.

36. Rimoldi, P., "The Future of Geonets and Geocomposites," *Proceedings of the GRI-13 Conference Geosynthetics in the Future: Year 2000 and Beyond*, Folsom, PA: Geosynthetic Information Institute, 1999, pp. 49-66.
37. Haxo, H. E. Jr., "Quality Assurance of Geomembranes Used as Linings for Hazardous Waste Containment," *Journal of Geotextiles and Geomembranes*, vol. 3, no. 4, 1986, pp. 225-248.
38. Schubert, W. R., "Bentonite Matting in Composite Lining Systems," *Proc. Geotechnical Practice for Waste Disposal '87*, R. D. Woods, Ed., Geotech Spec. Publ. 13, ASCE, pp. 784-796.
39. McRory, J. A.and Ashmawy, A.K., "Polymer Treatment of Bentonite Clay for Containment Resistant Barriers," *Proc. GeoFrontiers*, GSP 130 to 142, ASCE, 2005 (on CD).
40. Zanzinger, H., Koerner, R. M. and Touze-Foltz, N., Editors, *Proc. 3rd Intl. Symposium on Geosynthetic Clay Liners*, Wurzburg, Germany, 2010, 373 pgs.
41. Partos, A. M. and Kazaniwsky, P. M., "Geoboard Reduced Lateral Earth Pressure," *Proceedings of Geosynthetics '87*, IFAI, 1987, pp. 628-638.
42. Horvath, J. S., *Proceedings International Geotechnical Symposium on Polystyrene Foam in Below-Ground Applications*, New York, Manhattan College, May 1994.
43. Horvath, J. S., *Geofoam Geosynthetic*, Scarsdale, NY: Horvath Engineering P.C., 1995.
44. Negussey, D., *Properties and Applications of Geofoam*, Society of Plastics Industries, 1997.

PROBLEMS

1.1 This book deals exclusively with geosynthetics, i.e., synthetic materials placed in the ground. What materials would *not* be considered in this category. That is, what would be some *non*synthetic geotextiles, *non*synthetic geomembranes and *non*synthetic composites?

1.2 Regarding the unit prices of geosynthetics given in section 1.1.3, what factors do you think would influence these values?

1.3 Regarding the installation of geosynthetics, what labor group(s) would be involved in installing the following, considering both union and nonunion situations?

a. Geotextiles
b. Geogrids
c. Geonets
d. Geomembranes
e. Geosynthetic Clay Liners
f. Geofoam

1.4 What complications could you see arising in providing quality assurance geomembranes versus geotextiles? (Hint: Consider how many different firms are involved in the manufacture of each of the types mentioned.)

1.5 Name some major corporations that produce the following resins.

a. Polyethylene
b. Polypropylene
c. Polyester
d. Polyvinyl chloride

1.6 Identify the common polymer used to manufacture the following commonly used products.

a. Milk jugs
b. Soft drink bottles
c. Disposable coffee containers
d. Automobile dashboards
e. Automobile battery cases
f. Lightweight plastic canoes

1.7 The PVC curve in figure 1.3 was described as consisting of plasticizer/HCl, resin, carbon black/ash. What percentages of each were in this material?

1.8 Identify the components and their percentages in figure 1.3 of PP, PE, and PET.

1.9 The DSC traces shown in figure 1.5 list LDPE, MDPE, and HDPE. Identify the temperature window at which each type begins to melt, when melting is complete and the difference, i.e., the "melting window."

1.10 Using the OIT data of figure 1.6b, estimate the time for depletion of all the antioxidant in this particular geomembrane formulation at a service temperature of 20°C. Proceed using the following steps:

a. Replot the data on a semilong axis omitting the 115°C data.
b. Extrapolate each of the curves down to "zero" minutes.
c. Plot these values on an logarithmic y-axis against inverse temperature on the x-axis.
d. Determine the average slope of the resulting four points.
e. Extrapolate this slope down to an estimated service life temperature of 20°C to obtain the estimated time for antioxidant depletion.

1.11 Using the coefficients of thermal expansion shown in figure 1.7
a. What change in length is involved in 10 m of PET yarn if there is a 20°C change in temperature below the glass transition temperature (T_g)?
b. Recalculate part (a) for temperature changes above T_g?

1.12 To perform a forensic analysis of a failure involving a geosynthetic product in which the polymeric material itself was suspect, which tests in table 1.5 would you perform if
a. The material is a polypropylene geotextile?
b. The material is a polyester geogrid?
c. The material is a polyethylene geonet?
d. The material is a polyvinyl chloride geomembrane?

1.13 Name five major functions that geosynthetics perform and illustrate them by means of sketches.

1.14 In placing a geotextile beneath railroad ballast, the materials can serve in four different functions simultaneously. Describe and illustrate these functions.

1.15 The first person to describe the general use of geotextiles was Barrett in 1966 [19]. What major function did the geotextile serve? List the various uses he describes.

1.16 Shown on figure 1.10 is that two molecular weight curves are shown to essentially overlap one another. Why is this a significant finding?

1.17 Permeability of geotextiles refers to liquid moving through the voids created by the fibers and yarns that make up the fabric. This permeability is called "Darcian permeability."
a. Give the equation for Darcy's law?

b. Identify the various terms.
c. What are the major variables involved in variations of the *k* value?

1.18 Permeability of geomembranes refers to liquid vapor moving through the amorphous structure of the polymer. This permeability is called "diffusion permeability."
a. Give the equation for Fick's law.
b. Identify the various terms.
c. What are the major variables involved in variations of the *k* value?

1.19 If a hole is created in a geomembrane, the flow through the hole is governed by Bernoulli's Equation:
a. State Bernoulli's equation.
b. Identify the terms.
c. What kind of relationship do you think there would be between the diffusion permeability of an intact geomembrane and the Bernoulli flow through a hole(s) in a geomembrane?

1.20 What is a "corduroy" road, and how does it function?

1.21 Other than geosynthetics, what are some methods for strengthening soils (i.e., adding tensile strength to them)?

1.22 How would you estimate geosynthetic performance in severe climatic conditions such as; (Hint: How do plastics respond to cold and hot temperatures in general?)
a. In arctic conditions?
b. In desert conditions?

1.23 What two commonly used polymers in the manufacture of geosynthetic materials are in the polyolefin family?

1.24 Regarding the molecular structure of the polymers used to make geosynthetics:
a. What are typical lengths of the molecular chains?
b. What is meant by the "backbone" of the molecular chain?
c. Sketch the crystalline and amorphous molecular chains of polyethylene and show how they are linked together.

1.25 The molecular structure of high-molecular-weight polymeric materials has often been described as a "bowl of spaghetti." If this is the case,

a. What would be the length of a high molecular weight polymer like polyester, if the diameter was typical of spaghetti, e.g. 1.5 mm?
b. What happens to the "bowl of spaghetti" as the polymer structure is stressed?
c. What happens if it is stressed too high?

1.26 Degradation of polymeric materials involves "chain scission."
a. Describe this process.
b. What mechanisms can bring it about?

1.27 In general, all polymeric materials are susceptible to UV attack. Considering their chemical structure, why is this the case?

1.28 The usual processing step taken to avoid UV light degradation of polymers is the addition of carbon black. How does this material function? Are there other additives that can be used instead of carbon black such that the resulting geosynthetic is not black in color?

1.29 In the absence of ultraviolet light degradation, what would cause a polymer structure to *age*?

1.30 What are the major causes of degradation of the following polymers considering that they have been *timely* covered?
a. Polyethylene
b. Polypropylene
c. Polyester
d. Plasticized polyvinyl chloride
e. Polyamide

1.31 In considering the manufacturing of geomembranes as described in section 1.6.2, do you think that residual stresses could exist in the as-manufactured sheet? If so, how would you measure the magnitude and orientation?

1.32 Regarding the production cost of geotextiles, rank the following styles on the basis of an equivalent mass per unit area (i.e., consider that they are all the same in terms of g/m^2 weight and the same polymer type).
a. Woven monofilament
b. Woven slit film
c. Nonwoven heat bonded
d. Nonwoven resin bonded
e. Nonwoven needle punched

1.33 Regarding the material cost of geomembranes, rank the relative cost of the following styles on the basis of an equivalent thickness of material.
  a. High-density polyethylene (HDPE)
  b. Linear low-density polyethylene (LLDPE)
  c. Polyvinyl chloride (PVC)
  d. Flexible polypropylene—scrim reinforced (fPP-R)

1.34 Regarding geosynthetic clay liners (or GCLs):
  a. List some advantages and disadvantages over a geomembrane (GM).
  b. List some advantages and disadvantages over a compacted clay liner (CCL).

1.35 Regarding the use of a three-component liner consisting of a geomembrane over a GCL over a CCL, i.e., a GM/GCL/CCL composite:
  a. List some advantages and disadvantages over a GM/CCL composite.
  b. List some advantages and disadvantages over a GM/GCL composite.

1.36 Geofoam is listed in table 1.1 as having the primary function of separation
  a. What other function could it serve?
  b. If an additional primary function were added for geofoam, what would be the test descriptor?

1.37 List what you feel are the most relevant and unique properties of geofoam.

1.38 The term *geospacer* is sometimes referenced. What types of geosynthetic materials would it likely include?

Chapter 2

Designing with Geotextiles

2.0 INTRODUCTION

According to ASTM D4439, a *geotextile* is defined as follows:

> *Geotextile*: A permeable geosynthetic comprised solely of textiles. Geotextiles are used with foundation, soil, rock, earth, or any other geotechnical engineering-related material as an integral part of a human-made product, structure, or system.

The area of geotextiles is a well-established and exciting field, with new uses being developed regularly. As such, there are a number of possible applications and a large number of geotextiles to choose from. The vast majority of geotextiles are made from polypropylene (although some are polyester or polyethylene) polymers—formed into fibers or yarns (the choices being monofilament, multifilament, staple yarn, continuous yarn, slit-film monofilament or, silt film multifilament) and finally manufactured as a woven or nonwoven fabric. When placed in the ground, these textile fabrics are called *geotextiles*. In general, the words *fabric* and *geotextile* are used interchangeably as they are in this book. The following choices of fabric styles are available:

- Woven monofilament
- Woven multifilament
- Woven slit-film monofilament
- Woven slit-film multifilament
- Nonwoven continuous filament heat bonded
- Nonwoven continuous filament needle-punched,

- Nonwoven staple needle-punched
- Nonwoven heat bonded
- Nonwoven resin bonded (rare)
- Other woven or nonwoven combinations
- Knitted (very rare)

A complete description of the methods of manufacturing geotextiles is presented in section 1.3.

Due to the very wide range of applications and the tremendous variety of available geotextiles having widely different properties, the selection of a particular design method, or design philosophy, is a critical decision that must be made before the actual mechanics of the design process are initiated.

2.1 DESIGN METHODS

While many possible design methods or combinations of methods are available to the geotextile designer, the ultimate decision for a particular application usually takes one of three directions: design by cost and availability, design by specification, and design by function.

2.1.1 Design by Cost and Availability

Design by cost and availability is quite simple. The funds available divided by the area to be covered and a maximum available unit price that can be allocated for the geotextile are calculated. The geotextile with the best properties for the primary function intended is then selected within this unit price limit and according to its availability. The method is obviously weak technically but is one that is still sometimes practiced for temporary and/or noncritical applications.

2.1.2 Design by Specification

Design by specification is very common and is used almost exclusively when dealing with public agencies and many private owners as well. In this method several application categories are listed in association with various physical, mechanical, hydraulic, and/or endurance properties. The application areas are usually related to the intended primary function. A specification of this type used by the Pennsylvania Department of Transportation is shown in table 2.1. Subsurface filtration focuses on underdrain systems and related drainage applications. The combined category of separation/stabilization/reinforcement is distinguished by the condition of the soil subgrade, thickness of aggregate base course, and type of vehicular loading. Note the gradual increase in properties from type A to type C. Erosion control refers to geotextiles beneath rock riprap and articulated concrete mattress. The categories are distinguished by the site-specific installation survivability requirements. Sediment control refers to geotextile silt fences and is distinguished by having no geogrid support (type A) and having a geogrid support (type B). While this specification is typical in its format (listing the various common applications against minimum or maximum property values), it is not typical insofar as the numeric values of the various properties. Different agencies have different perspectives as to what properties and numeric values are important, and sometimes as to the test method of obtaining the numeric values.

A federal agency that has formulated a unified approach in the United States is the American Association of State Highway and Transportation Officials (AASHTO). In its M288 geotextile specifications (notes have been omitted), AASHTO provides for three different strength classifications (see table 2.2a). The classifications are essentially a list of strength properties meant to withstand varying degrees of installation survivability stresses. It is the first step in the process:

- *Class 1*: For severe or harsh survivability conditions where there is a greater potential for geotextile damage.
- *Class 2*: For typical survivability conditions; this is the default classification to be used in the absence of site-specific information.
- *Class 3*: For mild survivability conditions where there is little or no potential for geotextile damage.

TABLE 2.1 PENNSYLVANIA DEPARTMENT OF TRANSPORTATION GEOTEXTILE REQUIREMENTS BASED ON APPLICATION AREA[(1)]

Geotextile Properties	Test Method	Subsurface Filtration	Separation	Stabilization	Reinforcement	Erosion Control		Sediment Control	
			Type A	Type B	Type C	Type A	Type B	Type A	Type B
Grab tensile strength, N	ASTM D 4632	700	1200	1500[(5)]	1500[(5)]	900	400	900	400
Grab tensile elongation, %	ASTM D 4632	20 min	50 min	20	20	15-50	15 min	15-50	15 min
Burst strength, kPa	ASTM D 3786	1300	3000	-	-	2200	960	2200	960
Puncture, N (8.0 mm flat-end rod)	ASTM D 4833	250	450	620	900	360	180	360	180
Trapezoidal tear strength, N	ASTM D 4533	250	450	-	-	220	130	220	130
Apparent opening size (AOS) Sieve No.	ASTM D 4751	(2), (3)	(2), (3)	> No. 30	> No. 30	(2), (3)	(2), (3)	No. 20 max	No. 20 max
Permeability, K, cm/sec	ASTM D 4491	K fabric ≥ 10K soil[(3)]	K fabric ≥ 10K soil[(3)]	-	-	K fabric ≥ 10K soil[(3)]	K fabric ≥ 10K soil[(3)]	K fabric ≥ 10K soil[(3)]	K fabric ≥ 10K soil[(3)]
Permittivity, sec^{-1}	ASTM D 4491	0.2	-	-	-	-	-	0.01	0.01
Seam strength, N[(4)]	ASTM D 4632	310	1070	1600	2000	800	360	-	-
Ultraviolet resistance strength retention, %	ASTM D 4355	70 @ 150 hrs	70 @ 150 hrs	70 @ 150 hrs	70 @ 150 hrs	70 @ 150 hrs	70 @ 150 hrs	70 @ 150 hrs	70 @ 150 hrs

(1) The numerical values indicate minimum average roll value, or minimum to maximum range.
(2) Soil with 50% or less particles by weight passing No. 200 sieve, AOS ≥ No. 30 sieve.
Soil with more than 50% particles by weight passing No. 200 sieve, AOS > No. 50 sieve
(3) Design specified.
(4) Applies to both field and/or manufactured seams.
(5) Minimum grab tensile strength for the warp and fill direction at maximum elongation.

TABLE 2.2a AASHTO M288 GEOTEXTILE STRENGTH PROPERTY REQUIREMENTS

Property	Test Methods	Units	Geotextile Classification					
			Class 1		Class 2		Class 3	
			Elongation < 50 % i.e., woven	Elongation ≥ 50 % i.e., nonwoven	Elongation < 50 % i.e., woven	Elongation ≥ 50 % i.e., nonwoven	Elongation < 50 % i.e., woven	Elongation ≥ 50 % i.e., nonwoven
Grab Strength	ASTM D4632	N	1400	900	1100	700	800	500
Sewn Seam Strength	ASTM D4632	N	1260	810	990	630	720	450
Tear Strength	ASTM D4533	N	500	350	400	250	300	180
Puncture Strength	ASTM D4833	N	500	350	400	250	300	180
Permittivity	ASTM D4491	sec^{-1}	Minimum property requirements for permittivity, AOS and UV stability are based on geotextile application. Refer to Table 2.2b for subsurface filtration, Table 2.2c for separation, Table 2.2d for stabilization, or Table 2.2e for permanent erosion control.					
Apparent Opening Size	ASTM D4751	mm						
Ultraviolet Stability	ASTM D4355	%						

The second step is to select one of six different tables according to the specific application. These applications follow the intended primary function:

Table 2.2b: Filtration applications as in highway underdrains.
Table 2.2c: Separation when placed on firm strength subgrades.
Table 2.2d: Stabilization when placed on moderate strength subgrades.
Table 2.2e: Erosion control—for example, geotextiles serving as a filter beneath rock riprap.
Table 2.2f: Temporary silt fences for sediment control.
Table 2.2g: Geotextiles used for the prevention of reflective cracking in flexible pavement overlays.

TABLE 2.2b AASHTO M288 SUBSURFACE FILTRATION (CALLED "DRAINAGE" IN THE ACTUAL SPECIFICATION) GEOTEXTILE REQUIREMENTS

Property	Test Methods	Units	Requirements		
			Percent In-Situ Soil Passing 0.075 mm		
			< 15	15 to 50	> 50
Geotextile Class			Class 2 from Table 2.2a		
Permittivity	ASTM D4491	sec^{-1}	0.5	0.2	0.1
Apparent Opening Size	ASTM D4751	mm	0.43 max. avg. roll value	0.25 max. avg. roll value	0.22 max. avg. roll value
Ultraviolet Stability (Retained Strength)	ASTM D4355	%	50% after 500 hrs. of exposure		

TABLE 2.2c AASHTO M288 SEPARATION GEOTEXTILE PROPERTY REQUIREMENTS

Property	Test Methods	Units	Requirements
Geotextile Class			Class 2 from Table 2.2a
Permittivity	ASTM D4491	sec^{-1}	0.02
Apparent Opening Size	ASTM D4751	mm	0.60 max. avg. roll value
Ultraviolet Stability (Retained Strength)	ASTM D4355	%	50% after 500 hrs. of exposure

TABLE 2.2d AASHTO M288 STABILIZATION GEOTEXTILE PROPERTY REQUIREMENTS

Property	Test Methods	Units	Requirements
Geotextile Class			Class 1 from Table 2.2a
Permittivity	ASTM D4491	sec^{-1}	0.05
Apparent Opening Size	ASTM D4751	mm	0.43 max. avg. roll value
Ultraviolet Stability (Retained Strength)	ASTM D4355	%	50% after 500 hrs. of exposure

TABLE 2.2e AASHTO M288 PERMANENT EROSION CONTROL GEOTEXTILE REQUIREMENTS

Property	Test Methods	Units	Requirements		
			Percent In-Situ Soil Passing .075 mm		
			< 15	15 to 50	> 50
Geotextile Class Woven Monofilament Geotextiles All Other Geotextiles			Class 2 from Table 2.2a Class 1 from Table 2.2a		
Permittivity	ASTM D4491	sec^{-1}	0.7	0.2	0.1
Apparent Opening Size	ASTM D4751	mm	0.43 max. avg. roll value	0.25 max. avg. roll value	0.22 max. avg. roll value
Ultraviolet Stability (Retained Strength)	ASTM D4355	%	50% after 500 hrs. of exposure		

TABLE 2.2f AASHTO M288 TEMPORARY SILT FENCE PROPERTY REQUIREMENTS

Property	Test Methods	Units	Requirements		
			Supported Silt Fence	Unsupported Silt Fence	
				Elongation ≥ 50% i.e., nonwoven	Elongation < 50 i.e., woven
Maximum Post Spacing			1.2 m	1.2 m	2.0 m
Grab Strength	ASTM D4632	N			
Machine Direction			400	550	550
X-Machine Direction			400	450	450
Permittivity	ASTM D4491	sec^{-1}	0.05	0.05	0.05
Apparent Opening Size	ASTM D4751	mm	0.60 max. ave. roll value	0.60 max. avg. roll value	0.60 max.avg. roll value
Ultraviolet Stability (Retained Strength)	ASTM D4355	%	70% after 500 hrs of exposure	70% after 500 hrs of exposure	

TABLE 2.2g AASHTO M288 PREVENTION OF REFLECTIVE CRACKING, I.E., PAVING FABRICS, PROPERTY REQUIREMENTS

Property	Test Method	Units	Requirements
Grab Strength	ASTM D4632	N	450
Mass Per Unit Area	ASTM D5261	gm/m^2	140
Ultimate Elongation	ASTM D4632	%	≥ 50
Asphalt Retention	ASTM D6140	l/m^2	see M288
Melting Point	ASTM D276	deg. C	150

The following example illustrates the use of these AASHTO tables; the individual properties will be described later.

Example 2.1

Using the AASHTO M288 Specification of table 2.2, determine what geotextile properties (to be described in detail in section 2.3) are needed for the following applications:

(a) A nonwoven geotextile (ε > 50%) pavement underdrain filter adjacent to soil with 60% passing the 0.075 mm sieve and under typical installation survivability conditions.
(b) A woven geotextile (ε < 50%) pavement separator between firm soil subgrade and stone base course, under harsh survivability conditions according to the design engineer.

Solution: Tables 2.2b and 2.2g are used for the appropriate application properties and then table 2.2a is used for the required strength properties.

(a) First from table 2.2b and then from table 2.2a the required properties for the nonwoven geotextile filtration fabric are as follows:
- Permittivity ≥ 0.1 sec^{-1}
- AOS ≤ 0.22 mm
- Grab strength ≥ 700 N
- Sewn seam strength ≥ 630 N
- Tear strength ≥ 250 N
- Puncture strength ≥ 250 N
- UV stability ≥ 50% of 700 N, i.e., ≥ 350 N after 500 hrs.

(b) First from table 2.2c and then from table 2.2a the required properties for the woven geotextile separation fabric are as follows:
- Permittivity ≥ 0.02 sec^{-1}
- AOS ≤ 0.60 mm
- Grab strength ≥ 1100 N
- Sewn seam strength ≥ 990 N
- Tear strength ≥ 400 N
- Puncture strength ≥ 400 N
- UV stability ≥ 50% of 1100 N, i.e., ≥ 550 N after 500 hrs.

It must be cautioned that when using a design-by-specification method, the specifications sometime list *minimum* required fabric properties, whereas some manufacturers' literature may list either *average-lot* or *minimum-average-roll* property values. (In this regard, the word *lot* is defined as any unit of production taken for sampling or statistical testing, having one or more common properties and being readily separable from other similar units. Thus, a lot can be as large as an entire production run, or as small as a few rolls of fabric for a specific project. The point is that a lot is arbitrary and must be agreed on by the parties involved.) By comparing such a specification value to the manufacturer's listed values, you may be comparing different sets of numbers. This is so because average lot value is the mean value for the particular property in question from all the tests made on that lot of fabrics. This may be the compilation of thousands of tests made over many months or even years of production of that particular geotextile style. Thus, the average lot value is considerably higher than the minimum value (see figure 2.1). An intermediate value between these two extremes is the *m*inimum *a*verage *r*oll *v*alue (MARV)—the average of a representative number of tests made on selected rolls of the lot in question, which is limited in area to the particular site in question. This value is numerically equivalent to two standard deviations lower than the mean, or average, lot value. Thus, we can see that MARV is the minimum of a limited series of average roll values. These different values are shown in figure 2.1.

Note that in a true statistical sense, about 16% of all values will be lower than $\overline{X} - S$; 2.5% will be lower than $\overline{X} - 2S$; and 0.15% will be lower than $\overline{X} - 3S$, where

$\overline{X}$ = mean value, and

S = standard deviation.

Furthermore, the minimum average roll value (MARV), with 2.5% of the values falling below $\overline{X} - 2S$ is also called the 95% confidence level. (The other 2.5% is above $\overline{X} + 2S$, and is obviously not of concern since these values are well in excess of that required). One other consideration should be mentioned: the case where one is targeting a maximum value, such as a maximum AOS opening size value. Here you are considering the right side of the curve of figure 2.1 and the comparable value to MARV would logically become MaxARV.

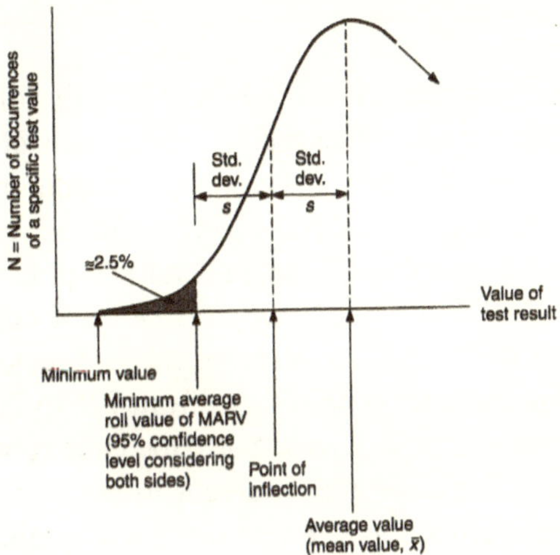

Figure 2.1 Relative relationships of different statistical values used in geotextile specifications and manufacturers' literature.

The mean value ($\overline{X}$) is calculated using $\Sigma X_i / N$ and the standard deviation using

$$S = \left[\frac{(X_1 - \overline{X})^2 + (X_2 - \overline{X})^2 + \ldots + (X_N - \overline{X})^2}{N-1} \right]^{1/2} \tag{2.1}$$

where

$\overline{X}$ = mean value
X_i = measured value, and
N = number of measurements.

These are, of course, standard statistical definitions. Still further, the coefficient of variation V, or simply "variance," is calculated using $(S/\overline{X})$ (100). The variance should be as low as possible, thereby indicating good-quality control during manufacturing. Most agencies, including AASHTO in its M288 specifications presented in tables 2.2 a to g, recommend the use of MARV or MaxARV for use in both the specification and the listing of manufacturers' product data. Example 2.2 illustrates the meaning of MARV insofar as field conformance testing is concerned.

Example 2.2

Consider a field construction site where 150 rolls of geotextile are delivered. This value then defines the *lot*. The quality control or quality assurance inspector would *sample*[*] a representative number of these rolls to determine MARV and see that it conforms with the value called for in the specification. Assume that the targeted value is grab tensile strength. A geotextile sample, full roll width and 1.0 m long, is cut from each of 6 randomly selected rolls in the lot. Note that according to ASTM D4354 on sampling technique, a lot consisting of from 126 to 216 rolls requires at least 6 rolls to be sampled. It is actually the cube root of the number of rolls in the lot. These samples are sent to an approved laboratory for testing. Within each sample, 8 test *specimens* are taken and tested according to ASTM D4632, the grab tensile test method. Given the test data shown in the table below, determine the MARV.

Solution: Assume that each of the six samples were cut into eight individual grab tensile specimens, were properly tested, and resulted in the following data set in units of newtons (*N*) at failure.

Specimen Test Number	Sample Number					
	1	2	3	4	5	6
1	643 N	627 N	637 N	642 N	652 N	637 N
2	627	615	643	646	641	624
3	652	621	628	658	639	631
4	629	616	662	641	657	620
5	632	619	646	635	642	618
6	641	621	633	642	651	633
7	662	622	619	658	641	641
8	635	628	636	662	645	625
Average =	640	621	638	648	646	629

* Throughout this book, a roll of geosynthetics is *sampled* by cutting a piece or swatch from it. This sample is then taken to a laboratory from which *specimens* are cut to exact size for subsequent testing according to a particular test protocol. In some cases, the sample will be cut into sections and incubated in an oven, or in liquid, or under light exposure, etc. It is then called a *coupon*, which is subsequently cut into specimens for actual testing purposes. Thus, the order of size hierarchy for all geosynthetics is a lot, roll, sample, coupon (sometimes), and specimen.

From this data set, it is seen that MARV is 621 N, which must equal, or exceed, the MARV value of grab-tensile strength required by the specification. Note that there are 6 individual test values in the entire data set which are numerically less than 621. These represent the statistical 2.5% of the values less than MARV as illustrated by the shaded portion in figure 2.1.

In summary, the design-by-specification method must compare like sets of numbers. If the intent of the specification is to list minimum average roll values (as it is with tables 2.1 and 2.2a-g), then manufacturer's listed mean or average values must be decreased by two standard deviations (approximately 5 to 20%) if average lot values are given. Only if minimum average roll values (MARV) are given by the manufacturer can they be directly compared to a MARV-based specification value on a like-set-of-number basis.

In closing, it is hoped that the current trend of both specifications and manufacturer's literature come together with a common unit that centers on MARV or, in a few cases, MaxARV. It is a concept that every one can live with, a value that can be field-verified, and a number that reflects the inherent variation in quality control of the manufacture of geotextiles. It is important to mention that only geotextiles (of all the geosynthetic materials) use the MARV concept. Other geosynthetics with smaller statistical variation work on an average or minimum average basis.

2.1.3 Design by Function

Design by function consists of assessing the primary function that the geotextile will be asked to serve and then calculating the required numerical value of a particular property for that function. By dividing this value into the candidate geotextile's allowable property value, a factor of safety (FS) will result.

$$Factor\ of\ Safety(FS) = \frac{\text{allowable (test) property}}{\text{required (design) property}} \tag{2.2a}$$

where

allowable property = a numeric value based on a laboratory test that models the actual situation or is adjusted accordingly,
required property = a numeric value obtained from a design method that models the actual situation, and
factor of safety (FS) = FS against unknown loads and/or uncertainties in the analytic or testing process; sometimes called a *global factor of safety.*

If the factor of safety is sufficiently greater than 1.0, the candidate geotextile is acceptable. The above process can be repeated for a number of available geotextiles; if others are acceptable, then the final choice becomes one of availability and least cost. The individual steps in this process are as follows:

1. Assess the particular application, considering not only the candidate geotextile but the material system on both sides of it.
2. Depending on the criticality of the situation (i.e., "If it fails, what are the consequences?"), decide on a minimum factor of safety. This value may be selected by the designer or imposed through regulations.
3. Decide on the geotextile's primary function.
4. Calculate numerically the required geotextile property value in question on the basis of its primary function.
5. Test for, or otherwise obtain, the candidate geotextile's allowable value of this particular property (recall the discussion in section 2.1.2 on the recommended use of MARV values).
6. Calculate the factor of safety on the basis of the allowable property (step 5) divided by required property (step 4) per equation 2.2a).
7. Compare this factor of safety to the required value decided on in step 2.
8. If not acceptable, repeat the process with a geotextile with more appropriate properties.
9. If it is then acceptable, determine whether any secondary function of the geotextile is more critical.

10. Repeat the process for other available geotextiles; if more than one satisfy the factor of safety requirement, select the geotextile on the basis of least cost and availability.

Note that the design-by-function process can also be used to solve for the required property value:

$$\text{required (design) property} = \frac{\text{allowable (test) property}}{\text{factor of safety}} \tag{2.2b}$$

Both calculation procedures will be illustrated later.

The design-by-function approach is the one that will be used throughout this book. This method obviously necessitates identifying the primary function that the geotextile is to serve; thus this chapter (and the subsequent chapters) has been laid out accordingly. A brief treatment of the major functions that a geotextile can serve is given in the next section.

2.2 GEOTEXTILE FUNCTIONS AND MECHANISMS

Section 1.3.3, which provided an overview of geotextiles, alluded to the many applications falling into categories vis-à-vis their major function (recall table 1.8). These categories—separation, reinforcement, filtration, drainage and containment—when properly identified, lead to design-by-function method. The purpose of this section is to demonstrate technically what these functions mean with respect to geotextiles and to elaborate on the actual mechanisms embodied within each type of function.

2.2.1 Separation

The concept of separation can perhaps be illustrated by the engineering adage, "10 kilograms of stone placed on 10 kilograms of mud results in 20 kilograms of mud." With this in mind, a geotextile serving in a separation function can be defined is as follows:

> *Geotextile separation:* The placement of a flexible porous textile between dissimilar materials so that the

integrity and functioning of both materials can remain intact or be improved.

When placing stone aggregate on fine-grained soils, there are two simultaneous mechanisms that tend to occur over time. One is that the fine soils attempt to enter into the voids of the stone aggregate, thereby ruining its drainage capability; the other is that the stone aggregate attempts to intrude into the fine soil, thereby ruining the stone aggregate's strength. When this occurs we have a situation that has been called *sacrificial aggregate*, which is all too often the case without the use of a proper separating geotextile. The two mechanisms are shown schematically in figure 2.2.

2.2.2 Reinforcement

Because geotextiles are materials possessing tensile strength, they can nicely complement those materials good in compression but weak in tension. Thus, low-strength, fine-grained silt and clay soils are prime targets for geotextile reinforcement. The following definition will clarify this point:

> *Geotextile reinforcement:* The synergistic improvement of a total system's strength created by the introduction of a geotextile (that is good in tension) into a soil (that is good in compression but poor in tension) or other disjointed and separated material.

Improvement in strength can be evaluated in a number of ways. The triaxial tests conducted by Broms [1] illustrate the beneficial effects of a geotextile when properly placed. Figure 2.3 shows two sets of triaxial tests on dense sand samples at confining pressures of 21 kPa and 210 kPa for different soil and geotextile configurations. In both parts, curves 1 represent the baseline shear strength data of the sand by itself; curves 2 have geotextiles on the extreme top and bottom of the soil and do not show improved shear strength behavior. Since these locations of the geotextiles are in the nonacting dead zones in conventional triaxial tests, this behavior is both logical and instructive. It is instructive because it is teaching us that if the geotextile is placed at the wrong location, it will have no beneficial effect. Upon placing

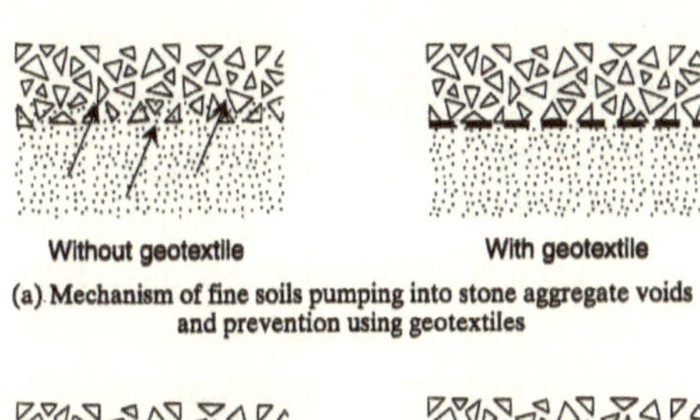

(a) Mechanism of fine soils pumping into stone aggregate voids and prevention using geotextiles

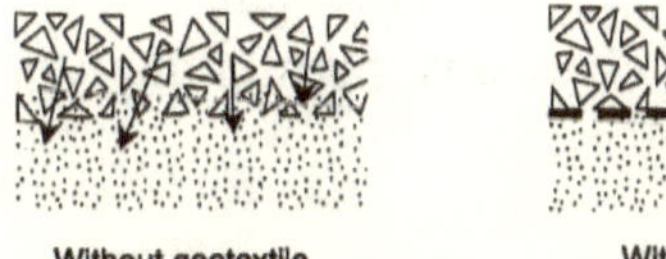

(b) Mechanism of stone aggregate intrusion into fine soil subgrade and prevention using geotextiles

Figure 2.2 Different physical mechanisms in the use of geotextiles involved in the separation function.

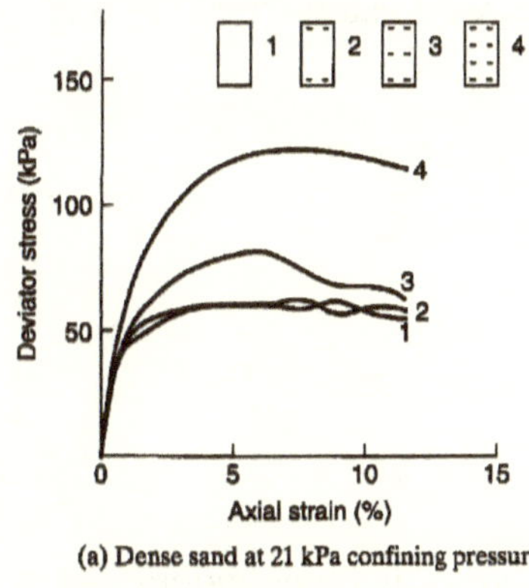

(a) Dense sand at 21 kPa confining pressure

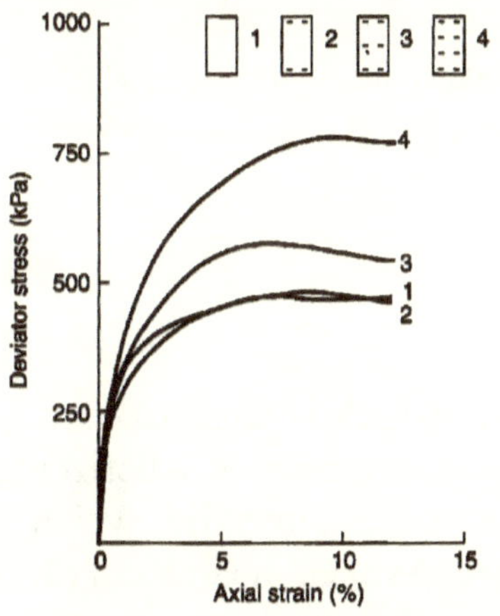

(b) Dense sand at 210 kPa confining pressure

Figure 2.3 Triaxial test results showing influence of geotextiles placed at various locations within soil specimen (After Broms [1]).

the geotextile in the center of the sample, as with curves 3, or at the one-third points, as with curves 4, however, beneficial effects are easily seen. Here the geotextile interrupts potential shear planes and has the influence of increasing the overall shear strength of the now-reinforced soil. As expected, the double layers placed at the one-third points (curves 4) are more beneficial than the single layer placed at the center of the sample (curves 3).

Within the general function of geotextile reinforcement of soils, there are three different reinforcement mechanisms: membrane type, shear type, and anchorage type.

Membrane Type. Membrane reinforcement occurs when a vertical force is applied to a geotextile that has been placed on a deformable subgrade. Depending on the depth at which the geotextile is placed from the force application, it is well established [2] that

$$\sigma_h = \frac{P}{2\pi z^2}\left[3\sin^2\theta\cos^3\theta - \frac{(1-2\mu)\cos^2\theta}{1+\cos\theta}\right] \tag{2.3}$$

where

σ_h = horizontal stress at depth z and angle θ,
P = applied vertical force,
z = depth beneath surface where σ_h is being calculated,
μ = Poisson's ratio, and
θ = angle from the vertical beneath the surface load P.

Note that directly beneath the load, where $\theta = 0$ deg.,

$$\sigma_h = -\frac{P}{2\pi z^2}\left(\frac{1}{2}-\mu\right) \tag{2.4}$$

Since μ is less than 0.5, σ_h is negative (which is tension); that is, the applied vertical downward force produces tension on a horizontal plane beneath it. Thus, tension results in the geotextile, which is precisely the objective of placing it there. As seen in equation (2.4), the larger the magnitude of P, the higher the tensile stress and the higher requirement of tensile strength of the geotextile. Also, the closer the geotextile is to the force (i.e., low values of z), the higher will be the applied stress on the geotextile. Many situations in which geotextiles are placed on soft soils or in a yielding situation use this particular reinforcement mechanism. When geotextiles are placed in closely spaced layers, as in walls and slopes, the situation is more complex but the principle is the same.

Shear Type. Shear reinforcement was illustrated by the triaxial tests of figure 2.3 but can be better visualized by means of direct shear tests. Here a geotextile placed on a soil is loaded in a normal direction, and then the two materials are sheared at their interface. The resulting geotextile-to-soil shear strength parameters (adhesion and friction angle) can be obtained as described in a traditional geotechnical manner using an adapted form of the Mohr-Coulomb failure criterion,

$$\tau = c_a + \sigma'_n \tan \delta \tag{2.5a}$$

where

τ = shear strength between the geotextile and soil,
σ'_n = effective normal stress on the shear plane,
c_a = adhesion of the geotextile to the soil (this is the complement to cohesion as occurs within clay soils), and
δ = friction angle between the geotextile and soil.

The shear strength parameters c_a and δ can be compared to the shear strength parameters of the soil by itself (i.e., soil against soil) as follows:

$$\tau = c + \sigma'_n \tan \varphi \tag{2.5b}$$

where

c = cohesion of soil-to-soil,
φ = friction angle of soil to soil.

Furthermore,

$$E_c = (c_d/c)\ 100 \quad (2.6)$$
$$E_\varphi = (\tan \delta/\tan \varphi)\ 100 \quad (2.7)$$

where

E_c = efficiency of cohesion mobilization, and
E_φ = efficiency of soil friction angle mobilization.

These efficiency ratios have limiting values of zero to unity. While a numeric value higher than unity is possible, such values cannot be mobilized since the failure plane would simply move into the soil itself and the situation reverts from equations 2.5a to 2.5b.

Anchorage Type. Anchorage reinforcement is similar to the shear type just described, but now the soil acts on both sides of the geotextile as a tensile force tends to pull the geotextile out of the soil. The laboratory modeling of this type of mechanism is similar to direct shear except that now the upper and lower soil is stationary in both halves of the test device and the geotextile extends out of the device at its center. It is gripped externally and pulled, while normal compressive stresses act on the soil and geotextile within the test box setup. The situation is readily described in terms of shear strength parameters by themselves and efficiencies as just discussed. Another approach could be to express the efficiency as a function of the amount of mobilized geotextile strength. Wide-width tensile values should be used in this case. Here anchorage efficiencies greater than unity can occur but are usually limited by the tensile strength of the geotextile. As with the other types of mechanisms of geotextile reinforcement, this category of geotextile anchorage is used quite often. The applications mentioned in section 1.3.3 illustrate the point.

For calculations, we will use the shear strength mobilized by the geotextile with the soil above and with the soil below and arithmetically sum the two values as the limiting anchorage value. In the absence of anchorage tests, we will use direct shear generated values for this purpose.

2.2.3 Filtration

The geotextile function of filtration involves the movement of liquid through the geotextile itself—i.e., across its manufactured plane. At the same time, the geotextile most serve the purpose of retaining the soil on its upstream side. Both adequate permeability requiring an open fabric structure and soil retention requiring a tight fabric structure are required simultaneously. A third factor is also involved: a long-term soil-to-geotextile flow compatibility that will not excessively clog the fabric during the lifetime of the system. Thus, geotextile filtration can be defined as follows:

> *Geotextile filtration:* The equilibrium soil-to-geotextile system that allows for adequate liquid flow with limited soil loss across the plane of the geotextile over a service lifetime compatible with the application under consideration.

This function of filtration is a major one for the geotextile industry, recall the application areas presented in section 1.3.3. Geotextiles, when properly designed and constructed, offer a practical remedy to many problems involving the flow of liquids.

Permeability. This particular discussion of geotextile permeability refers to cross-plane permeability when liquid flow is perpendicular to the plane of the fabric. Some of the geotextiles used for this purpose are relatively thick and compressible. For this reason the thickness is associated with the permeability coefficient and is used as *permittivity*, which is defined as follows:

$$\psi = \frac{k_n}{t} \tag{2.8}$$

where

ψ = permittivity,
k_n = cross-plane permeability coefficient (the subscript n is often omitted), and
t = thickness at a specified normal pressure.

The testing for geotextile permittivity follows similar lines as used for testing soil permeability. It should be noted that some designers prefer to work directly with permeability and require the geotextile's permeability to be some multiple of the adjacent soil's permeability—e.g., 0.1, 1.0 or 10.0 (see Christopher and Fisher [3]).

Soil Retention. For the required flow of liquid to be allowed through the geotextile, the void spaces in it must be sufficiently large. There is, however, a limit—that being when the upstream soil particles start to pass through the geotextile voids along with the flowing liquid. This can lead to an unacceptable situation called *soil piping*, in which soil particles are carried through the geotextile, leaving unstable soil voids behind. The velocity of the liquid then increases, accelerating the process, until the upstream soil structure begins to collapse. This collapse often leads to small sinkhole-type patterns that grow larger with time.

This process is prevented by making the geotextile voids tight enough to retain the soil on the upstream side of the fabric. It is the coarser soil fraction that must be initially retained and that is the targeted soil size in the design process. These coarser-sized particles eventually block the finer-sized particles from moving and build up a stable upstream soil structure. In a sense, the geotextile is acting as a catalyst to make the upstream soil do its own filtration. Fortunately, filtration concepts are well established in the design of soil filters, and those same ideas will be used to design an adequate geotextile filter.

There are many formulas that can be applied to soil-retention design, most of which use the soil particle size characteristics and compare them to the 95% opening size of the geotextile, which is defined as the O_{95} value. The test method used in the United States to determine this value is called the *apparent opening size* (AOS) and is obtained using a dry-sieving method. In Europe and Canada, the test method is called *filtration opening size* (FOS) and is accomplished by wet or hydrodynamic sieving. Both of these latter methods are preferable to the dry-sieving method used in the United States, but there seems to be a reluctance to change.

The simplest of the design procedures examines the percentage of soil passing the no. 200 sieve, whose openings are 0.074 mm. According to AASHTO [4], the following is recommended:

- For soil with ≤ 50% passing the no. 200 sieve: $O_{95} < 0.60$ mm—i.e., AOS of the fabric ≥ no. 30 sieve.
- For soil > 50% passing the no. 200 sieve: $O_{95} < 0.30$ mm—i.e., AOS of the fabric ≥ no. 50 sieve

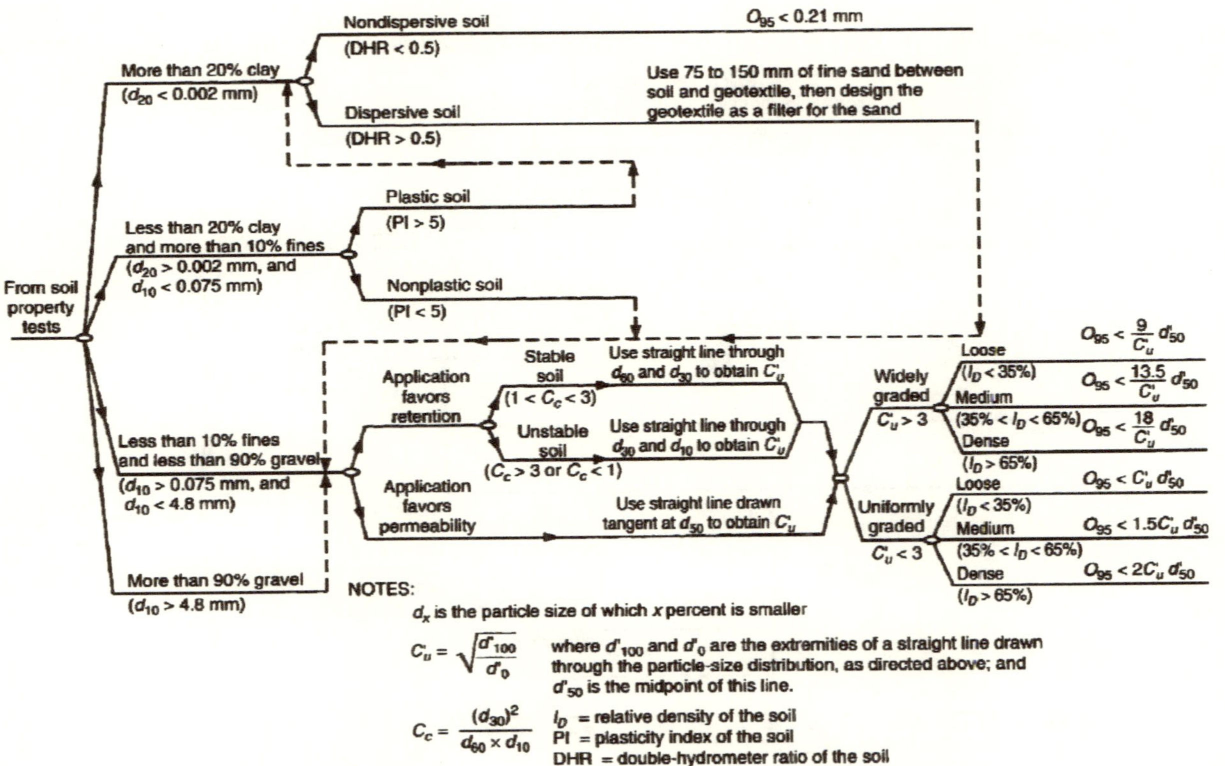

Figure 2.4a Soil retention criteria for geotextile filter design using steady-state flow conditions (After Luettich, et al. [6]).

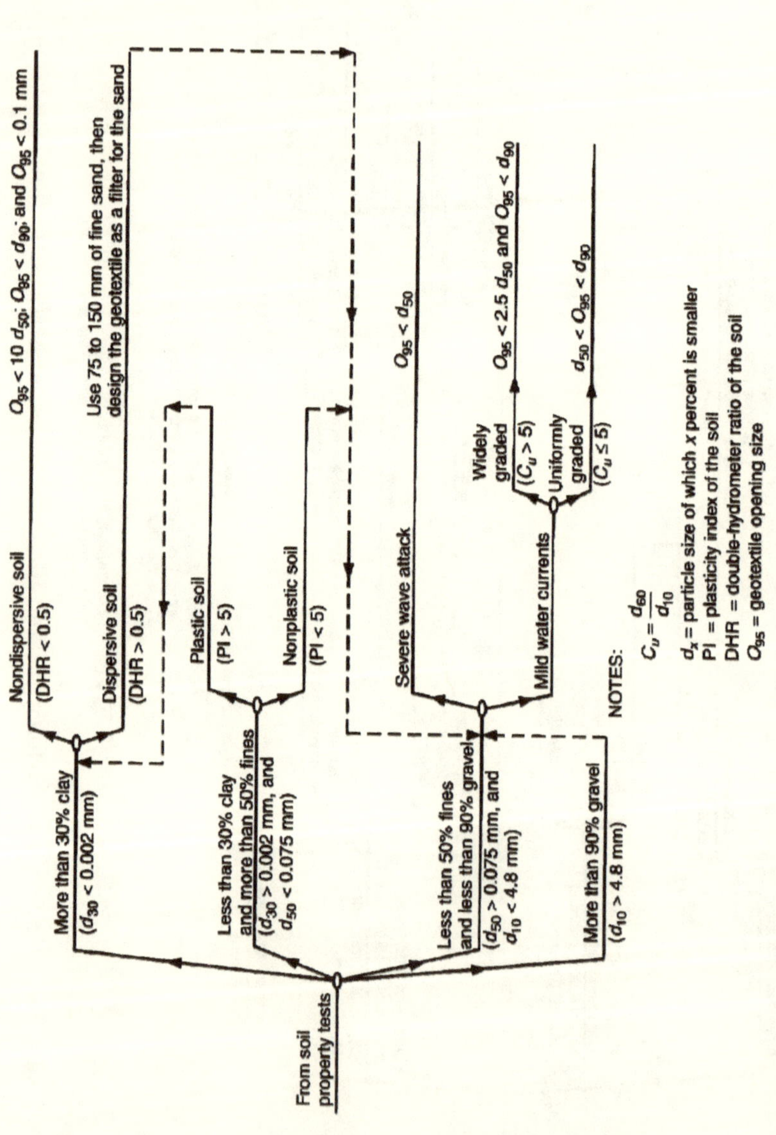

Figure 2.4b Soil retention criteria for geotextile filter design under dynamic flow conditions. (After Luettich, et al. [6])

To extend this further, a series of direct comparisons of geotextile-opening size (O_{95}, O_{50} or O_{15}) was made in ratio form to some soil particle size to be retained (d_{90}, d_{85}, d_{50} or d_{15}; see Christopher and Fischer [3]). The numeric value of the ratio depends on the geotextile type, the soil type, the flow regime, etc. For example, Carroll [5] recommends the following:

$$O_{95} < (2 \text{ or } 3)\, d_{85} \tag{2.9}$$

where d_{85} is the soil particle size in mm, for which 85% of the total soil is finer.

In contrast to the simplified methods above, a more comprehensive approach toward soil-retention criteria is given in figure 2.4a and b, for steady-state and dynamic flow conditions respectively (Luettich et al. [6]). To utilize the figures, we must first characterize the upstream soil. A grain-size distribution, along with Atterberg limits and dispersivity characteristics for the fine fraction, is necessary. Two examples illustrate the use of each of the figures.

Example 2.3

What is the appropriate formula to obtain the required O_{95} of a geotextile filter under steady-state flow conditions if the upstream soil is 25% less than 0.002 mm and the fine fraction is nondispersive?

Solution: From figure 2.4a, we see that $O_{95} < 0.21$ mm, which is equivalent to a no. 70 sieve or tighter.

Example 2.4

What is the appropriate formula to obtain the required O_{95} of a geotextile filter under dynamic flow conditions if the upstream soil is less than 50% fines and less than 90% gravel, and the situation is one of severe wave attack.

Solution: From figure 2.4b, we see that $O_{95} < d_{50}$, where d_{50} is the median particle size of the upstream soil

Long-Term Flow Compatibility. Perhaps the most frequently asked question regarding the use of geotextiles in hydraulic-related systems is, "Will it eventually clog?" Obviously, some soil particles will embed themselves on, or within, the geotextile structure and an understandable reduction in the as-manufactured permeability or permittivity will almost always occur. This type of *partial clogging* is expected. A more perceptive question asks if the geotextile will *excessively clog*, such that the flow of liquid through it will be decreased to the point where the system will not adequately perform its function. There are guidelines available for noncritical, nonsevere cases [3], but the question can be answered directly by taking a soil sample and the candidate geotextile(s) and testing them in the laboratory. We can perform one of three techniques: the gradient ratio (GR) test [7] such that the $GR \leq 3.0$, the long-term flow (LTF) test [8] such that the terminal slope of the flow rate versus time curve is adequate for site-specific conditions, or the hydraulic conductivity ratio (HCR) test [9], such that resulting HCR values are between 0.7 and 0.3. These tests will be described later.

A different approach to the answer of the excessive clogging question is simply to avoid situations that have been known to lead to such problems. Experience shows that the following conditions give rise to concerns about geotextile filter applications; Koerner and Koerner [10].

- Poorly graded (i.e., all-uniform size) fine, cohesionless, soils such as loess, rock flour and stone quarry fines.
- Cohesionless soils consisting of gap-graded particle size distributions and functioning under high hydraulic gradients.
- Dispersive clays which separate into individual fine particles over time. (Note, however, that nondispersive clays possessing true cohesion are generally not troublesome due to these same cohesive forces keeping the upstream soil structure intact.)
- High-suspended solids in the permeating liquid, as found in turbid river water or dredged water that can build up on, or within the geotextile.
- High-suspended solids coupled with high-microorganisms content, as in landfill leachates and agricultural wastes, that combine to build up on or within the geotextile.
- High-alkalinity groundwater where the slowing of the liquid as it flows through the geotextile can cause a calcium, sodium, or magnesium precipitate to be deposited.

- Atypical fluids like oily wastes and sludges of high viscosity.
- Excessive downstream coverage of the geotextile filter by paving blocks, concrete amoring, or soil-cement protection layers.

For most of these cases one could use a very open geotextile and allow for fine particles, sediment, microorganisms, or viscous fluids to pass through into the downstream drain. Recognize, however, that whatever the downstream drain is (gravel, drainage core, perforated pipe, etc.), it must be designed to adequately accept and transport such particulate or viscous matter without itself excessively clogging.

This discussion of soil-to-geotextile compatibility assumes the establishment of a set of mechanisms that are in equilibrium with the flow regime being imposed on the system. Numerous attempts at insight into these phenomena have been attempted (see McGown [11], Heerten [12], and Giroud [13]). There exists a number of possibilities, including upstream soil filter formation, blocking, arching, partial clogging, and depth filtration. These are shown schematically in figure 2.5. With respect to how these mechanisms interact, it has been mentioned that the geotextile serves as a catalyst to promote the upstream soil and the now-soil-modified geotextile likely to generate its own internal filter system. Obviously, a number of phenomena are working together simultaneously, and just what mechanism dominates under what conditions of soil type, geotextile type, and flow regime (reversing flow conditions are most troublesome; see Maiser and Myles [14]) is still an issue that deserves further investigation.

2.2.4 Drainage

Fabrics placed in such a way as to transmit liquid within the plane of their structure provide a drainage function. Thus, a definition of drainage is as follows:

> *Geotextile drainage:* The equilibrium soil-to-geotextile system that allows for adequate liquid flow with limited soil loss within the plane of the geotextile over a service lifetime compatible with the application under consideration.

All geotextiles can provide such a function but to widely varying degrees [15]. For example, thin woven geotextiles, by virtue of their fibers crossing over and under one another, can transmit liquid within the spaces created at these crossover points, but to an extremely low degree. Conversely, thick, nonwoven needle-punched geotextiles have considerable void space in their structure, and this space is available for liquid transmission. Furthermore (to preview the discussion in chapters 4 and 8), geonets and drainage geocomposites can transmit much more liquid than can geotextiles—even thick, bulky ones. Obviously, proper design will dictate just what type of geosynthetic drainage material is necessary.

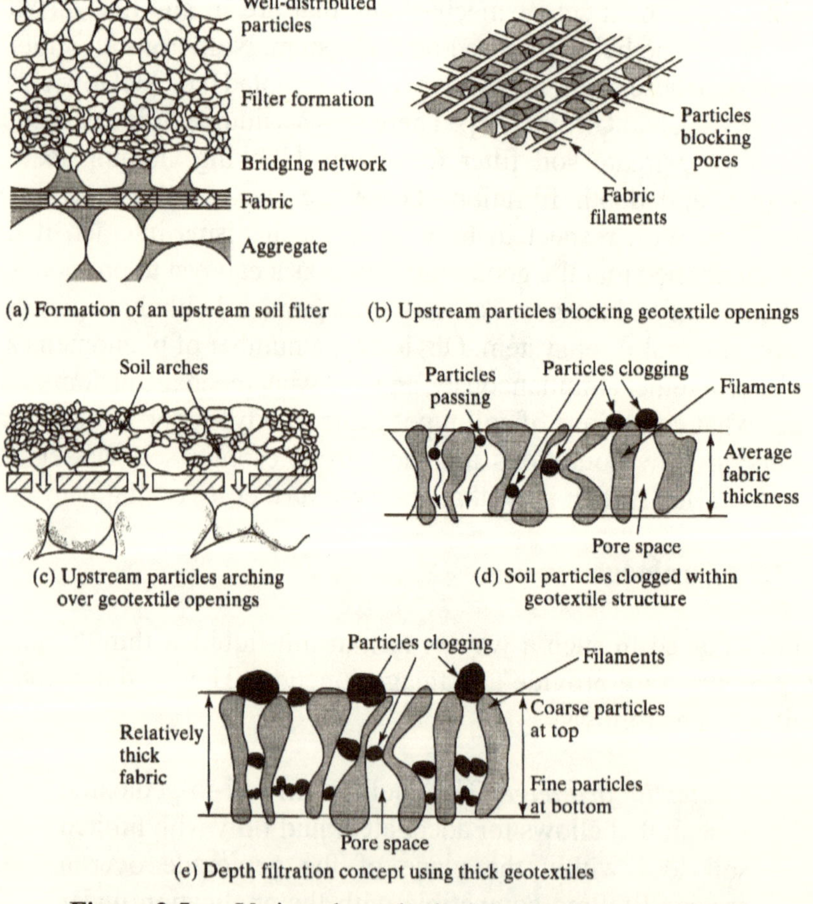

Figure 2.5 Various hypothetical mechanisms involved in long-term soil-to-fabric flow compatibility. (Parts [a-d] after McGown [11]; Part [e] after Heerten [12]

Note that this discussion on drainage overlaps considerably the preceding section on filtration. For these two functions (except for the consideration of flow direction), the soil retention and long-term compatibility concepts are the same.

Permeability. Referring now to in-plane permeability for the drainage function, we must recognize that the geotextile's thickness will decrease with increasing normal stress on it. For this reason we will define a term called *transmissivity* as follows:

$$\theta = k_p t \tag{2.10}$$

where

θ = transmissivity,
k_p = in-plane permeability coefficient (the subscript p is often omitted), and
t = thickness at a specified normal pressure.

The testing method for geotextile transmissivity will be covered later.

Soil Retention. The criteria used to design the opening spaces of a geotextile so that it retains the adjacent soil were covered in section 2.2.3. The concepts and design guides are precisely the same for the drainage function as they were for filtration.

Long-Term Flow Compatibility. As with the filtration function, we must ensure the compatibility of the soil with the geotextile over the lifetime of the system being built. The criteria discussed in section 2.2.3 hold for the drainage function the same as they do for filtration.

2.2.5 Containment

By virtue of their inherent porosity, geotextiles rarely serve in the containment function. The notable exception is when the geotextile is purposely impregnated with bitumen or polymer. These few applications will be noted accordingly.

2.2.6 Combined Functions

The introduction to this chapter described design-by-function. The procedure as outlined identified the geotextile's primary function and set the design accordingly. Where geotextiles are used for a single function, this can indeed be done. However, geotextiles often serve multiple or combined functions. Some examples are as follows:

- Beneath railroad ballast, where separation, reinforcement, filtration, and drainage can all be involved.
- In flexible-forming systems to contain concrete, grout or soil, where separation, reinforcement, and filtration are involved.
- Prevention of crack reflection in asphalt pavement overlays, where both reinforcement and waterproofing functions are involved.

In these situations all functions—primary, secondary, tertiary, and so on—must all be evaluated. They must all satisfy the required factor of safety (FS). If the situation is properly assessed, the calculated FS should increase progressively as we proceed through the primary, secondary, tertiary, etc., functions. If not (i.e., if the factors of safety jump around as we proceed through the calculations), it means that the critical functions were not properly assessed to begin with. Thus, the minimum FS will always indicate the primary function, the next highest value of FS will indicate the secondary function, and so on. This approach, of course, assumes that a reasonably accurate quantitative analysis can be developed for each of the functions described. Before we discuss this, however, we will treat a very important aspect of the subject dealing with specific geotextile properties and how they are obtained. That subject is a quantification of geotextile properties via their current test methods and procedures that follows.

2.3 GEOTEXTILE PROPERTIES AND TEST METHODS

This section presents the necessary test methods, relevant details, and selected data for the design-by-function procedure to be developed in the remainder of the chapter. (It also applies to the design-by-specification procedure). The reader should refer back to this section continuously as the various design methods are developed, since results from these test methods become the numerator of the design-by-function equation (recall equations 2.2a and 2.2b).

2.3.1 General Comments

In a growing and ever-changing area such as geotextiles, it should come as no surprise that a completely unified set of worldwide standards and test methods is currently not available. Yet the activity toward such an ultimate goal is very intense. Organizations that are involved in this activity are spread across the entire spectrum of potential users: raw material suppliers, manufacturers, manufacturers representatives, contractors and installers, testing organizations, design engineering firms, owners, regulators, research institutes, and universities.

Within these groups one will often hear reference to either *index* or *performance* tests. This terminology is somewhat unfortunate, since a particular index test to one group might be (and usually is) very much a performance test to another group. For example, a geotextile puncture test using a steel probe may be of an index variety to a geotechnical engineer, but to the manufacturer it is instead a measure of the quality control performance of the particular manufacturing process. Thus, this book does not make continual reference to a test method as being either index or performance related, but when it does so, the test method is identified from the perspective of the design engineer.

In the review of available geotextile test methods to follow, it should be recognized that many of the test methods are not fully harmonized between countries as far as their test procedures are concerned. The main groups developing and promoting test methods are American Society for Testing and Materials (ASTM), the International Standards Organization (ISO), and the Geosynthetic Research Institute (GRI). Between the three organizations, there are currently about 300 geosynthetic standards. The standards (test methods, guides, practices, and specifications) from these three groups will be constantly referenced in this book.

It should come as no surprise that many physical and mechanical test methods for geotextiles are partially, or completely, taken from existing textile standards [16-18]. The tests that differ between textiles and geotextiles are those which involve hydraulic, endurance, and environmental properties. These are generally new tests oriented completely toward geotextiles.

The section will be subdivided into the following major categories: physical properties, mechanical properties, hydraulic properties, endurance properties, and degradation considerations.

2.3.2 Physical Properties

The properties discussed in this subsection all refer to the geotextile in its manufactured or as-received condition. These tests are often referred to as being index tests.

Specific Gravity. The specific gravity of the fibers from which geotextiles are made is actually the specific gravity of the polymer formulation (see ASTM D792 or D1505). As customary, specific gravity is defined as the ratio of the material's unit volume weight (without any voids) to that of distilled, de-aired water at 4°C. Some typical values of the specific gravity of commonly used polymers made into geotextiles are listed below (steel, soil, glass, and cotton are added for comparison).

- Steel = 7.87
- Soil/Rock = 2.9 to 2.4
- Glass = 2.54
- Polyvinyl Chloride = 1.69
- Cotton = 1.55
- Polyester = 1.38 to 1.22
- Nylon = 1.14 to 1.05
- Polyethylene = 0.96 to 0.90
- Polypropylene = 0.91

Note that the specific gravity of the polyolefins is less than 1.0, which must be considered when working in water, i.e., they will float.

Mass per Unit Area (Weight). *Mass per unit area* is the proper term for what most people mean when they state or ask for the *weight* of the geotextile. It is also sometimes called *basis weight*, but this is equally incorrect, since neither weight nor basis weight explicitly considers area. Geotextile mass per unit area (the proper term) is given in units of grams per square meter (g/m^2). Unfortunately, still other values are listed in the literature, such as grams per linear meter for a geotextile of given width. Sometimes the latter value is given inversely as linear meters per kilogram. The point here is that one must clearly state what value is being communicated. Methods for the test are ASTM D5261 and ISO 9864.

Testwise, the mass (or weight) should be measured to the nearest 0.01% of the total specimen mass, and length and width should be measured under zero geotextile tension. The range of typical values for most geotextiles is from 150 to 750 g/m^2 although geotextiles in excess of 2000 g/m^2 have been used. Since fabric cost (and, in general, mechanical properties) is directly related to mass per unit area, it is an important property.

Thickness. The thickness of geotextiles is sometimes mentioned in specifications, but this is really more of a descriptive property than design-oriented property. It is measured as the distance between the upper and lower surface of the fabric, measured at a specified pressure. ASTM D5199 stipulates that the thickness of a geotextile is to be measured to an accuracy of at least 0.02 mm under a pressure of 2.0 kPa. The comparable ISO 9863 test method allows for the specifier to select the pressure. The thicknesses of commonly used geotextiles range from 0.25 to 3.5 mm.

Stiffness. Stiffness, or flexibility, of a geotextile should not be confused with its modulus (which is determined as the initial portion of the stress-versus-strain curve). In this test, stiffness is a measure of the interaction between the geotextile mass and its bending stiffness as shown by the manner in which the geotextile gravitationally bends under its own weight; the test method is designated as ASTM D1388. It is more appropriately called *flex stiffness*. The method takes a 25 mm wide strip of geotextile specimen and slides it out lengthwise over the edge of a horizontal surface. The length of overhang is measured when the tip of the geotextile bends under its own weight and just touches an inclined plane making an angle of 41.5° with the horizontal. One-half of this length is the bending length of the specimen. The cube of this quantity multiplied by the mass per unit area of the geotextile is its flexural rigidity or stiffness. The value is expressed in mg-cm units.

The property is indicative of the geotextile's inherent capability of providing a suitable working surface for installation. In placing a geotextile on extremely soft soils, a high-geotextile stiffness is very desirable. Haliburton et al. [19] have related this property to various soil subgrade strength values as given in table 2.3.

2.3.3 Mechanical Properties

The mechanical properties to be discussed here indicate the geotextile's resistance to tensile stresses mobilized from applied loads and/or installationconditions. Some are performed with the geotextile by itself (i.e., often called *index*, or *in-isolation tests*) while others are associated with a standard soil or with the site-specific soil (i.e., often called *performance tests*).

Compressibility. The compressibility of a geotextile is its thickness at varying applied normal stresses. For most geotextiles, the compressibility is relatively low and of little direct consequence as far as design is concerned (e.g., with woven fabrics and with nonwoven heat-bonded and heavily calendered geotextiles). For nonwoven needle-punched or bulky resin-bonded geotextiles, however, compressibility is important. This is because such geotextiles are

TABLE 2.3 RECOMMENDED GEOTEXTILE STIFFNESS VALUES FOR VARYING DEGREES OF REQUIRED WORKABILITY, AFTER HALIBURTON, ET AL. [19]

Subgrade CBR* (%)	Workability Benefit of Vegetative Cover**	Field Workability Requirements	Minimum Fabric Stiffness*** (mg · cm)
CBR ≤ 0.5	Poor	Very high	25,000
	Good	High	15,000
0.5 < CBR ≤ 1.0	Poor	High	15,000
	Good	Moderate	10,000
1.0 < CBR ≤ 2.0	Poor	Moderate	10,000
	Good	Low	5,000
CBR > 2.0	Poor	None	1,000
	Good	None	1,000

*CBR refers to *soaked* California Bearing Ratio, which is a test routinely used in geotechnical engineering to evaluate soil subgrade strength. The values for *unsoaked* CBR are considerably higher. The test, both soaked and unsoaked alternatives, is standardized as ASTM D1883.

**Medium to dense root system will probably exhibit some inherent workability benefits, whereas little to no root system will be of no benefit.

***Test conforms to ASTM D1388, except uses 300 mm long by 50 mm wide test specimens.

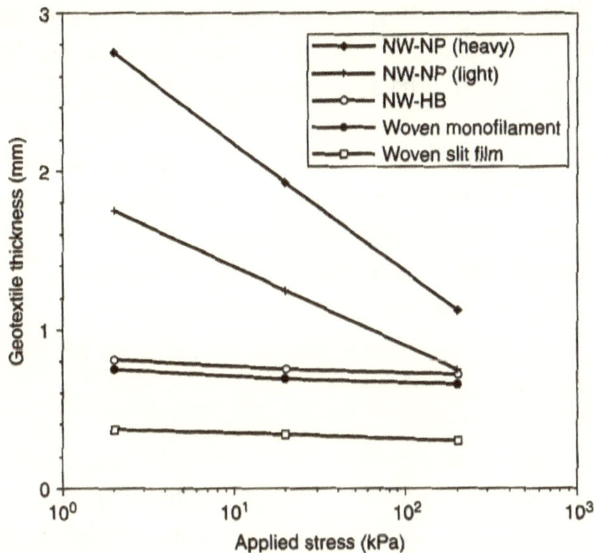

Figure 2.6 Compressibility of different types of geotextiles, including nonwoven needle-punched (NW-NP), nonwoven heat-bonded (NW-HB), and two types of woven.

often used to convey liquid within the plane of their structure. The more a fabric compresses under load, the lower its transmissivity. Figure 2.6 illustrates the compressibility of several geotextile types, where the influence of normal stress on thickness is clearly seen. The nonwoven needle-punched geotextiles are the most compressible, and this, in turn, is directly related to their mass per unit area.

Tensile Strength. Perhaps the single most important property of a geotextile is its tensile strength. In this regard, *strength* is defined as the maximum tensile stress that the test specimen can sustain at the point of failure. Invariably, all geotextile applications rely on this property either as the primary function (as in reinforcement applications) or as a secondary function (as in separation, filtration or drainage). The actual performance of the test contains the geotextile test specimen within a set of clamps or grips, then places this assembly in a constant rate of extension (CRE) testing machine and then stretches the geotextile in tension at a uniform strain rate until failure occurs. Fabric failure is generally easy to identify and often it is even audible. During the extension process, it is customary to measure both load and deformation in such a way that a stress-versus-strain curve can be generated. Stress is usually given as force per unit width and strain is calculated as deformation divided by original specimen length. From the stress-versus-strain curve, four values are obtained:

1. Maximum tensile stress (referred to as the geotextile's *strength*)
2. Strain at failure (generally referred to as *maximum elongation*, or simply *elongation*)
3. Toughness (work done per unit volume before failure, usually taken as the area under the stress-strain curve)
4. Modulus of elasticity (which is the slope of the initial portion of the stress-versus-strain curve)

Typical responses of geotextiles made from different manufacturing processes are given in figure 2.7. Note that the vertical axis is in units of force per unit width of fabric (i.e., kN/m), which is not a bona fide stress unit. To obtain true stress units, this value would have to

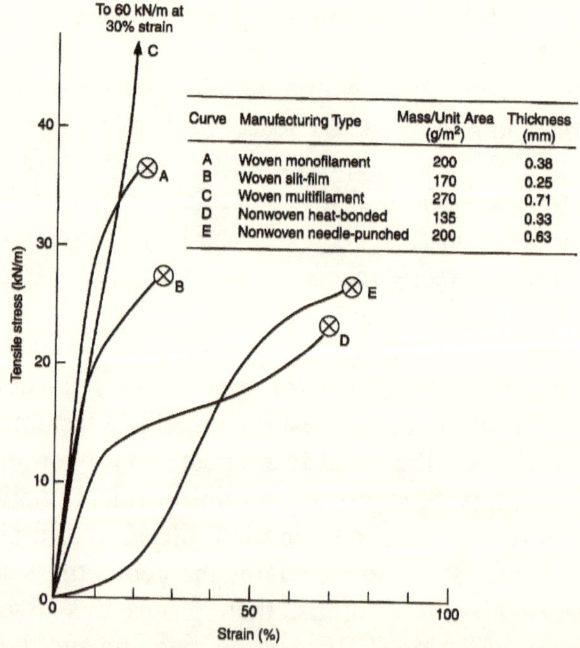

Figure 2.7 Tensile test response of various geotextiles manufactured by different processes. All are polypropylene fabrics; specimens were initially 200 mm wide by 100 mm high and tested according to ASTM D4595.

be divided by the geotextile's thickness. This is not conventionally done, since the thickness varies greatly under load and during the extension process. This, of course, has implications in the toughness and modulus values as well, since they too would have to be divided by thickness to obtain conventional engineering units. The example that follows illustrates these features.

Example 2.5

For the nonwoven heat-bonded geotextile illustrated in figure 2.7 (curve D), determine the strength, elongation, toughness, and modulus in common geotextile units and (on the basis of a nominal thickness of 0.33 mm) in standard engineering units.

Solution: By observation, the strength is

$$T_{\text{max}} = 23\text{kN / m}$$

and for 0.33 mm thickness,

$$\sigma_{\text{max}} = 69{,}700 \text{ kN / m}^2 = 69{,}700 \text{ kPa}$$

The elongation, i.e., maximum strain, is also determined by observation:

$$\varepsilon_f = 69\%$$

The toughness (U) is then calculated as 1/2 ($T_{max} \times \varepsilon_f$) (actually this is an approximation since it should be the actual area under the curve):

$$U_g = \frac{1}{2}(23 \times 0.69)$$

$$= 7.9 kN / m$$

and for 0.33 mm thickness,

$$U = 24{,}000 \text{ kN/m}^2 = 24{,}000 \text{ kPa}$$

Finally, the modulus is taken from the initial slope of the curve as

$$E = \frac{12}{0.10} = 120 \ kN/m$$

and for 0.33 mm thickness,

$$E = 364{,}000 \text{ kN/m}^2 = 364{,}000 \text{ kPa}$$
$$= 364 \text{ mPa}$$

There are several features of the tensile test that require further discussion, since they have implications for subsequent design procedures, the major ones being the modulus and the specimen size. Regarding the modulus, several choices are available for measuring the initial slope of the curve. These are the following:

- *Initial tangent modulus:* This is straightforward for many woven geotextiles in both their warp and weft directions and for nonwoven heat-bonded geotextiles. Here the initial slope is quite linear (as in conventional soil testing) and a reasonably accurate modulus value can be obtained.
- *Offset tangent modulus:* This concept is sometimes used when the initial slope is very low and is typical of nonwoven needle-punched geotextiles (see figure 2.7, curve “E”). To obtain the relevant modulus, we avoid the initial portion of the curve and essentially shifts the *y*-axis to the right, where it meets the downward extension of the linear portion of the response curve. The slope is then taken from this adjusted axis location.
- *Secant modulus:* To avoid the some arbitrariness of the above mentioned methods, we could stipulate the procedure of obtaining a modulus value—e.g., a secant modulus at 10% strain. Here, we draw a line from the axes’ origin to the designated curve at 10% strain and measure its slope from the origin irrespective of the actual curve to this point. Example 2.6 illustrates these various procedures.

Example 2.6

For the nonwoven needle-punched fabric E shown in figure 2.7, determine the initial tangent modulus, offset tangent modulus, and secant moduli at 10% and 35% strain in units of kN/m and kN/m² (kPa) based on an initial thickness of 0.63 mm.

Solution: Scaling directly from the curve:

$$E_T = \frac{4.3}{0.50} = 8.6\text{kN / m} \text{ or } 13{,}600\text{kN / m}^2$$

$$E_{OT} = \frac{20}{0.46-0.20} = 77\text{kN / m} \text{ or } 122{,}000\text{kN / m}^2$$

$$E_{S10} = \frac{1.1}{0.10} = 11\text{kN / m} \text{ or } 17{,}500\text{kN / m}^2$$

$$E_{S35} = \frac{11.6}{0.35} = 33\text{kN / m} \text{ or } 52{,}600\text{kN / m}^2$$

Regarding the test specimen size (length, width, aspect ratio, and shape), much has been written. ASTM standards D1682, D751, D4632, and D4595 along with ISO 10319 allow for a number of variations. Figure 2.8 illustrates the current most popular test specimen sizes. The grab tensile test D4632 is a very widely used and reported test. The geotextile specimen dimensions are 100 mm wide and 150 mm long, but the jaws of the clamps grip only the central 25 mm of the test specimen. Almost all geotextile manufacturers and geotextile specifications use this value, recall tables 2.1 and 2.2. Narrow strip tests (usually 25 or 50 mm wide) are used in many research and development studies since they use a minimum amount of geotextile. The reason wide-width specimens are necessary is that geotextiles (particularly nonwovens) when tensioned tend to have a severe necking effect under increasing stress, and they rope up, giving artificially high values. Thus, the tendency for design-related tests is to use wide-width specimens. The most common wide-width tests

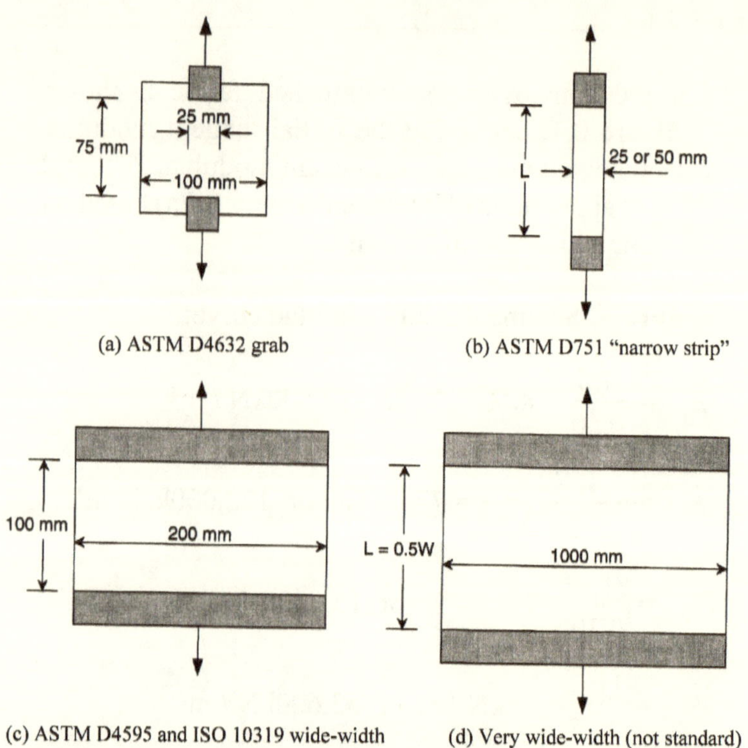

(a) ASTM D4632 grab

(b) ASTM D751 "narrow strip"

(c) ASTM D4595 and ISO 10319 wide-width

(d) Very wide-width (not standard)

Figure 2.8 Various tensile test specimen patterns used to obtained fabric strength properties.

are ASTM D4595 and ISO 10319, both of which use a 200 mm wide specimen, which is 100 mm long between the faces of the opposing grips. Such a test is not intended to be a routine or index test. The grab specimen should continue to be used in this regard (e.g., as a manufacturers' quality-control or conformance test). There are no universal relationships between the different test specimen sizes or shapes, and therefore, the choice of specimen size depends on the intended use of the data. Proper identification of the specimen size on the test data is always necessary. Regarding other features of tensile testing of geotextiles (effect of conditioning, load rate, load method, etc.), the applicable standard(s) should be consulted.

As the strength of the geotextile being tested increases, a number of operational problems arise. The most obvious need is for a higher-capacity tensile testing machine than is used for conventional geotextiles. This is straightforward. The types of devices used to grip the geotextile, however, is another matter. Figure 2.9 illustrates the

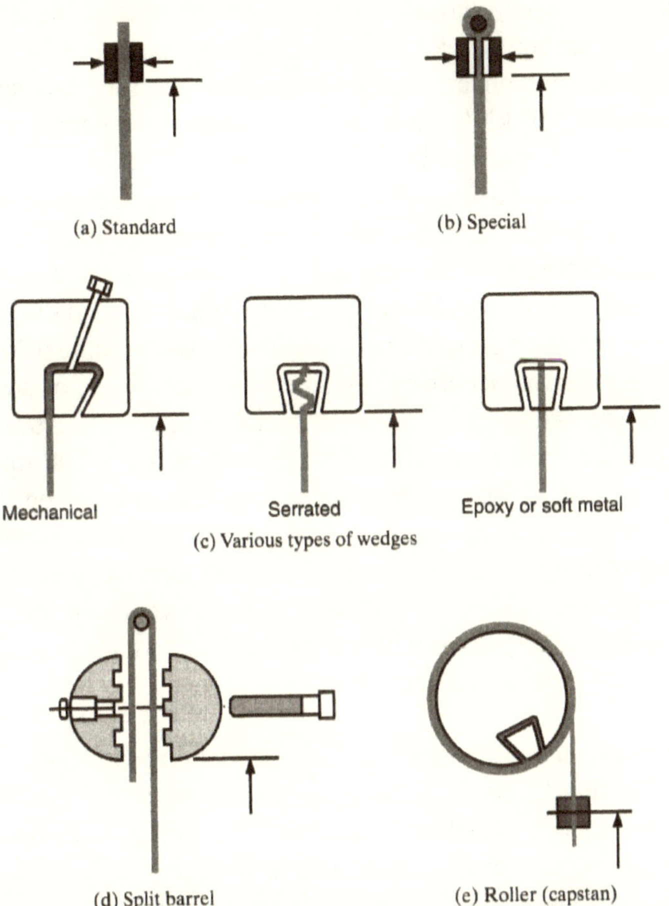

Figure 2.9 Various grip types for testing geotextiles and geogrids. (Adapted from Myles and Carswell [20])

various grip types recommended for use according to the ultimate strength of the geotextile (see GRI-GT9 for details). With geotextile strengths greater than approximately 50 kN/m, standard clamping jaws (figure 2.9a) are not satisfactory. This is due to slippage within the conventional grips or stress concentrations at the face of the grips leading to erroneous values of stress and strain. Note that slippage gives erroneously high values of elongation and grip stress concentration gives erroneously low values of strength. Standard grips can be made adaptable up to approximately 90 kN/m (see figure 2.9b). At higher strengths, some type of wedge grips become necessary (see figure 2.9c). Wedge grips can be made in a number of styles, using mechanical wedges, serrated wedges, or cast metal wedges. Split barrel types (see figure 2.9d) have also been attempted. However, even these grips become unacceptable with geotextile strengths greater than approximately 180 kN/m, due mainly to stress concentration failures at the edge of the upper or lower grips. Here stresses are very high and can only be avoided using roller, capstan on Demgen type of grips (see figure 2.9d). In this case, the geotextile tightens on itself around the rollers, and failure is within the test specimen between the opposing set of rollers. Elongation, however, can no longer be read directly from the testing machine's crosshead movement, since geotextile take-up around the rollers is occurring. This necessitates the use of an external measuring device such as a linear variable differential transformer (LVDT), laser sensor, or infrared sensor. Most use a 100 mm gauge distance located in the center of the test specimen between the roller grips. This output is fed into an *x*-*y* recorder for the *x*-axis elongation (or strain) reading. The *y*-axis or load is taken directly from the tensile testing machine. Thus, we obtain the stress-versus-strain diagram similar to figure 2.7, albeit with considerable effort and the associated added cost of specialized equipment.

Confined Tensile Strength. Before finishing the topic of tensile strength, it should be cautioned that all the tests just described are performed without lateral confining pressure (i.e., they are in-isolation tests). With lateral confinement, which obviously is how geotextiles are eventually used, results can be different. McGown et al. [21,22] have pioneered this test variation using a boxlike chamber separated in two halves where the geotextile test specimen is sandwiched between lubricated membranes and thin soil layers that have been pressurized by rubber bellows. It is important to allow the test specimen to elongate freely without friction being mobilized by the confining soil adjacent to the lubricated membranes. A friction mobilized pull-out test will be described later. The confinement pressure within the bellows simulates the in-situ pressure. The test specimen is 200 mm wide and 100 mm long in the test zone. Although this process is tedious and relatively complex, it is the best attempt at obtaining a true tensile strength/elongation response known to the author. Wilson-Fahmy et al. [23] present data on a number of different geotextiles. It is important to note from this study that *only* the nonwoven needle-punched geotextiles show significantly improved stress-versus-strain behavior under confinement. There was no measurable improved strength behavior noted with woven geotextiles, heat-bonded geotextiles, geonets, geomembranes, or GCLs when they were placed under confinement.

Seam Strength. Often the ends or sides of rolls of geotextiles have to be joined together for the purpose of transferring tensile stress. By far, the most common method is by sewing. Various styles of sewn seams will be described later, but whatever the type, they must be laboratory-evaluated for their load-transfer capability from one geotextile roll to another. ASTM D4884 and ISO 13426 test methods call for the following requirements.

- The shape of the seamed test specimen is 200 mm wide except at the seam itself. Here an additional 25 mm of seamed material is allowed to protrude from both sides; that is, at the seam, the test specimen is 250 mm wide. This accounts for a certain amount of loss of seam strength when the seaming yarns are cut during specimen preparation, but to what degree is quite uncertain.
- The resulting ultimate load is divided by a 200 mm width and reported in units of kN/m. The appropriateness of this computational step is questionable.
- The rate of extension is 10%/min.
- Elongation across the seam is not required to be measured—i.e., the test measures only tensile strength.

Test results from the evaluation of well-made sewn seams of geotextiles having wide-width strengths up to approximately 20 kN/m usually result in sewn-strength efficiencies above 85%; i.e.,

$$E(\%) = \frac{T_{\text{seam}}}{T_{\text{geotextile}}} \times 100 \tag{2.11}$$

where

E = seam efficiency (percent),
T_{seam} = wide-width seam strength, and
$T_{\text{geotextile}}$ = wide-width (unseamed) geotextile strength.

As the geotextile strength becomes higher, seam strengths become progressively less efficient (see figure 2.10, where the upper bound is typical of good factory seams and lower bound is typical of poor field seams). Above 50 kN/m, most seam strengths fall beneath 75% efficiency, and beyond 200 to 250 kN/m, the best one can do is approximately 50% seam efficiency. Note that by this point poorly made

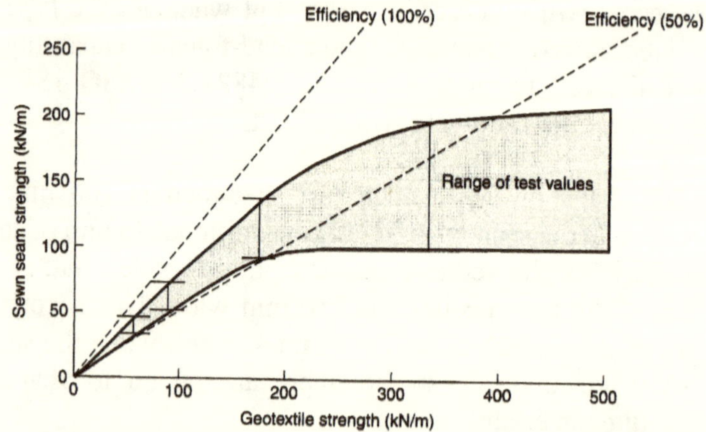

Figure 2.10 Behavior of sewn geotextile strength in comparison to the parent (unseamed) geotextile strength.

seams become extremely low in their load-transfer capabilities. The seaming of high-strength geotextiles simply begs for better joining or bonding methods than sewing. Other possibilities that are available are the use of epoxy resins [24] or mechanical joining.

Fatigue Strength. Fatigue strength, or fatigue resistance, is the ability of a geotextile to withstand repetitive loading before undergoing failure. A tensile test specimen, usually of a wide-width variety, is stressed longitudinally at a constant rate of extension to a predetermined load (less than failure) and then back to a lower, or zero, load. This cycling is repeated until failure occurs. The resulting cyclic stress-versus-strain response (i.e., the hysteresis loops) can be used to calculate a cyclic modulus that becomes evident after a number of load cycles are applied. Also important is the number of cycles required to bring the geotextile to failure and the respective loads that were applied. The load resulting in failure is converted to stress, and this value is usually expressed as a fraction of the quasi-statically applied failure stress (strength) described previously. As expected, the lower the stress level, the larger the number of cycles required before failure.

Although many variables remain to be defined (primarily, the decision as to what loads to apply during testing), the test reasonably simulates in-situ conditions for applications such as seismic and railroad loadings,and wave or tidal action. Research in this area seems justified (see Ashmawy and Bourdeau [25]).

Burst Strength. There are two test methods that stress geotextiles out of plane, thereby mobilizing tension until failure occurs. The most common is the Mullen burst test, which was covered in ASTM D3786 but is now depreciated. In this test, an inflatable rubber membrane is used to distort the geotextile into the shape of a hemisphere of 30 mm diameter. Bursting of the geotextile occurs when no further deformation is possible. The test has been used for quality control but is seeing less and less use in manufacturer's literature and specifications. The trend in this regard is to use puncture tests of the 50 mm probe variety.

An alternative test uses a large rectangular test specimen and deforms it by an underlying rubber membrane. Called a diaphragm test by Raumann [26], the central portion of the geotextile (along the minor axis) is very close to plane strain conditions. As such, the pressure-versus-strain response yields a very accurate modulus. It is a difficult test to setup and perform, and the current tendency is to utilize wide-width tensile tests of the type described previously.

The most recent type of burst test is a multi-axial test where the specimen is gripped in both machine and cross-machine directions and then is inflated by an underlying diaphragm in an out-of-plane mode (see Anderjack and Wartman [27]).

Tear Tests. During installation, geotextiles are subjected to tearing stresses. Although a test simulating such situations is important, it will be seen that the methods developed to date can vary widely in their response. There are three tear tests commonly used: trapezoidal, tongue, and Elmendorf.

Trapezoidal Tear Test. The trapezoidal tearing load is the force required to break individual yarns in a fabric. One such test was originally developed to test automotive fabrics from failure by screw drivers in the back pocket of people sitting down. Since then, it has been discontinued for that use, but has been revised and modified for geotextiles. The current trapezoidal tear tests are ASTM D4533 and ISO 13434. In these tests, the geotextile is inserted into a tensile testing machine on the bias, which causes the yarns to tear progressively. An initial 15 mm cut is made to start the process. The load actually stresses the individual yarns gripped in the clamps rather than stressing the fabric structure. The value—commonly referred as *trap tear*—is reported by all manufacturers and used in most specifications.

Tongue Tear Test. As indicated in ASTM D751, the tongue tear test, uses a 75 mm by 200 mm geotextile specimen with a 75 mm long initiation cut. The geotextile is placed in a testing machine with the cut ends in the grips of the machine. An increasing tensile force is applied to make the geotextile tear along the initiation cut. The test configuration permits the yarns to rope up and work together to resist tear propagation. Thus, the values resulting from tongue tear tests are usually much higher than those from trapezoidal tear tests.

Elmendorf Tear Test. The Elmendorf tear test is covered in ASTM D1424 and involves a procedure for the determination of the average force required to dynamically propagate a single-rip tongue-type tear starting from a premade cut in a woven geotextile. The cut is then continued by means of a rotating pendulum apparatus. The tearing force is the force required to continue the tear previously started in the test specimen. The strength is calculated as the work done in tearing the specimen divided by twice the length of the tear. The test is often used in Europe to measure tear strength. It is generally not used for nonwoven geotextiles.

Impact Tests. Since falling objects such as rocks, tools, other construction items can readily create punctures and tears in geotextiles, a number of tests have been developed to assess the impact resistance of geotextiles. One such test that measures impact resistance directly in energy units (Joules) has been developed for an Elmendorf tear apparatus. The impacting cone is attached to the pendulum arm of the Elmendorf tear tester and penetrates through the geotextile specimen, which is fixed on the end of the device. The fixture holding the geotextile specimen is called a *Spencer impact attachment*. Impact resistance units are read directly from the device. Unfortunately, the limit of most commercially available systems is about 25 J, which is too low for many geotextiles. Thus, it is necessary to use impact pendulum devices developed for other materials, such as metals, which have energies of up to 300 J. Such devices are covered under ASTM A370 and ASTM D256. The test specimen holder, however, must be converted to hold geotextiles rather than notched metal bars. See Koerner et al. [28] for details and results from this type of test procedure.

Another dynamic impact test that punctures then tears the geotextile test specimen is a drop-cone test initiated by Alfheim and Sorlie [29] and subsequently standardized as ISO 13433. The tapered cone is marked, and the amount of penetration into the fabric is indicative of a number of properties and, in general, its robustness to harsh installation and in situ conditions. This is an important test method and is seen referenced in most European specifications and manufacturers' literature.

Puncture Tests. In addition to the dynamic tests just described for impact resistance, there is need for an assessment of geotextile resistance to objects such as stones and stumps under quasi-static conditions. Such a test is described under ASTM D4833. This test uses a penetrating steel rod of 8.0 mm in diameter. The geotextile test specimen is firmly clamped in an empty cylinder of 45 mm inside diameter and the rod pushed through it via a compression testing machine at a prescribed rate. Resistance to puncture is measured in force units.

This test is a popular one due to its simplicity and ability to be automated. It is reported by all manufacturers and listed in most specifications. A considerably large database exists using this test method (e.g., see Koerner et al. [28]). It is important to note the exact shape of the end of the metal rod. Three types are in current use: hemispherical, flat, and beveled flat. The interrelationships and differences between these types have not been identified. The latter type, with a 0.8 mm, 45° bevel around its circumference is preferred and covered in ASTM D4833.

The small size of the above described device is also of concern. For example, a lightweight nonwoven geotextile can selectively be chosen in a low-density fiber region or in a high-density fiber region. The differences in puncture resistance will be very large.

With such a concern in mind, a larger-size puncture test has been developed [30]. It uses a conventional soil-testing CBR plunger and mold. The penetrating steel rod is 50 mm in diameter, and the geotextile is firmly clamped in an empty mold of 150 mm inside diameter. The circumference of the plunger should be beveled 0.80 mm on a 45° angle so as not to cut the yarns at the edge of the penetrating rod. This test is formalized as ASTM D6241 and ISO 12236.

There is a direct relationship between the CBR puncture-resistance value and the wide-width tensile strength of geotextiles. This is because the geotextile between the inner edge of the specimen holder, and the outer edge of the puncturing rod is in a state of pure axi-symmetric tension. Cazzuffi and Venesia [31] propose the following empirical equation as a correlation between the puncture breaking force of the CBR test and the wide-width tensile strength for isotropic, nonwoven geotextiles

$$Tf = Fp/2\pi r \tag{2.12}$$

where

T_f = tensile force per unit width of fabric (kN/m),
F_p = puncture breaking force (kN), and
r = radius of the puncturing rod (m).

Both the German and Italian standards have correlations between the CBR test results and the wide-width tensile elongation of the geotextile. According to the German (DIN) standard, the tensile elongation at failure (ε_f) is calculated as follows:

$$\varepsilon_f = \frac{(x - a)}{a} \times 100 \tag{2.13}$$

where

x = diagonal elongation of the geosynthetic at failure (m), and
a = horizontal distance between the outer edge of the plunger and the inner edge of the mold (m).

The Italian (ENEL) standard uses the following equation to calculate the tensile elongation at failure:

$$\varepsilon_f = \frac{\left[\pi(R+r)\,x + \pi r^2 - \pi R^2\right]}{\pi R^2} \times 100 \tag{2.14}$$

where

R = radius of the mold (m), and
r = radius of the puncturing rod (m).

It appears that the strength predictions are reasonable [28], with the nonwovens being more accurate than the wovens. The variation in predicted elongation at failure has more scatter, but is still reasonable. Clearly, the CBR test for puncture strength or as a form of axi-symmetric tensile strength has considerable merit.

Friction Behavior

In many design problems, it is necessary to know the soil-to-geotextile friction behavior. The generally accepted test setup is an adaptation of the direct shear test used in geotechnical engineering [32]. As shown in figure 2.11a, the geotextile is firmly fixed to one-half of the test device with soil (or another geosynthetic) in the other half. After normal stress is applied and equilibrates, a shear force is mobilized until sliding occurs between the geotextile and the soil with no further increase in required shear force. When the test is repeated at different normal stresses, the data are plotted and trends are established as shown in figure 2.11b. From these trend curves, limiting data (peak and residual stresses) can be obtained and then the curves of figure 2.11c can be drawn. These curves result in the establishment of the Mohr-Coulomb failure parameters (adhesion and friction angle) of the interface being tested (recall equation 2.5a). From a comparison of the geotextile-to-soil response versus the soil-to-soil response, the shear strength efficiencies on the soil's cohesion and friction angle can be obtained (recall equations 2.5b, 2.6 and 2.7). Note that the soil's shear strength parameters are the upper limit—that is, an efficiency of 100%. This implies that if the soil-to-geotextile interface is stronger than the shear strength of the soil itself, failure will occur entirely in the soil either above or beneath the geotextile inclusion.

Also shown in figure 2.11 b and c is a residual strength lower than the peak strength. This is not uncommon in carrying out

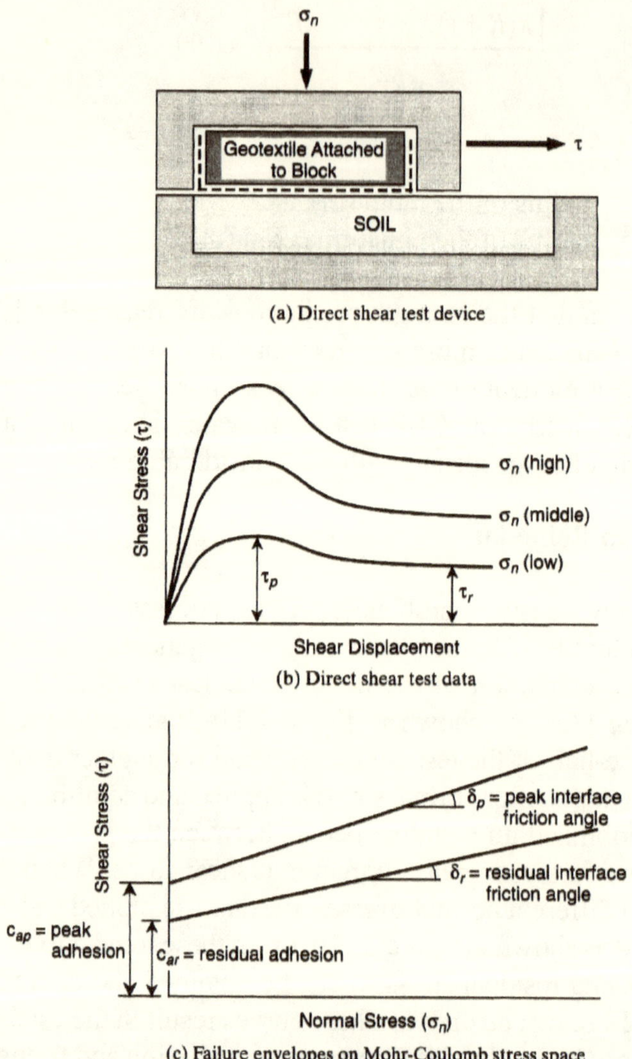

Figure 2.11 Test setup and procedure to assess interface shear strengths involving geotextiles.

geosynthetic-to-soil direct shear tests, including those tests which have geotextiles. Such data present to the design engineer a major decision as to what value to select. If peak values are used, traditional factor of safety values can generally be used, the assumption being that the movement of the interface in the field cannot go beyond the deformation needed to mobilize peak strength. If the interface deformations in the field go beyond peak, then a lower shear strength must be used and, in the limit, the residual shear strength. Using less than peak strength, the acceptable factor of safety value can probably be lower, but how much so is not clear. This issue of peak, residual, or somewhere in-between, is very significant when considering multi-geosynthetic lined slopes as typical in landfill liners and covers [33,34]. The issue will be further discussed in chapter 5 on geomembranes.

The results from such a test setup by Martin et al. [35] are presented in table 2.4 for four geotextile types against three different cohesionless soils. *Peak* soil-to-geotextile friction angles are given (in all cases the adhesion was zero), as well as the geotextile efficiency versus the peak soil friction angle by itself as per equation (2.7). Here it is seen that most geotextiles can mobilize a high percentage of the soil's friction and can be used to advantage in situations requiring this feature.

Both ASTM D5321 and ISO 12957 direct shear tests call for a shear box of 300 mm × 300 mm in size. While such a large test box is appropriate for geonets, geogrids, many geocomposites, and large particle-sized soils, this author considers it to be excessive for geotextiles (and certainly for geomembranes) against sands, silts, and clays and against one another. Standard geotechnical engineering laboratory shear boxes (e.g., 100 mm × 100 mm) are felt to be satisfactory for geotextile testing and focus should be on more relevant shear strength testing parameters than size, such as the following:

- Use of site-specific soil types as well as product-specific geotextile types
- Control of density and moisture content of the as-placed soil
- Geotextile fixity conditions to the end platen(s)
- Saturation conditions during consolidation and shear testing
- Specific type of saturating fluid (e.g., leachate)
- Use of field anticipated strain rates
- Adequate shear box deformation to achieve residual shear strength

Pullout (Anchorage) Tests. Geotextiles are often called on to provide anchorage for many applications within the reinforcement function. Such anchorage usually has the geotextile sandwiched between soil on either side. The resistance can be modeled in the laboratory using a pullout test, which will be detailed in section 3.1.2 on geogrids. The pullout resistance of the geotextile is obviously

TABLE 2.4 PEAK SOIL-TO-GEOTEXTILE FRICTION ANGLES AND EFFICIENCIES (IN PARENTHESES) FOR SELECTED COHESIONLESS SOILS

Geotextile type	Concrete sand $\varphi = 30$ deg.	Rounded sand $\varphi = 28$ deg.	Silty sand $\varphi = 26$ deg.
Woven, monofilament	26 deg. (84%)	–	–
Woven, slit film	24 deg. (77%)	24 deg. (84%)	23 deg. (87%)
Nonwoven, heat bonded	26 deg. (84%)	–	–
Nonwoven, needle punched	30 deg. (100%)	26 deg. (92%)	25 deg. (96%)

Source: After Martin, et al. [35]

Note: Literature values such as the above should *not* be used in critical designs. Site specific geotextiles and soils must be individually tested and evaluated in accordance with the particular project conditions, e.g., saturation, type of liquid, normal stress, consolidation time, shear rate, displacement amount, etc.

dependent on the normal force applied to the soil, which mobilizes shearing resistances on both surfaces of the geotextile.

Since the test greatly resembles a direct shear test, albeit with stationary soil on both sides of the tensioned geotextile, a possible design strategy is to take direct shear test results (for both sides of the geotextile) and use these values for pullout design purposes. However, this may not be a conservative practice.

Test results by Collios et al. [36] show a relationship of pullout test results to shear test results with some notable exceptions. For pullout testing, if the soil particles are smaller than the geotextile openings, efficiencies are high; if not, they can be low. In all cases, however, pullout test resistances are less than the sum of the direct shear test resistances. This is due to the fact that the geotextile is taut in the pullout test and exhibits large deformations. This, in turn, causes the soil particles to reorient themselves into a reduced shear strength mode at the soil-to-geotextile interfaces, resulting in lower pullout resistance. The stress state mobilized in this test is both interesting and complex as evidenced by the approximate one-hundred technical references on this topic.

2.3.4 Hydraulic Properties

Unlike the physical and mechanical properties just discussed, traditional tests on textile materials rarely have hydraulic applications;

that is, the garment and industrial fabrics industries obviously do not test for liquid flow. As a result, hydraulic testing of geotextiles has required completely new and original test concepts, methods, devices, interpretation, and databases. Both geotextile tests in-isolation and with soil will be described in this section.

Porosity. As conventionally defined with soils in geotechnical engineering, the porosity of a geotextile is the ratio of void volume to total volume. It is related to the ability of liquid to flow through or within the geotextile but is rarely measured directly. Instead, it is calculated from other properties of the geotextile.

$$n = 1 - \frac{m}{\rho t} \tag{2.15}$$

where

n = porosity (dimensionless),
m = mass per unit area i.e., weight (g/m^2),
ρ = density (g/m^3), and
t = thickness (m).

It is seen in equation (2.15) that for a given geotextile's weight and density, the porosity is directly related to thickness. Thickness in turn is related to the applied normal stress (see again figure 2.6).

Pore size can be measured by careful sieving with controlled-size glass beads (see the AOS test later in this section), by the use of image analyzers [37], or by the use of mercury intrusion [38]. Bhatia et al. [39] have compared these different measurement techniques on a variety of geotextiles illustrating behavioral trends and comparisons. The image analyzer results presented by ICI Fibers [40] for their various weights of geotextiles are instructive in showing that the pore size shifts gradually lower as the geotextile weight increases (see figure 2.12a). McGown [41] has provided information of the same type comparing different geotextile manufacturing styles (see figure 2.12b). These results are for the as-manufactured geotextile. Recognize, however, that geotextiles have pore sizes that are sensitive to changes in geotextile thickness due to applied normal stresses and adjacent soil gradations as would be typical of in situ conditions.

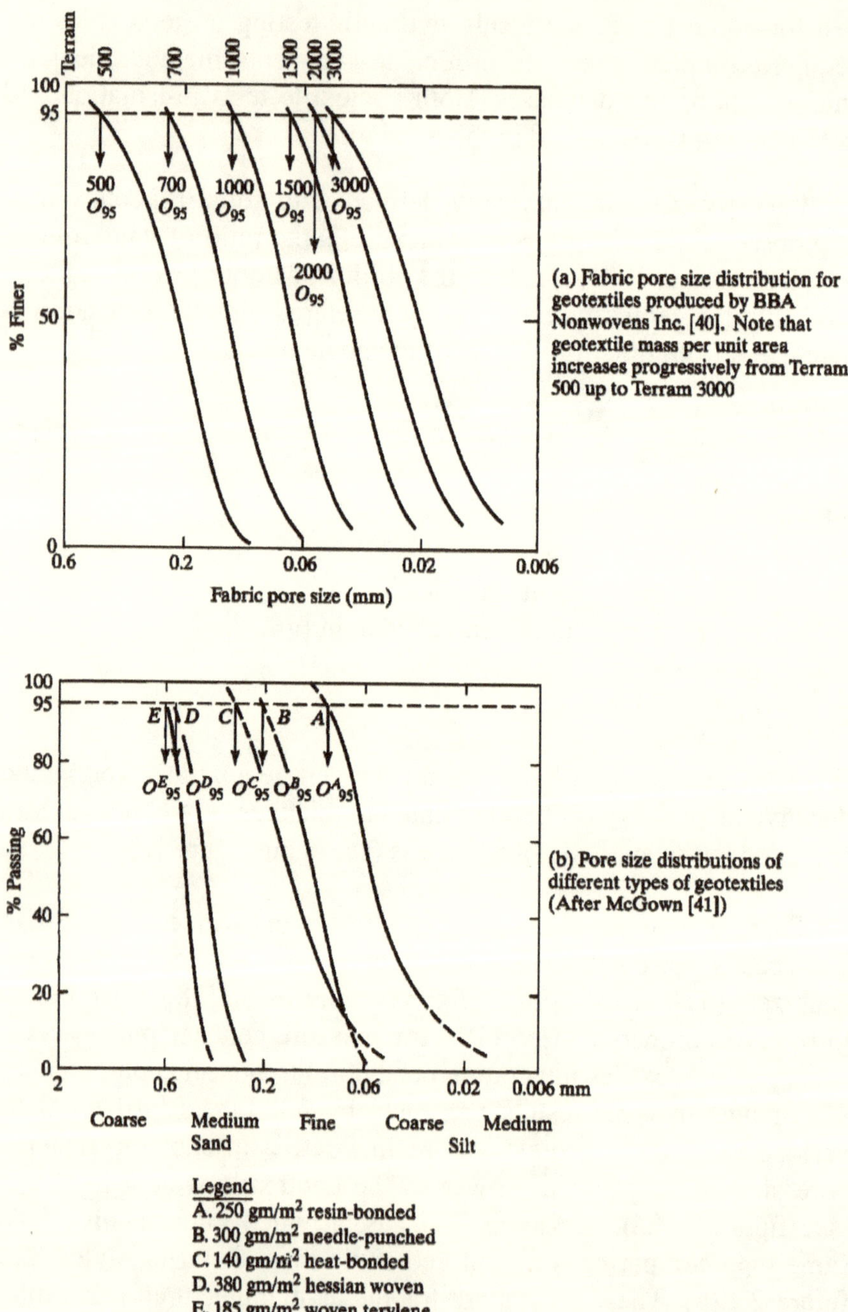

Figure 2.12 Various fabric pore size distribution curves.

Percent Open Area. Percent open area (POA) is a geotextile property that has applicability only for woven geotextiles, and even then only for woven monofilament geotextiles. POA is a comparison of the total open area (the void areas between adjacent yarns) to the total specimen area. A convenient way to measure the open area is to project a light through the geotextile onto a large poster-sized piece of cardboard. The magnified open spaces (resembling a window screen) can be mapped by a planimeter. Alternatively, a cardboard background that is crosshatched like graph paper can be used. Here, the squares are counted and summed up for the open area. The total area (yarns plus voids) must be measured at the same magnification as the voids measurement. Woven monofilament geotextiles vary from essentially a closed structure (POA ≅ 0%) to one that is extremely open (POA = 36%), with many commercial woven monofilament geotextiles being in the range of 6 to 12%.

The test is not applicable to nonwovens since the overlapping yarns block any light from passing directly through the geotextile. Thus, a different test method is required to measure void sizes in nonwovens.

Apparent Opening Size (or Equivalent Opening Size). A test for measuring the apparent opening size was developed in the 1970s by the US Army Corps of Engineers to evaluate woven geotextiles. The test has since been extended to cover all geotextiles, including the nonwoven types. The apparent opening size (AOS) or equivalent opening size (EOS)—AOS and EOS are essentially equivalent terms—are defined in CW-02215 as the US standard sieve number that has openings closest in size to the openings in the geotextile. The subsequent ASTM test is designated D4751. The test uses known-diameter glass beads and determines the O_{95} size by standard *dry* sieving. Sieving is done using beads of successively larger diameter until the weight of beads passing through the test specimen is 5%. This defines the O_{95} size of the geotextile's openings in millimeters. Values of O_{95} are indicated on the curves of figure 2.12. Note, however, that the O_{95} value only defines one particular opening size of the geotextile, not the total pore-size distribution. A conversion of the O_{95} size in millimeters can then be made using table 2.5 to obtain the closest US sieve size, and its number defines the AOS (or EOS) value. Thus, AOS, EOS, and O_{95} all refer to the same specific

TABLE 2.5 CONVERSION OF U.S. STANDARD SIEVE SIZES TO EQUIVALENT SQUARE OPENING SIZES (in mm).

Sieve Size (No.)	Opening (mm)
4	4.750
6	3.350
8	2.360
10	2.000
16	1.180
20	0.850
30	0.600
40	0.425
50	0.300
60	0.250
70	0.210
80	0.180
100	0.150
140	0.106
170	0.088
200	0.075
270	0.053
400	0.037

pore size, the difference being that AOS and EOS are sieve numbers, while O_{95} is the corresponding sieve-opening size in millimeters. It should also be noted in the conversion on table 2.5 that as the AOS sieve number increases, the O_{95} particle size value decreases; that is, the numbers are inversely related to one another. In this book we will generally use the O_{95} value since it is the target value for design purposes.

The AOS test per ASTM D4751 as just described is a poor test, having many problems, but simplicity of the test and its inertia seem to sustain its use in the United States. Some of the problems associated with the test are as follows:

- The test is conducted dry, whereas filtration and drainage always involve liquids.
- The glass beads can easily get trapped in the geotextile itself (particularly in thick nonwovens) and not pass through at all.
- Electrostatic charges often result in the finer glass beads clinging to the inside of the sieve and not participating in the test at all.

- Yarns in some geotextiles easily move with respect to one another (as they do in woven slit-film geotextiles), thereby allowing the beads to pass through an enlarged void not representative of the total geotextile test specimen.
- Slight changes in fabric structure do not result in different O_{95} values. This is perplexing since structure, temperature, humidity, bead size variation, and test duration all potentially influence the test results.
- The test is directed only at the 5% size (equivalent to the 95% passing size), which allows for determination of the O_{95} size. The remainder of the pore size curve is not defined.

Alternatives to the dry-sieving test just described include the following of wet-sieving methods [42].

- In Canada (CGSB-148.1) and France, a frame containing the geotextile specimen has well-graded glass beads placed on it and is repeatedly submerged in water. The bead fraction that passes is calculated and a O_{95} equivalent particle size is obtained.
- In Germany, the setup is similar but a water spray is used. The soil fraction that passes as well as an effective opening diameter is calculated.
- The ISO 12956 test is also a wet-sieving test and will hopefully be seeing greater use than dry sieving in the future.

In general, these wet sieving tests avoid many of the problems of dry sieving and are felt to be more representative of site conditions. In addition, even more sophisticated measurement techniques are emerging, including bubble point, mercury intrusion, and image analysis. Figure 2.13 illustrates that the differences in pore size are quite pronounced.

Permittivity (Cross-Plane Permeability). One of the major functions that geotextiles perform is that of filtration. (Note that most transportation agency specifications and some manufacturers' literature incorrectly call this "drainage.") In filtration, the liquid flows perpendicularly through the geotextile into crushed stone, a perforated pipe, a geosynthetic drainage core, or some other drainage material.

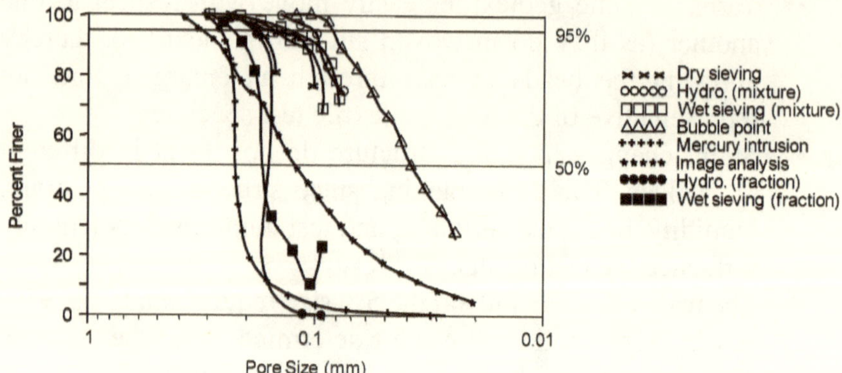

Figure 2.13 Complete pore size distribution curves for a continuous filament needle-punched nonwoven geotextile using different test methods. (After Bhatia et al. [39])

It is important that the geotextile allow for this flow to occur and not be impeded. Hence, the geotextile's cross-plane permeability must be quantified. As we discussed in the compressibility section, however, fabrics deform under load (recall figure 2.6). Thus, a new term, permittivity (Ψ) as was previously defined as equation (2.8), is repeated here.

$$\Psi = \frac{k_n}{t} \tag{2.8}$$

where

Ψ = permittivity (sec^{-1}),
k_n = permeability (properly called *hydraulic conductivity*) normal to the geotextile where the subscript n is often omitted (m/sec), and
t = thickness of the geotextile (m).

The above equation is used in Darcy's formula as follows:

$$q = k_n i A$$

$$q = k_n \frac{\Delta h}{t} A$$

$$\frac{k_n}{t} = \Psi = \frac{q}{(\Delta h)(A)} \tag{2.16}$$

where

q = flow rate (m^3/sec),
i = hydraulic gradient (dimensionless),
Δh = total head lost (m), and
A = total area of geotextile test specimen (m^2).

The formulation above is used for constant head tests in an identical manner as with soil permeability testing. Typically, the flow rate (q) is measured at one value of Δh, and then the test is repeated at different values of Δh. These different values of Δh produce correspondingly different values of q. When plotted as (ΔhA) on the horizontal axis and (q) on the vertical axis, the slope of the resulting straight line yields the desired value of Ψ.

The test can also be conducted using a falling (variable) head procedure as is also performed on soils. In this case, Darcy's formula is integrated over the head drop in an interval of time and used in the following equation:

$$\frac{k_n}{t} = \Psi = 2.3\frac{a}{A\Delta t}\log_{10}\frac{h_0}{h_f} \tag{2.17}$$

where

Ψ = permittivity (sec^{-1}),
a = area of water supply standpipe (m^2),
A = total area of geotextile test specimen (m^2),
Δt = time change between h_0 and h_f (sec),
h_0 = head at beginning of test (m), and
h_f = head at end of test (m).

In either case, the resulting permittivity value can be multiplied by the geotextile thickness to obtain the traditional permeability value, if so desired.

If the permeating fluid is not water (e.g., leachate, sludge or waste oil) compensation for differences in density and viscosity must be made (Hausmann [43]). This is done by using the following conversion:

$$\Psi_f = \Psi_w \frac{\rho_f}{\rho_w} \frac{\mu_w}{\mu_f} \tag{2.18}$$

where

Ψ_f = permittivity of the fluid under consideration,
Ψ_w = permittivity using water,
ρ_f = density of the fluid,
ρ_w = density of water,
μ_w = viscosity of water, and
μ_f = viscosity of the fluid.

ASTM D4491 uses a device as shown in figure 2.14 to measure the permittivity of geotextile test specimens. It is similar to ISO 11058. Either constant head or falling head can be used, although the standard is written around the constant head test, at a head of 50 mm. As with the permeability of soils, geotextile values of permittivity (and permeability) range over several orders of magnitude:

- Permittivity, ψ: from 0.02 to 2.2 s^{-1}
- Permeability, k_n: from 8×10^{-6} to 2×10^{-3} m/s

Some important test considerations are preconditioning of the test specimen, temperature, and the use of de-aired water. ASTM D4491 requires a dissolved oxygen content of the permeating water to be less than 6.0 mg/l. Tap water is allowed unless disputes arise, in which case deionized water should be used. Note that conventional soil-testing permeameters cannot be used to test geotextiles, since the size of their water outlets is rarely large enough to handle the flow coming through most geotextiles. The testing of soil and geotextile systems for long-term flow compatibility will be treated later under endurance properties in section 2.3.5.

Permittivity Under Load. The previously described permittivity test had the geotextile test specimen under zero normal stress, a situation rarely encountered in the field. To make the test more performance oriented, numerous attempts to construct a permittivity-under-load device have been made. Generally, a number of layers of geotextile (two

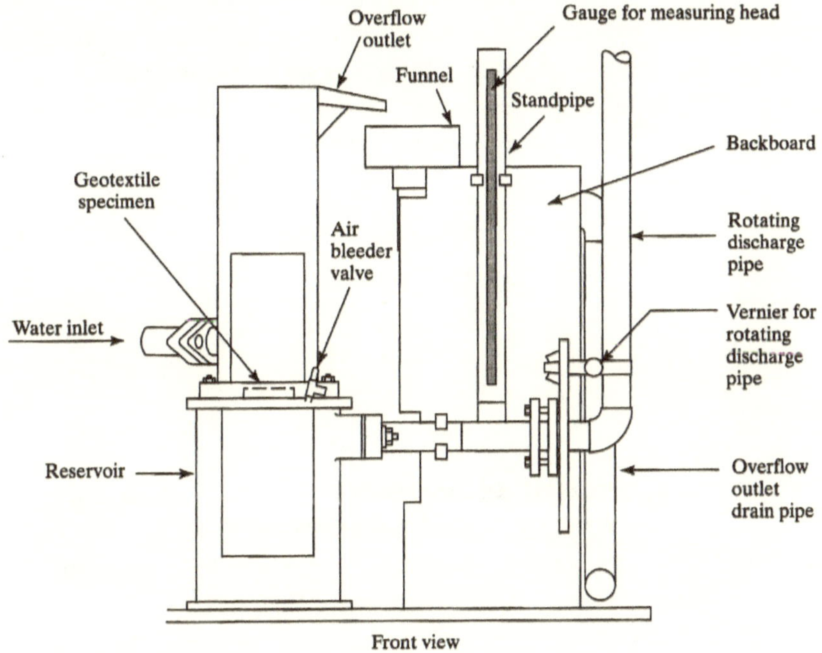

Figure 2.14 Permeability device for measuring geotextile permittivity (cross-plane flow).

to five) are placed on one another with an open-mesh stainless steel grid on top and bottom. This assembly is placed inside a permeameter and loaded normally via ceramic balls of approximately 12 mm diameter. Thus, normal stress is imposed on the geotextile, but flow is only nominally restricted. Loading by soil itself (which would definitely affect flow) is completed avoided. The test has been standardized as ASTM D5493, and results seem to indicate the following trends between standard permittivity and permittivity-under-load;

- Woven monofilament geotextiles: no change to a slight increase when under load.
- Woven slit-film geotextiles: data scatter is too large to establish trends.
- Nonwoven heat-bonded geotextiles: no change to slight decrease when under load.
- Nonwoven needle-punched geotextiles: slight decrease to moderate decrease depending on magnitude of load and mass/ unit area of geotextile.

Transmissivity (In-Plane Permeability). For flow of water within the plane of the geotextile (e.g., in the utilization of the drainage function), the variation of geotextile thickness (its compressibility under load) is again a major issue. Thus, transmissivity (θ) was introduced in equation (2.10); it is used in Darcy's formula as follows:

$$q = k_p iA \tag{2.19}$$

$$q = k_p i(W \times t)$$

$$k_p t = \theta = \frac{q}{iW} \tag{2.20}$$

where

- θ = transmissivity of the geotextile (m^2/sec or m^3/sec-m),
- k_p = permeability (properly called hydraulic conductivity) in the plane of the geotextile where the subscript p is often omitted (m/sec),
- t = thickness of the geotextile (m),
- q = flow rate (m^3/sec),
- W = width of the geotextile test specimen (m),
- i = hydraulic gradient (dimensionless) = $\Delta h/L$,
- Δh = total head lost (m), and
- L = length of the geotextile (m).

If the permeating fluid is other than water (e.g., a turbid water sludge or leachate), the density and viscosity can be accommodated for by using equation (2.18). Also, note in equation (2.20) that θ and q/W carry the same units, but they are numerically equal *only* at a hydraulic gradient i, of unity.

A number of test devices are configured to model the above formulation, where liquid (usually water) flows in the plane of the geotextile (of dimensions L × W × t) in a *parallel* flow trajectory; ASTM D4716 and ISO 12958 use such a device. Koerner and Bove [44] provide a review of such devices, and table 2.6, after Gerry and Raymond [15], gives typical values. These test devices are necessary for high-flow geonets and drainage geocomposites (discussed in chapters 4 and 8 respectively), but they are somewhat unwieldy for geotextiles. Such devices are large, time-consuming to setup and difficult to seal against sidewall leaks. This last item is particularly important for geosynthetics

TABLE 2.6 TYPICAL VALUES OF TRANSMISSIVITY AND IN-PLANE PERMEABILITY OF GEOTEXTILES, DATA AFTER GERRY AND RAYMOND [14]*

Type of Geotextile	Transmissivity m^2/s	Permeability Coefficient m/s
nonwoven, heat bonded	3.0×10^{-9}	6×10^{-6}
woven, slit film	1.2×10^{-8}	2×10^{-5}
woven, monofilament	3.0×10^{-8}	4×10^{-5}
nonwoven, needle punched	2.0×10^{-6}	4×10^{-4}

*Values are measured at an applied normal stress of 40 kPa .

with relatively low transmissivity values such as geotextiles. A low flow-rate measurement, with a high (and unknown) potential for leakage, results in a relatively uncertain value of transmissivity.

Instead, a variation of this concept is recommended for geotextile testing whereby *radial* drainage is achieved. Schematically such a device is shown in figure 2.15, where liquid enters into the inside of the upper load bonnet, then flows radially through the geotextile and is collected around the outer perimeter of the stationary reaction section of the device. The transmissivity concept is adapted accordingly as follows:

$$q = k_p iA$$

$$q = k_p \frac{dh}{dr}(2\pi rt) \tag{2.19}$$

$$2\pi(k_p t)\int dh = q\int \frac{dr}{r}$$

$$(k_p t) = \theta = \frac{q\,ln(r_2 / r_1)}{2\pi\Delta h} \tag{2.21}$$

where

r_2 = outer radius of the geotextile test specimen, and
r_1 = inner radius of the geotextile test specimen.

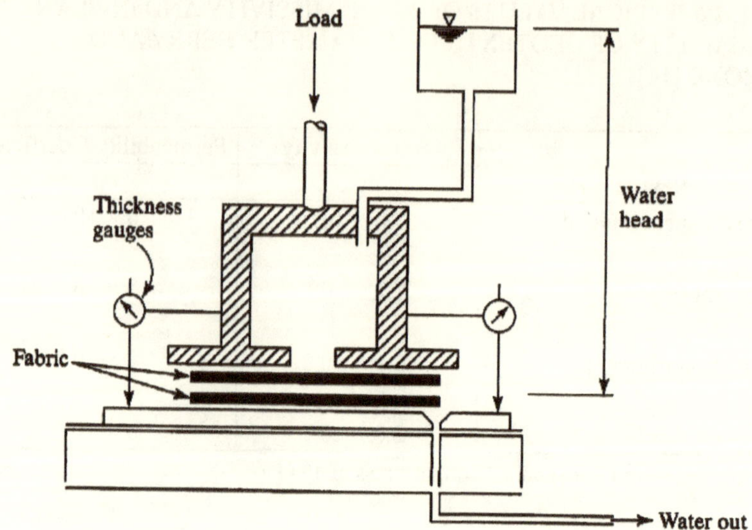

Figure 2.15 Permeability device for measuring geotextile transmissivity (radial in-plane flow).

The thicker, nonwoven geotextiles are best suited to convey water in the drainage function, but these are the same geotextiles that are subject to relatively high compression under load. Thus, the exponential decrease in transmissivity of the geotextiles shown in figure 2.16 should come as no surprise. Fortunately, most geotextiles reach constant values after approximately 100 kPa, beyond which the yarn structure is sufficiently tight and dense to hold the load and still convey liquid to the extent shown. Note also the increase in transmissivity with increasing mass per unit area (or number of layers).

Lastly, it should be noted that the radial device shown in figure 2.15 can readily be adapted to measure the in-plane flow of gases (e.g., air, methane, radon) by placing a shroud around the outside of the load bonnet. The gas is introduced under controlled pressure and measured at the outlet for its flow rate. Typical air transmissivity data is given in figure 2.17a. The same device can be used under combined airflow through partially saturated geotextiles to assess *permselectivity*, as shown in figure 2.17b.

Soil Retention: Underwater Turbidity Curtains. One variation of a soil-retention test is directed primarily toward the use of geotextiles as *underwater turbidity (or silt) curtains*. The test device consists of two rectangular tanks that are placed end to end with slide gates facing one another. Between these two slide gates is the geotextile test specimen. The upstream tank (with the slide gate closed) is now filled with water that has a known amount of uniformly mixed silt in it. The gate valve at the exit end of the downstream tank is opened. The test begins by lifting up the slide gates on each side of the geotextile and allowing the (turbid) water to pass through the geotextile, which is acting as a submerged soil filter. Clear water is continually added to the upstream tank to maintain a constant head. Two values are generated:

1. The flow rate and velocity through the geotextile, which is indicative of its void space and the amount of clogging that occurs.
2. The percentage of solids passing through the geotextile during the test process, which is indicative of the geotextile's retention capability.

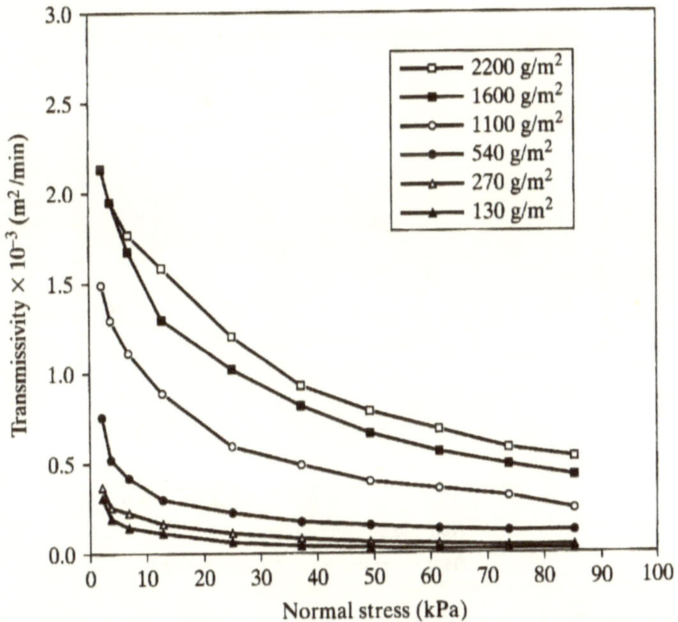

Figure 2.16 Transmissivity test results for different mass per unit area of nonwoven needle- punched geotextiles.

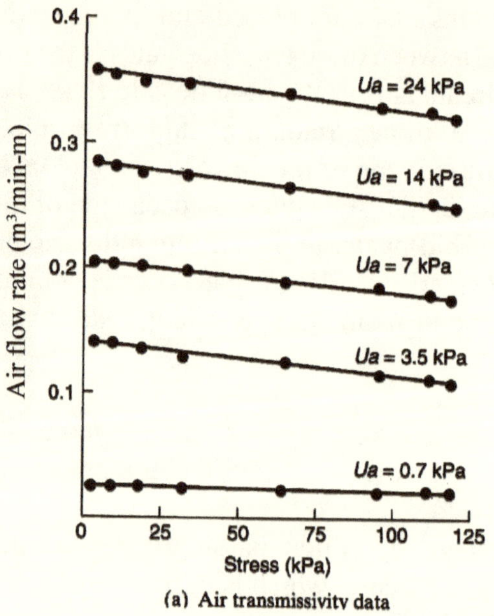

(a) Air transmissivity data

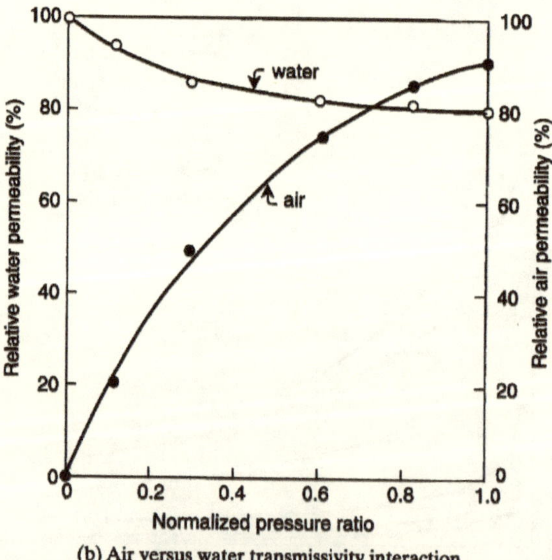

(b) Air versus water transmissivity interaction under 120 kPa normal pressure

Figure 2.17 Radial transmissivity data for air and air/water mixtures on a 550 g/m² needle- punched nonwoven geotextile. (After Koerner, et al. [45])

This test, developed by the New York Department of Transportation, is aimed at an assessment of turbidity of rivers and streams during adjacent construction activities.

Soil Retention: Above Ground Silt Fences. The second variation of a soil-retention test is directed toward the use of geotextiles as *aboveground silt fences*. The test protocol and setup is covered in ASTM D5141. Here the soil (usually a silty sand) is slurried in water and poured into a flume box measuring 1200 mm long × 800 mm wide × 300 mm high. The candidate geotextile, measuring 800 mm × 300 mm, forms the downstream end of the box, which is set at an 8% slope. The flow rate of the soil-water mixture is monitored with time, and the amount of fines passing through the geotextile is measured to determine the soil-retention capability. The process is repeated at least three times to determine the degree of clogging that has occurred. Two values are generally reported:

1. The slurry flow rate (l/min-m)
2. The retention efficiency (percent)

The recommended procedure, developed by the Virginia Department of Transportation under designation VTM-51, also includes a field method with the same objective of determining the filtering efficiency of the geotextile.

2.3.5 Endurance Properties

Thus far, the testing of geotextiles has concentrated on the short-term material behavior of the as-manufactured fabric. Yet the question remains regarding behavior during service conditions over the design lifetime of the system; in other words, *endurance* is of concern. This section addresses some of the tests that focus on this question. The reader should also see ASTM D5819 and ISO 13429, which are guides for selecting various endurance test methods.

Installation Damage. It should be obvious that harsh installation stresses can cause geotextile damage. In some cases installation stresses might be more severe than the actual design stresses for which the geotextile is intended. There are a number of references available, but most involve removal of the geotextile after considerable time, usually years. In this section, focus is on the immediate damage that can be caused by the contractors operations during installation.

In order to assess this situation, one hundred field sites were evaluated by removing approximately 1 m^2 of the geotextile within hours after placement. The procedure followed closely the ISO 13437 protocol. Most of the geotextiles were used for highway base separation, but some were for embankments, walls, underdrains, erosion control, staging areas, access ways, and so on. The entire exhumed sample was brought into the laboratory along with an equal size of the unused and uninstalled geotextile for comparison purposes. Test specimens were taken from the exhumed and unused geotextiles and were tested. The percentage of strength reduction was calculated. The following mechanical tests were performed:

- Grab tensile (3 to 6 tests per site)
- Puncture resistance (3 to 6 tests per site)
- Trapezoidal-tear resistance (3 to 6 tests per site)
- Burst resistance (3 to 6 tests per site)
- Wide-width strength in machine direction (1 to 2 tests per site)
- Wide-width strength in cross-machine direction (1 to 2 tests per site)

A hole count was also made of the exhumed geotextiles. In figure 2.18a, the relationship between retained strength (the weighted average of the above tests) and number of holes per square meter is observed. The entire data set was arbitrarily divided into three groups: acceptable, questionable, and nonacceptable (regions A, B, and C respectively in figure 2.18b). While loss of strength can be accommodated via a suitable installation-damage reduction factor (the inverse of the numeric strength retained value shown in figure 2.18 that will be given later), we could well wonder how the separation and filtration functions can properly work with the occurrence of so many holes. It should be noted that the recommendations of this study by Koerner and Koerner [46] suggests that no geotextile less than 270 g/m^2 should be used unless special precautions are taken, such as a sand cushioning layer along with lightweight construction equipment, to avoid installation damage.

Creep (Constant Stress) Tests. *Creep* is the common name applied to the elongation of a geotextile under constant load. Since polymers are generally considered creep-sensitive materials, it is an important property to evaluate. The test specimens should be of the wide-width variety (recall figure 2.8c) and are usually stressed by means of stationary hanging weights. Since the test duration should be long, a number of tests are often conducted simultaneously by cascading the test specimens and their respective loads. The setup can also be horizontal with a number of specimens connected to one another.

Selection of the load is important, and it is usually based on a percentage of the geotextile's strength as determined from a conventional test, as described in section 2.3.3. If such a value is considered to be 100%, creep test stresses of 20%, 40%, and 60% are sometimes evaluated. Stresses are commonly applied for one thousand to ten thousand hours (depending on the particular application) and deformation readings are taken at progressively longer time increments from the beginning of the test. Obviously, continuous deformation readings can be taken by LVDTs or other electronic monitoring equipment. The elongation or percent strain (deformation divided by original length) versus time should be plotted for each stress increment. Both ASTM D5262 and ISO 13431 describe details of the testing procedure. Although completely arbitrary, the slope of the terminal portion of the curve (in units of mm/hr.) is then reported. The resulting values can then be empirically compared to maximum allowable values or used directly in a predictive procedure, as described by Shrestha and Bell [47]. Creep rates can also be calculated from theseresponse curves, plotted and compared to some agreed on limiting value of strain,—for example, at 10%.

An accelerated creep test based on the concept of time-temperature-superposition has recently been used with success. It is known as the stepped isothermal method, or SIM, see Thornton and Baker [49]. The test is conducted per ASTM D6992 wherein a test specimen within an environmental chamber is maintained at a given tensile stress level and time; deformation is monitored continuously. The stress is then ramped higher and held accordingly. This process is continued producing a series of incremental deformation (strain) curves. Computer software is then used to obtain a master curve that represents simulated times out to a hundred, or more, years.

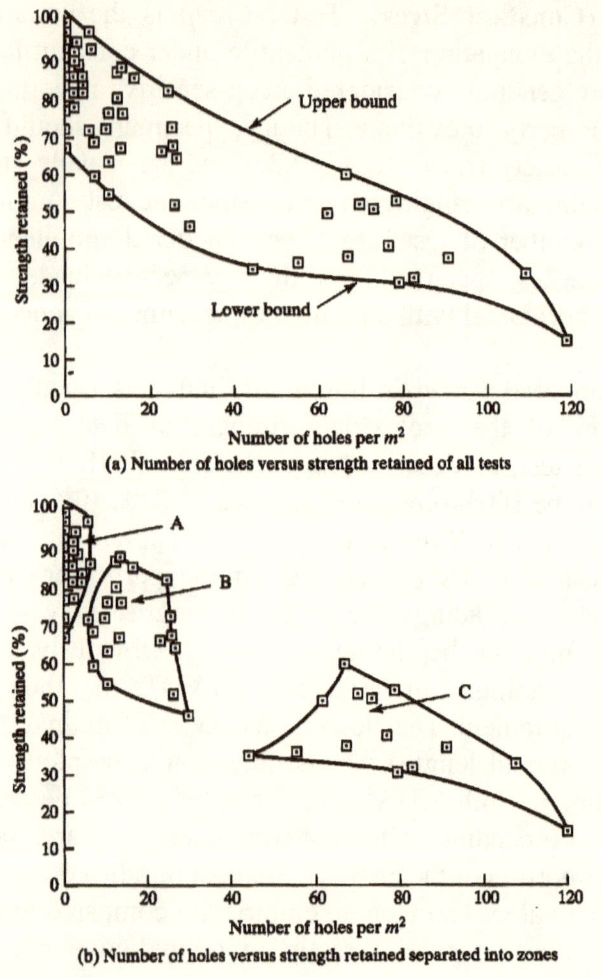

Figure 2.18 Results of field exhuming of geotextiles immediately after installation to assess installation damage. (After Koerner and Koerner [46])

Numerous references are available on the creep behavior of geotextile-forming yarns and fabrics. Perhaps the greatest sensitivity is due to stress level and polymer type (see figure 2.19). Such information is very important in design, since the inverse of the quasi-static strength at which no (or a minimum) creep occurs is used as a value for the reduction factor necessary to avoid objectionable creep deformation. Trends from den Hoedt [48] are shown in figure 2.19 and creep reduction factors will be given later.

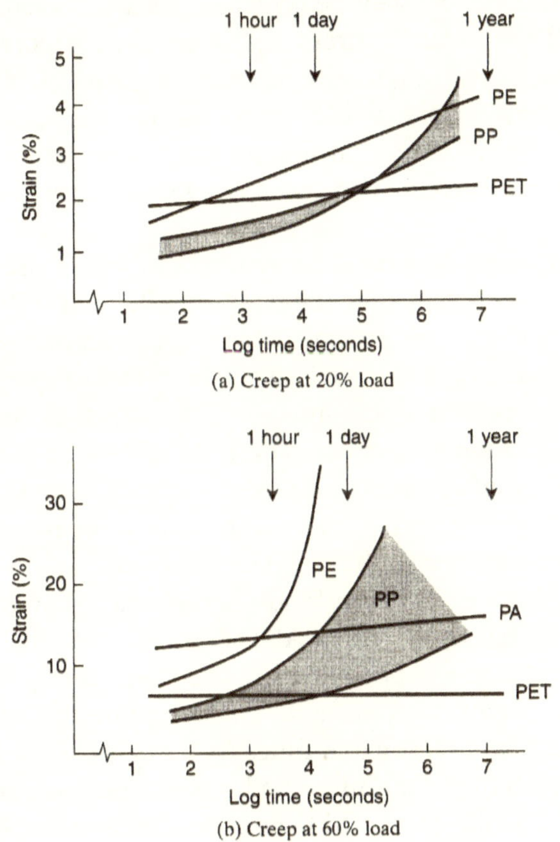

Figure 2.19 Results of creep tests on various yarns of different polymers. (After den Hoedt [48])

Confined Creep Tests. As with the confined wide-width tensile test mentioned previously, McGown et al. [21,22] have also performed confined creep tests. The general tendency of these tests is to show that the creep behavior can be improved with soil confinement. As with the short-term confined tensile tests of section 2.3.3, the major improvement is with nonwoven needle-punched geotextiles, followed by other nonwovens, and then by woven geotextiles, the last appearing to show little if any improvement. While expensive and time-consuming to perform, such tests are important to set realistic creep-reduction factors.

Stress Relaxation (Constant Strain) Tests. Stress relaxation is the common name given to the reduction in stress of a material while it is maintained under a constant deformation. As with creep (to which it is mathematically related), stress relaxation is an important property. Unfortunately, the test setup is difficult, requiring load cells and electronic strain measurements and, as with creep, taking considerable time. The literature is notably absent on this property in regard to geotextiles, however, there are geogrid and geomembrane data available [50].

Stress (Creep) Rupture. In general, polymers used in-service above their glass transition temperature (PE and PP) are in the "rubbery" state and are governed by a creep-limited strain that is sometimes used as a 10% maximum. The previous creep and stress relaxation tests pertain to this situation. Polymers that function below their glass transition temperature (PET, PA, PVA) are in the "glassy" state. Hence, the creep strains are generally quite low and stress (creep) rupture will likely occur before a given strain value is reached. Thus, the need for a variation in the creep test method.

Fortunately, the test setup and configuration for stress rupture is identical to that described in ASTM D5262 and ISO 13431. The difference is that the applied stresses are significantly higher. They must be sufficiently high so as to cause rupture, or failure, of the test specimens in a reasonable time. As many as 10 to 15 replicate specimens are brought to failure under different loads and analyzed accordingly. Ingold et al. [51] describe the process that will be addressed in detail in chapter 3, which deals with geogrids that are always used in the reinforcement function. It should be mentioned that rupture at low loads is difficult to accomplish at typical laboratory testing temperatures and elevated temperatures are used accordingly. As a result, time-temperature superposition (TTS) per ASTM D5292 and stepped isothermal testing (SIM) per ASTM D6992 along with the necessary curve shifting is necessary. Both methods significantly decrease testing time while the latter has the added advantage of using a single test specimen.

Abrasion Tests. The abrasion of geotextiles when in service can be the cause of failure of soil-geotextile systems. The ASTM test methods for abrasion resistance of textile fabrics are designated D1175 and cover six different procedures: inflated diaphragm, flexing and abrasion, oscillatory cylinder, rotary platform-double head, uniform abrasion, and impeller tumble.

In all cases, abrasion is defined as "the wearing away of any part of a material by rubbing against another surface." There are, however, a large number of variables to be considered in such a test. Results are reported as the percent weight loss or strength/elongation retained under the specified test and its particular conditions.

One of the tests for evaluating the abrasion resistance of geotextiles is the rotary platform-double head (Taber Test, Model 503) method. In the test, both heads are weighted and consist of vitrified (CS-17) abrasion wheels. The test specimen is disk-shaped with a 90 mm outer diameter and a 60 mm inner diameter. The specimen is placed on a rubber base on the platform that is rotated and abraded by the stationary abrasion wheels for up to 1000 cycles. Two strip tensile specimens are then cut from the abraded geotextile and tested for their tensile strengths. The average value is then compared to the tensile strength of nonabraded geotextile and the results reported as a percentage of strength retained by the geotextile after abrasion. The percentage of elongation retained after abrasion can also be reported.

Although this particular test is straightforward to perform and many devices are commercially available, the ASTM and ISO preference is for a uniform (sandpaper) abrasion test. The designations are D4886 and 13427 respectively. Data presentation and analyses are similar to that just described.

All these tests, however, are questionable simulators of field abrasion conditions. In many cases it would be better to use some sort of tumble test, such as the German Test Standard DIN 5385, which is a large test using basalt-stone aggregate abrading a geotextile test specimen within a one meter diameter rotating drum.

Long-Term Flow (Clogging) Test. One of the greatest hydraulic endurance concerns is that of the long-term flow capability of a geotextile with respect to the flow coming from the upstream soil. Tests are needed to assess the potential of *excessive* geotextile clogging.

The most direct testing approach is to take a sample of the soil at the site and place it above the candidate geotextile, which is fixed in position in a test cylinder. It is then evaluated under constant head flow over a long period of time. A set of such test units can be built and a series of candidate geotextiles tested simultaneously (see figure 2.20a). The general response is piecewise linear, with the initial portion due largely to densification of the soil and not of direct interest. At a transition time of approximately 10 hr. (for granular soils) to 200 hr. (for fine-grained soils), the soil-geotextile system will enter its field-simulated behavior. If the slope of the response curve becomes essentially zero after this transition time, the geotextile is compatible with the soil at least under the imposed test conditions (hydraulic gradient, temperature, water, etc; see figure 2.20b). Assuming that the flow-rate value at equilibrium is adequate for the situation, the candidate geotextile(s) should be appropriate. If the slope continues to be negative, however, increased clogging is indicated and eventually excessive or complete clogging could occur. In such a case, the geotextile is not suited for this type of soil and these test conditions. The database for this particular test has been extended for both clear and turbid water using a number of soil-to-geotextile conditions [53]. If the slope reverses and goes positive, upstream soil loss (also called "piping") is indicated. This is equally unsatisfactory as is excessive clogging, but for a different reason. Piping indicates that the geotextile is too open for the upstream soil, while excessive clogging indicates that the voids are too tight. The test is a good discriminator in this regard.

Although seemingly straightforward in its approach to answering the clogging question, there are drawbacks to the test, the major one being time. The test should normally run for 1000 hr. (≅ 40 days) to clearly establish the slope of the curve beyond the transition time. (Note the time axis in figure 2.20b is logarithmic.) This is unfortunately too long for many real-time situations, where an answer regarding potential clogging is usually needed within a few days. Also, the test chamber can develop bacteria growth in a warm laboratory environment over the required test time, and periodic flushing with a detergent is necessary; this could cause changes in the soil-geotextile system. Finally, the question of de-aired and/or de-ionized water must always be addressed in hydraulic tests of this type.

Gradient Ratio (Clogging) Test. A test that may be performed in a considerably shorter time than the long-term flow test, and that is aimed at determining the hydraulic compatibility of a soil-geotextile system is the US Army Corps of Engineer's Gradient Ratio Test CW-02215. It has been adopted (with slight variations) by ASTM as the D5101 Test Method. In this test, the flow configuration is set up similarly to the long-term flow test just described. Now, however, instead of measuring flow rates, the hydraulic head at various locations in the soil-geotextile column is measured.

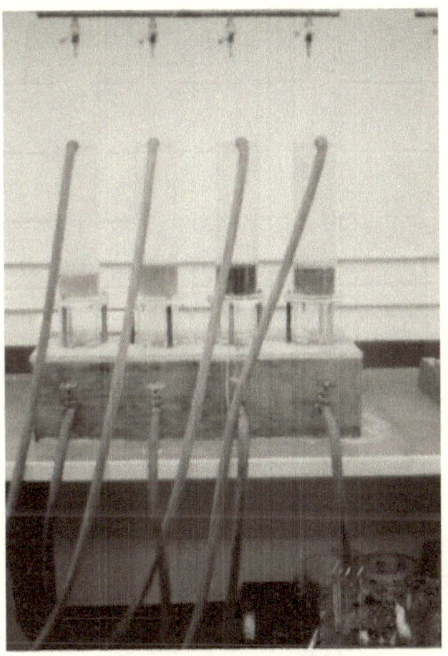

(a) Photograph of setup

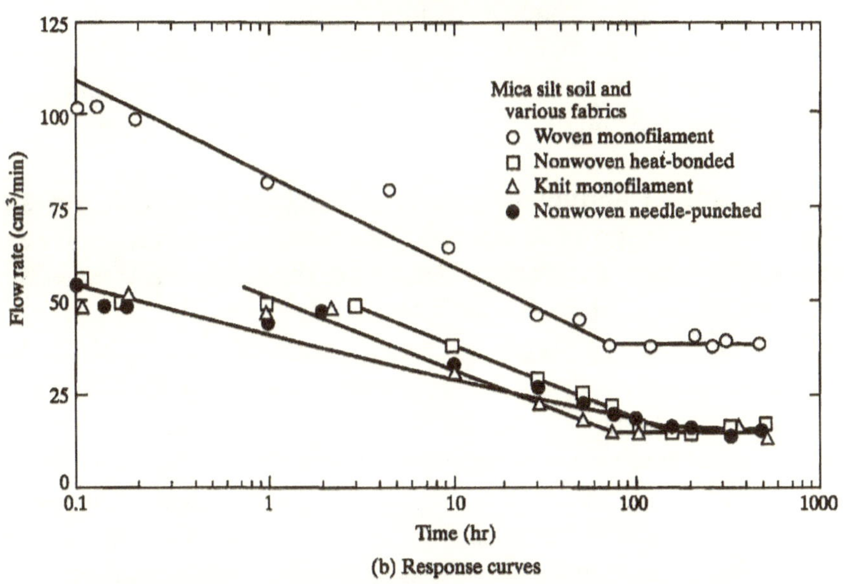

(b) Response curves

Figure 2.20 Long-term flow tests on soil-geotextile systems and typical response curves (After Koerner and Ko [52])

Head differences are then converted to hydraulic gradients and finally the gradient ratio (GR) value, as defined below, is calculated.

$$\textit{Gradient Ratio} = \frac{\Delta h_{GT+25S} / t_{GT+25}}{\Delta h_{50S} / t_{50}} \tag{2.22}$$

where

Δh_{GT+25S} = head change (mm) from the bottom of the geotextile to 25 mm of soil above the geotextile,

t_{GT+25} = geotextile thickness (mm) plus 25 mm of soil,

Δh_{50S} = head change (mm) between 50 mm of soil above the geotextile,

t_{50} = 50 mm of soil.

The Corps of Engineers suggests that gradient ratio values greater than three indicate nonacceptable geotextiles for the type of soil under test.

The test is not without its share of problems and complications, including long-term stability of the gradient ratio value [8], piping along the test cylinder walls, use of de-aired or de-ionized water, and air pockets in the soil, geotextile, and head monitoring system.

Hydraulic Conductivity Ratio (Clogging) Test. Williams and Abouzakhm [54] propose the use of a flexible wall permeameter test to assess not only excessive clogging conditions but also excessive soil loss and equilibrium conditions. In the hydraulic conductivity ratio (HCR) test, one prepares the soil column as per ASTM D5084, the customary flexible wall permeameter test for soils albeit with the candidate geotextile on the top of the soil column. Complete geotechnical engineering procedures should be deployed—that is, back-pressure saturation, stress state loading conditions, site-specific soil, and site-specific permeating liquid. A hydraulic conductivity (permeability) test on the sample is now performed in two separate modes according to ASTM D5567.

1. The permeant flows down through the *clean* geotextile and through the soil column, resulting in the soil permeability, k_s.

2. Flow is then reversed and the permeant flows up through the soil column and the covering (now downgradient) geotextile, thereby challenging its behavior and ultimately resulting in the soil/geotextile permeability, k_{sg}.

Using such data, a HCR value is calculated using the equilibrium values of k_s and k_{sg} as follows.

$$HCR = k_{sg}/k_s \tag{2.23}$$

Values of HCR are then plotted against pore values passed through the system. An interpretation of these curves is suggested by Luettich and Williams [55] whereby high values of HCR suggest soil loss, low values of HCR suggest excessive clogging, and intermediate values of HCR (for example, between 0.4 and 0.8) suggest soil-to-geotextile equilibrium. The author feels that this test is the premier laboratory test to assess the potential of excessive geotextile clogging and/or soil-retention concerns.

2.3.6 Degradation Considerations

"How long will the geotextile last?" This important question for permanent and/or critical applications is asked more frequently than any other in geosynthetics. It will be addressed through a description of different degradation mechanisms and testing procedures in this subsection. It should be noted that all the degradation mechanisms to be described result in some form of molecular chain scission, bond breaking, cross-linking, or the extraction of formulated components. Thus, there is a fundamental change (albeit it extremely slow in a buried environment) in the polymer at the molecular level from its as-manufactured state. While many chemical fingerprinting methods can be used to detect these changes (recall table 1.5) they are time-consuming, expensive, and tedious to perform. At the macroscopic level, the mechanical properties will eventually change from the as-manufactured state. For example, a stress-versus-strain curve will gradually transition from a plastic to brittle behavior in that; the modulus will increase, the elongation at failure will decrease, and the strength will often temporarily increase, but will eventually decrease. The reader should also see ASTM D5819 and ISO 13429, which are guides for selecting various degradation test methods.

Sunlight (Ultraviolet) Degradation. Sunlight is an important cause of degradation to all organic materials, including the polymers from which geosynthetics are made. For geosynthetic purposes, energy from the sun is divided into three parts:

- Infrared, with wavelengths longer than 760 nm
- Visible, with wavelengths between 760 and 400 nm
- Ultraviolet or UV, with wavelengths shorter 400 nm

The UV region is further subdivided into UV-A (400 to 315 nm), which causes some polymer damage; UV-B (315 to 280 nm), which causes severe polymer damage; and UV-C (280 to 100 nm), which is only found in outer space.

From summer to winter there are changes in both the intensity and the spectrum of sunlight, most significant being the loss of shorter-wavelengths of UV radiation during the winter months. Other factors in the UV degradation process of polymers are geographic location, temperature, cloud cover, wind, moisture, and atmospheric pollution. These should be considered in any test method. Laboratory simulations are approximate but nonetheless very important.

For laboratory simulation of sunlight, artificial light sources (lamps) are generally compared with worst-case conditions, or the "solar maximum condition." The actual degradation is caused by photons of light breaking the polymer's chemical bonds. For each type of bond, there is a threshold wavelength for bond scission; above the threshold the bonds will not break, below it they will. Thus, the short wavelengths are critical. The literature [56] shows that polyethylene is most sensitive to UV degradation around 300 nm, polyester around 325 nm, polypropylene around 370 nm, polyvinyl chloride around 312 nm, and polystyrene around 315 nm; that said, they are all within the UV-A or UV-B range of the wavelength spectrum.

Of the available laboratory weathering devices, the two most common are the Xenon arc (ASTM D4355) and the ultraviolet fluorescent devices (ASTM D7238). Both produce spectra in the UV-A and UV-B regions quite accurately (where degradation occurs), but the Xenon arc continues into visible light while the UV-fluorescent drops off. That said, essentially everything else favors the UV-fluorescent device as table 2.7 indicates. The author feels that for long-term weathering evaluations to full degradation of the geosynthetic material, the geosynthetics industry should favor the UV-fluorescent method and write specifications accordingly.

TABLE 2.7 APPROXIMATE COSTS FOR DIFFERENT LABORATORY WEATHERING DEVICES

Item	Xenon Arc	UV-Fluorescent
initial cost	\$70-80,000	\$10-15,000
tubes/bulbs	\$15,000/year	\$300/year
power cost	\$5000/year	\$400/year
water cost	\$3000/year	none
sewer cost	\$1500/year	none

Using either exposure device the test coupons are removed at designated times, cut into strip-tension test specimens, and evaluated for their retained strength and elongation. The results are then compared to the unexposed geotextile for percent retained values [57]. Alternatively, one can use the information obtained in a predictive manner to determine the equivalent field behavior at a specific location.

Example 2.7

Assume that a geotextile reaches its half-life elongation in laboratory weathering device (Energy = 517.8 W/m^2) in 2000 hours. What is the equivalent lifetime in Philadelphia with a known average exposure energy of 5021 MJ/m^2-yr? *Note*: Joule (J) = watts (W) × seconds (sec).

Solution:

$$E_{test} = (517.8)(2000)(3600)(1\times10^{-6})$$
$$= 3728 MJ/m^2$$
$$E_{Phila} = (5021 MJ/m^2 - yr)(1/4 \; sun \; time)$$
$$= 1255 MJ/m^2 - year$$
$$T_{Phila} = \frac{3728}{1255} = 2.97 \; years$$

Thus, the acceleration factor (AF) of the exposure device over real-time exposure is as follows:

$$AF = \frac{(2.97)(365)}{(2000)(1/24)} = 13$$

Ultraviolet degradation is also covered by ASTM under the title "Outdoor Weathering of Plastics," designated D1435. This procedure is intended to define site-specific conditions for the exposure of plastic materials to light and weather. It is a comparative test depending on a defined climate, time of year, atmospheric conditions, and so on. Racks are constructed with the geotextile coupons to be evaluated fixed to them. Samples can be placed at 0, 45°, or 90° to the horizontal and in different solar orientations. Exposure test samples should simulate service conditions of the end-use application as much as practical. A specific version of this test is available as ASTM D5970. Clearly, if a test of site-specific UV degradation is desired at a critical field site where the geotextile is to be exposed for months or years, outdoor weathering tests of this type should be considered.

Whatever the test method used to produce UV-degradation results, it is clear that geotextiles must be shielded from prolonged ultraviolet light exposure. Geotextile rolls are always shipped with a protective plastic covering and only when the material is ready for use should it be unrolled and exposed. The manufacturer's recommendations for "timely cover" (backfilling) must be rigidly met. Cover placement (soil or another geosynthetic) for polypropylene geotextiles should generally be within fourteen days (per AASHTO M288), with polyesters being allowed longer exposure times. For long-term exposed applications, either outdoor or laboratory testing is necessary; both of which can be performed on a product-specific and actual or simulated site-specific basis.

Temperature Degradation. Extremely high temperature causes all polymer degradation mechanisms to occur at an accelerated rate. In fact, at the heart of *time-temperature superposition* lifetime prediction techniques used in Arrhenius modeling is to test laboratory specimens at high temperatures (e.g., from 50°C to 100°C) and extrapolate the accelerated degradation down to field-anticipated temperatures. Thus, high temperature is an acceleration phenomenon acting with other degradation mechanisms like sunlight, oxidation, hydrolysis, chemical, radiation, biological, and so on. As such, laboratory temperature testing (per se) as an individual degradation mechanism will not be discussed separately.

Regarding the mechanical behavior of plastics (insofar as engineering properties are concerned), hot and cold temperatures cause a softening and stiffening respectively, as would be expected. For geotextiles, high temperatures slightly increase flexibility, and ASTM D1388 can be used to quantify the behavior (recall section 2.3.2). ASTM Test Method D746 addresses the effect of cold temperatures on plastics and, in particular, on brittleness and impact strength. At severely cold temperatures, specimens are tested by a specified impact device (recall section 2.3.3).

For geotextiles, however, neither hot nor cold *ambient* temperatures are generally important topics or issues, except in extreme environmental situations.

Oxidation Degradation. While all types of polymers react with oxygen causing degradation, the polyolefins (polypropylene and polyethylene) are generally considered to be the most susceptible to this phenomenon. Hsuan et al. [58] describe the chemical mechanism. ASTM Recommended Practice D794 describes high-temperature oxidation testing for plastics. Only the incubation procedure for heat exposure is specified, the test method(s) for assessment being governed by the potential end use. Heat is applied using a forced-air oven with controlled airflow and with substantial fresh-air intake. Two types of incubation are described: continuous and cyclic heat. In the former, heat is gradually increased until failure occurs. Failure is defined as a change in appearance, weight, dimension, or other properties that alter the material to a degree that it is no longer serviceable for the purpose. The test may be very short or require months, depending on the rate of temperature increase. The cyclic heat test repeatedly applies heat up to a constant value until failure.

A number of research efforts are ongoing to assess geotextile oxidative behavior using forced-air ovens at (constant) elevated temperatures; the higher the temperature, the greater the rate of oxidative degradation. Changes in tensile strength, elongation and modulus are tracked over time. Properly plotted, these trends are back-extrapolated to a site-specific (i.e., lower) temperature to arrive at a predicted lifetime. The procedure is the essence of time-temperature superposition, followed by Arrhenius modeling. It will be described in chapter 5 on geomembranes.

Caution should be exercised in the incubation of geotextiles at extremely high temperature. Polypropylene melts at 165°C and polyethylene melts at 125°C. Such high temperatures should obviously be avoided and incubation should be at significantly lower temperatures.

Hydrolytic Degradation. Hydrolysis can cause degradation via either internal or external fiber or yarn reactions. Geosynthetics manufactured using polyester resins are particularly affected when the immersion liquid has a very high (pH > 10) or very low (pH < 3) alkalinity.

Trends in degradation behavior insofar as loss of strength is concerned is provided by Hsuan et al. [59]. Extremely high pH values can affect some polyesters, while extremely low pH values can be harsh on some polyesters and polyamides. It is important that the polyester resin used for permanent geotextile applications has a high molecular weight (e.g., > 25,000) and a low carboxyl end group (CEG) concentration (e.g., < 30). These effects are further described by Hsuan et al. [58] and Hsieh et al. [60].

In cases of concern, the candidate geotextile should be incubated in water having the prevailing pH levels at 20°C and 50°C for at least 120 days and then tested for changes in strength and elongation. For a base-line the comparison, it is important to have a complete parallel set of samples incubated in distilled water (pH = 7) at the same temperatures.

Chemical Degradation. ASTM Method D543 covers chemical degradation under the title "Resistance of Plastics to Chemical Reagents." The test method includes provisions for reporting changes in weight, dimensions, appearance, and strength. Provisions are also made for various exposure times and exposure to reagents at elevated temperatures. A list of fifty standard reagents is supplied in order to attempt some sort of standardization.

The DuPont Company has evaluated most of its yarns (including acetate, dacron, nylon, orlon, rayon, cotton, wool, and silk) under a wide range of chemicals (sulfuric acid, hydrochloric acid, nitric acid, hydrofluoric acid, phosphoric acid, organic acids, sodium hydroxide, bleaching agents, scouring, and laundering agents, salt solutions, and organic and miscellaneous chemicals), many of which were at different concentrations and at different temperatures. After the specified exposure, the coupons were rinsed, air-dried, and then conditioned at 21°C and 65% relative humidity for 16 hours. Data on test specimen strength, elongation, and toughness of the exposed yarns were compared to control specimens of the yarns that were not exposed to the chemical. Similar information is available from most raw material suppliers and some geotextile manufacturers for their particular products in many standard chemical environments.

Notable exceptions to the use of literature values using standard incubation liquids are geotextiles used at landfill sites, heap leach pads, and agriculture waste facilities. Here the liquid can be very aggressive and uniquely site specific. In such cases, the actual leachate, or a synthetized version thereof, is to be used as the incubation liquid. ASTM D5322 presents the laboratory incubation methodology and ASTM D5496 gives an alternative field incubation methodology. Upon removal of the geotextile coupons (typically 120 days at 50°C), they are tested according to ASTM D6389 to assess the percent change from the nonincubated samples. Decisions as to the geotextile's compatibility or noncompatibility to the site-specific leachate are made accordingly.

Biological Degradation. In order for microorganisms such as bacteria or fungi to degrade polymers, the organisms must attach themselves to the fiber or yarn surfaces and use the polymer as a feedstock. This is highly unlikely. All the resins used for geosynthetics are very high in molecular weight with relatively few chain endings for the biodegradation process to be initiated. Irrespective of the above comment, ISO 12961 is focused on microbiological degradation.

The additives to the polymer, however, might be somewhat more vulnerable than the resin itself. Plasticizers or processing aids could be vulnerable, although there is no authoritative research on the subject to my knowledge.

Converse to biological *degradation*, biological *clogging* is clearly a concern depending on the characteristics of the permeating liquid. For those liquids high in microorganism content, exacerbated by high suspended solids content, excessive filter clogging has occurred. Clearly, landfill leachates have been problematic in this regard [61, 62], and animal waste runoff from large farms promises to be likewise.

Radioactive Degradation. While there are no references in the open literature on the radioactive degradation of geotextiles, the subject is generally discussed when dealing with radioactive waste disposal. It is assumed, though clearly not proven, that low-level radioactive exposure (by hospital garments, nuclear power plant tools and clothing, ground emanating radon, etc.) is orders of magnitude too low to cause chain scission of geosynthetic-related polymers (see NRC Workshop [63]). Conversely, high-level radioactive waste (e.g., spent nuclear fuel rods) is suspect and, if in the proximity of geotextiles, could cause radiation degradation. The situation must be experimentally evaluated if radioactive conditions are anticipated, although such testing will undoubtedly involve major health and safety issues and be extremely expensive.

Other Degradation (and Damage) Processes. Other processes that may degrade geotextiles are ozone attack (only in certain climates and when exposed) and rodent or muskrat attack (which is site specific and has happened when the animal was trapped). Also, seagulls, hawks, and vultures have been known to pull apart exposed geotextiles at many landfill and reservoir sites. In spite of these degradation and/or damage mechanisms, the performance record for geotextile durability has been quite good as will be described in the next section.

Geotextile Aging. While specific test standards are not available to measure the aging of geotextiles (due to the complexity of the many mechanisms involved), field-exhuming work is continually being reported. Perhaps the longest functioning geotextile that is periodically exhumed is at the Volcros Dam in France (Gourc and Faure [64]). Both mechanical and hydraulic properties are examined and compared to original properties. Losses were generally nominal with maximum reductions (perhaps installation-related) being 30%.

Numerous studies on the exhuming of geotextiles have been reported in the literature. Tests on these recovered samples show that geotextiles remain in good to excellent condition. This indicates to the author that the proper polymers and formulations are at hand and are being utilized in the manufacture of geotextiles used in the applications to be discussed. For extremely long service lifetimes a laboratory procedure using time-temperature-superposition and Arrhenius modeling will be presented in chapter 5.

2.3.7 Summary

This section on geotextile properties is very important and could have been dealt with in greater detail than we have. The subject deserves this attention, for any quantifiable design method will result in numbers to be compared to the candidate geotextile's actual properties. This section dealt with the relevant properties and subsequent test values, how they are obtained, their authenticity, and their reliability. In some cases, additional inquiry and research was recommended.

The section included a mixture of in-isolation (or index) properties and soil-to-geotextile related (or performance) properties. Eventually, the tests for these different properties will sort out into their respective categories and uses, but most organizations are looking at the complete collection of tests as they were presented here.

Regarding a summary of geotextile properties, the rapidly changing market and its demands make it difficult to give precise values. However, for many commercially available geotextiles, the annual *Specifier's Guide of Geosynthetics Magazine* is an excellent source [65]. For specific values of a particular type of geotextile the respective manufacturers should be consulted; all have up-to-date websites that can be freely accessed.

2.4 ALLOWABLE VERSUS ULTIMATE GEOTEXTILE PROPERTIES

It is important to recognize that the preceding geotextile test properties represent idealized conditions and, therefore, result in the maximum possible numeric values when used directly in design; that is, they result in upper-bound values. In the design-by-function concept described in section 2.1.3, the factor of safety was formulated around an allowable test value (equations 2.2a and 2.2b). Generally, most laboratory test values cannot be used directly; they must be suitably modified for the anticipated in-situ conditions. This could be done directly in the test procedure, for example by conducting a completely simulated performance test; but in most cases, this simply is not possible. Simulating installation damage, performing long-term creep testing, using site-specific liquids, reproducing in-situ pore water stresses, providing complete stress state modeling, and so on are generally not feasible. To account for such differences between the laboratory measured test value and the desired performance value, two approaches can be taken:

1. Require an extremely high factor of safety at the end of a problem.
2. Use reduction factors on the laboratory-generated test value to make it into a site-specific allowable value.

At the suggestion of Voskamp and Risseeuw [66], the latter alternative of *reduction factors* (RF values) will be used in this book. By doing this, the usual value of factor of safety can be assessed in the final analysis. Our approach will be to refer to the general laboratory-obtained value as an "ultimate" value and to modify it by reduction factors to obtain an "allowable" value. See [67] for additional rationale and numeric values.

2.4.1 Strength Related Problems

For problems dealing with geotextile strength, such as in separation and reinforcement applications, the formulation of the allowable values takes the form of equation 2.24a. Typical ranges of values for reduction factors for different applications are given in table 2.8a. These values must be tempered by the site-specific considerations. Also, note that if the laboratory test includes the mechanism listed, it would appear in the equation as a value of 1.0.

$$T_{allow} = T_{ult}\left(\frac{1}{RF_{ID} \times RF_{CR} \times RF_{CBD}}\right) \tag{2.24a}$$

$$T_{allow} = T_{ult}\left(\frac{1}{\Pi RF}\right) \tag{2.24b}$$

TABLE 2.8a RECOMMENDED STRENGTH REDUCTION FACTOR VALUES FOR USE IN EQ. 2.24a

Area	Range of Reduction Factors		
	Installation Damage	Creep*	Chemical/Biological Degradation**
Separation	1.1 to 2.5	1.5 to 2.5	1.0 to 1.5
Cushioning	1.1 to 2.0	1.2 to 1.5	1.0 to 2.0
Unpaved roads	1.1 to 2.0	1.5 to 2.5	1.0 to 1.5
Walls	1.1 to 2.0	2.0 to 4.0	1.0 to 1.5
Embankments	1.1 to 2.0	2.0 to 3.5	1.0 to 1.5
Bearing and foundations	1.1 to 2.0	2.0 to 4.0	1.0 to 1.5
Slope stabilization	1.1 to 1.5	2.0 to 3.0	1.0 to 1.5
Pavement overlays	1.1 to 1.5	1.0 to 2.0	1.0 to 1.5
Railroads (filter/sep.)	1.5 to 3.0	1.0 to 1.5	1.5 to 2.0
Flexible forms	1.1 to 1.5	1.5 to 3.0	1.0 to 1.5
Silt fences	1.1 to 1.5	1.5 to 2.5	1.0 to 1.5

*The low end of the range refers to applications which have relatively short service lifetimes and/or situations where creep deformations are not critical to the overall system performance.
**Previous editions of this book have listed biological degradation as a separate reduction factor. As described in Section 2.3.6, however, there is no evidence of such degradation for the polymers used to manufacture geotextiles.

where

T_{allow} = allowable tensile strength,
T_{ult} = ultimate tensile strength,
RF_{ID} = reduction factor for installation damage (≥ 1.0),
RF_{CR} = reduction factor for creep (≥ 1.0),
RF_{CBD} = reduction factor for chemical and biological degradation (≥ 1.0), and
ΠRF = value of cumulative reduction factors (≥ 1.0).

Note that equation 2.24a could have included additional site-specific terms, such as reduction factors for seams and for intentionally made holes. It also could have been formulated with fractional multipliers (values ≤ 1.0) and placed in the numerator of the equation or on the opposite side of the equation as with the *load and reduction factor design method* (LRFD). While the equation indicates tensile strength, it can be applied to burst strength, tear strength, puncture strength, impact strength, and so on.

2.4.2 Flow Related Problems

For problems dealing with flow through or within a geotextile, such as filtration and drainage applications, the formulation of the allowable values takes the form of equation 2.25a. Typical values for reduction factors are given in table 2.8b. Note that these values must be tempered by the site-specific conditions as in section 2.4.1. If the laboratory test includes the mechanism listed, it would appear in the equation as a value of 1.0.

$$q_{allow} = q_{ult}\left(\frac{1}{RF_{SCB} \times RF_{CR} \times RF_{IN} \times RF_{CC} \times RF_{BC}}\right) \tag{2.25a}$$

$$q_{allow} = q_{ult}\left(\frac{1}{\Pi RF}\right) \tag{2.25b}$$

where

q_{allow} = allowable flow rate,
q_{ult} = ultimate flow rate,
RF_{SCB} = reduction factor for soil clogging and blinding (≥ 1.0),
RF_{CR} = reduction factor for creep reduction of void space (≥ 1.0),
RF_{IN} = reduction factor for adjacent materials intruding into geotextile's void space (≥ 1.0),
RF_{CC} = reduction factor for chemical clogging (≥ 1.0),
RF_{BC} = reduction factor for biological clogging (≥ 1.0), and
ΠRF = value of cumulative reduction factors (≥ 1.0).

TABLE 2.8b RECOMMENDED FLOW REDUCTION FACTOR VALUES FOR USE IN EQUATION 2.25a

Application	Range of Reduction Factors				
	Soil Clogging and Blinding*	Creep Reduction of Voids	Intrusion into Voids	Chemical Clogging**	Biological Clogging***
Retaining wall filters	2.0 to 4.0	1.5 to 2.0	1.0 to 1.2	1.0 to 1.2	1.0 to 1.3
Underdrain filters	2.0 to 10	1.0 to 1.5	1.0 to 1.2	1.2 to 1.5	2.0 to 4.0***
Erosion control filters	2.0 to 10	1.0 to 1.5	1.0 to 1.2	1.0 to 1.2	2.0 to 4.0
Landfill filters	2.0 to 10	1.5 to 2.0	1.0 to 1.2	1.2 to 1.5	2.0 to 5.0***
Gravity drainage	2.0 to 4.0	2.0 to 3.0	1.0 to 1.2	1.2 to 1.5	1.2 to 1.5
Pressure drainage	2.0 to 3.0	2.0 to 3.0	1.0 to 1.2	1.1 to 1.3	1.1 to 1.3

*If stone rip-rap or concrete blocks cover the surface of the geotextile use either the upper values, orinclude an addition reduction factor.

**Values can be higher particularly for high alkalinity groundwater.

***Values can be higher for extremely high microorganism content and/or growth of organisms and plant/vegetation roots.

As with the equation 2.24 for strength reduction, this flow reduction equation could also have additional site-specific terms included, e.g., blocking of a portion of the geotextile's surface by rip-rap or concrete blocks.

2.5 DESIGNING FOR SEPARATION

Application areas for geotextiles used for the separation function were given in section 1.3.3. There are indeed many specific applications, and it could said, in a general sense, that geotextiles always serve a separation function. If they do not serve this function, any other function, including the primary function, will not be served properly. This should not give the impression that the geotextile function of separation always play a secondary role. Many situations call for separation only, and in such cases the geotextiles do serve a significant and worthwhile purpose.

2.5.1 Separation Applications

There are many geotextile applications where separation is the major function. Certainly, the use of thick and heavy geotextiles used to cushion geomembranes against stone puncture falls into this category. Since it is geomembrane-related it will be addressed in chapter 5. A much more common application that best illustrates the use of geotextiles as separators is its placement between a reasonably firm soil subgrade (beneath) and a stone base course, aggregate, or ballast (above). We say "reasonably firm" because it is assumed that the subgrade deformation is not sufficiently large to mobilize uniformly high-tensile stress in the geotextile. (The application of geotextiles in unpaved roads on soft soils with membrane-type reinforcement is treated later in section 2.6.1.) Thus, for a separation function to occur, the geotextile has only to be placed on the soil subgrade and then have stone placed, spread, and compacted on top of it. The subsequent deformations are very localized and occur around each individual stone particle. A number of design scenarios can be developed showing what geotextile properties are required for this type of situation.

2.5.2 Burst Resistance

Consider a geotextile on a soil subgrade with stone of average particle diameter (d_a) placed above it. If the stone is uniformly sized, there will be voids within it that will be available for the geotextile to enter. This entry is caused by the simultaneous action of the traffic loads being transmitted to the stone, through the geotextile, and into the underlying soil. The stressed soil then tries to push the geotextile up into the voids within the stone. The situation is shown schematically in figure 2.21. Giroud [68] provides a formulation for the required geotextile strength that can be adopted for this application.

$$T_{reqd} = \frac{1}{2} p' d_v [f(\varepsilon)] \qquad (2.26)$$

where

T_{reqd} = required geotextile strength,
p' = stress on the geotextile that is slightly less than, p, the tire inflation pressure at the ground surface,
d_v = maximum void diameter of the stone $\cong 0.33 d_a$,
d_a = the average stone diameter,
$f(\varepsilon)$ = strain function of the deformed geotextile
$= \frac{1}{4}\left(\frac{2y}{b} + \frac{b}{2y}\right)$, in which
b = width of opening (or void)
y = deformation into the opening (or void)

The field situation is analogous to the ASTM D3786 (Mullen) burst test, which has the geotextile being stressed into a gradually increasing hemispherical shape until it fails (recall section 2.3.3). Thus, the adapted form of equation 2.26 is as follows:

$$T_{ult} = \frac{1}{2} p_{test} d_{test} [f(\varepsilon)] \qquad (2.27)$$

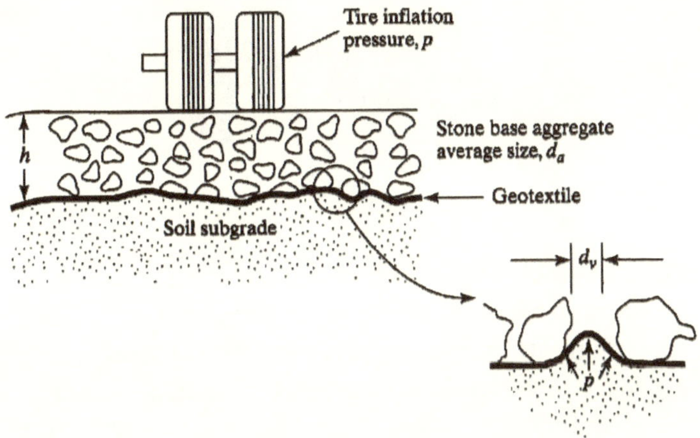

Figure 2.21 Geotextile being forced up into voids of stone base by traffic tire loads.

where

T_{ult} = ultimate geotextile strength,
p_{test} = burst pressure of the geotextile at failure (its strength), and
d_{test} = diameter of the burst test device (= 30 mm).

Knowing that $T_{allow} = T_{ult}/(\Pi RF)$, where ΠRF = cumulative reduction factors, we can formulate an expression for the factor-of-safety (FS) as follows:

$$FS = \frac{T_{allow}}{T_{reqd}}$$
$$= \frac{(p_{test}\, d_{test})}{(\Pi RF)p' d_v}$$

For example, if d_{test} = 30 mm; $d_v = 0.33\, d_a$; and $\Pi RF = 1.5$ (which is reasonable since creep is not an issue with this application), then the FS value is the following, with d_a in units of mm.

$$FS = \frac{p_{test}(30)}{(1.5)p'(0.33d_a)}$$

$$FS = \frac{60.6 p_{test}}{p' d_a} \qquad (2.28)$$

Example 2.8:

Given a truck with 700 kPa tire inflation pressure on a poorly graded aggregate layer consisting of 50 mm maximum-size stone. What is the factor of safety using a geotextile beneath the aggregate having an ultimate burst strength of 2000 kPa and cumulative reduction factors of 1.5.

Solution: Assuming that the tire inflation pressure is not significantly reduced through the thickness of the stone base, we can solve equation 2.28 as follows:

$$FS = \frac{60.6(2000)}{700(50)}$$

$$FS = 3.5$$

Note that with the reduction factors of 1.5 already included, the resulting factor of safety value is acceptable.

For a range of stone base particle diameters (d_a), values of tire inflation pressure (p'), and cumulative reduction factors of 1.5, along with a factor of safety of 2.0, we can generate the design guide in figure 2.22. Here it can be seen that stone size is quite significant insofar as the required burst pressure values are concerned. Note also that these are poorly graded aggregates and that the presence of fines will lessen the severity of the design; hence this approach should be considered to be a worst-case design.

2.5.3 Tensile Strength

Continuing the discussion of geotextile roadway separation, there are other processes acting on the geotextile simultaneous as its tendency to burst in an out-of-plane mode. One of these is tensile stress being mobilized by in-plane deformation. This occurs as the geotextile is locked into position by stone-base aggregate above it and soil subgrade below it. A lateral, or in-plane, tensile stress in the geotextile is mobilized when an upper piece of aggregate is forced between two lower pieces that are in contact with the geotextile. The analogy to the grab tensile test can be readily visualized, as illustrated in figure 2.23. Here we can estimate the maximum strain that the geotextile will undergo as the upper stone wedges itself down to the level of the geotextile. Using the dimensions shown (where $S \simeq d/2$ and l_f = deformed geotextile length), the maximum strain with no slippage nor stone breakage can be calculated.

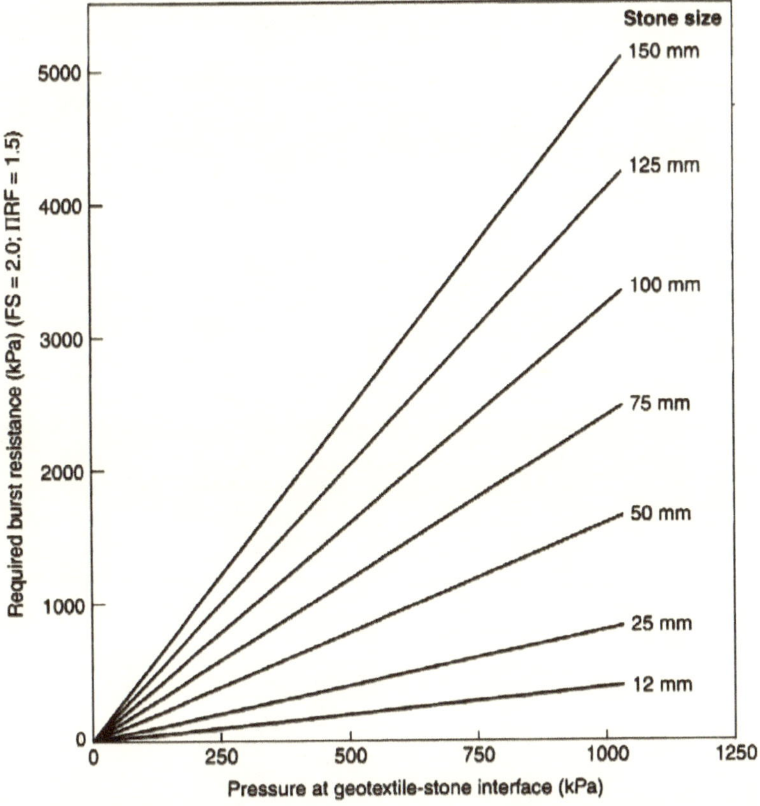

Figure 2.22 Design guide for burst analysis of geotextile used in a separation function based on cumulative reduction factors of 1.5 and a factor of safety of 2.0

$$\varepsilon = \frac{l_f - l_o}{l_o}(100)$$
$$= \frac{[d + 2(d/2)] - 3(d/2)}{3(d/2)}(100)$$
$$= \frac{4(d/2) - 3(d/2)}{3(d/2)}(100)$$
$$\varepsilon = 33\%$$

Note that the preceding assumptions result in a strain that is independent of particle size. Thus, the strain in the geotextile could be as high as 33% given the idealized (upper bound) assumptions stated above. Now the tensile force being mobilized is related to the pressure exerting on the stone as follows [68]:

$$T_{reqd} = p' \, (d_v)^2 \, [f(\varepsilon)] \tag{2.29}$$

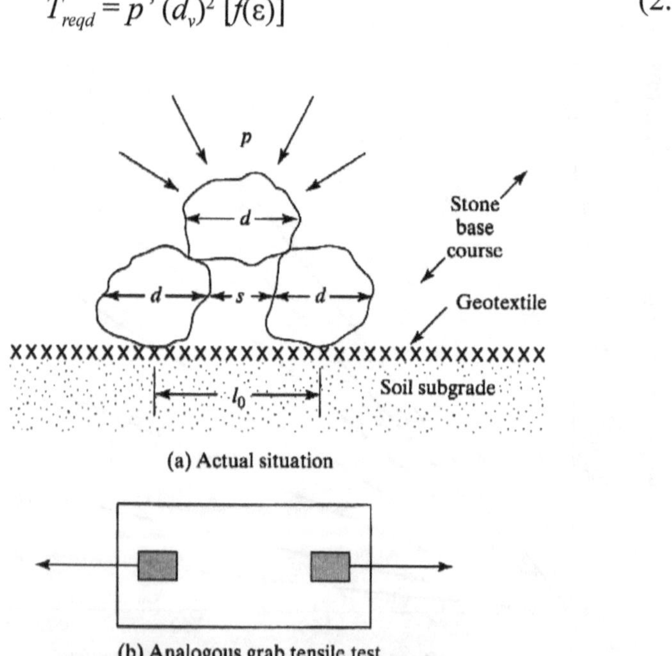

Figure 2.23 Geotextile being subjected to tensile stress as surface pressure is applied and stone base attempts to spread laterally.

where

T_{reqd} = required grab tensile force,
p' = applied pressure,
d_v = maximum void diameter $\simeq 0.33\ d_a$, where
d_a = average stone diameter, and
$f(\varepsilon)$ = strain function of the deformed geotextile

$$= \frac{1}{4}\left(\frac{2y}{b} + \frac{b}{2y}\right)\text{, where}$$

b = width of stone void, and
y = deformation into stone void.

Example 2.9:

Given a truck with 700 kPa tire inflation pressure on a stone base course consisting of 50 mm maximum-size stone with a geotextile beneath it. Calculate (a) the required grab tensile stress on the geotextile, and (b) the factor of safety for a geotextile whose grab strength at 33% is 500 N with cumulative reduction factors of 2.5. Use a value of f(ε) = 0.52.

Solution: (a) Using an empirical relationship that d_v = 0.33 d_a and the value of $f(\varepsilon) = 0.52$, the required grab tensile strength is as follows:

$$\begin{aligned} T_{reqd} &= p'(d_v)^2(0.52) \\ &= p'(0.33d_a)^2(0.52) \\ &= 0.057p'd_a^2 \\ &= 0.057(700)(1000)(0.050)^2 \\ T_{reqd} &= 100N \end{aligned}$$

(b) The factor of safety on a 500 N grab tensile geotextile at 33% strain with reduction factors of 2.5 is as follows:

$$FS = \frac{T_{\text{allow}}}{T_{\text{reqd}}}$$

$$= \frac{500/2.5}{100}$$

$$FS = 2.0, \textit{ which is acceptable.}$$

2.5.4 Puncture Resistance

The geotextile must always survive the installation process. This is not just related to the roadway separation function; indeed, fabric survivability is critical in all types of applications; without it the best of designs are futile (recall figure 2.18). In this regard, sharp stones, tree stumps, roots, miscellaneous debris, and other items, either on the ground surface beneath the geotextile or placed above it, could puncture through the geotextile during backfilling and when traffic loads are imposed. The design method suggested for this situation is shown schematically in figure 2.24. For these conditions, the vertical force exerted on the geotextile (which is gradually tightening around the protruding object) is as follows:

$$F_{reqd} = p' \, d_a^2 S_1 S_2 S_3 \tag{2.30}$$

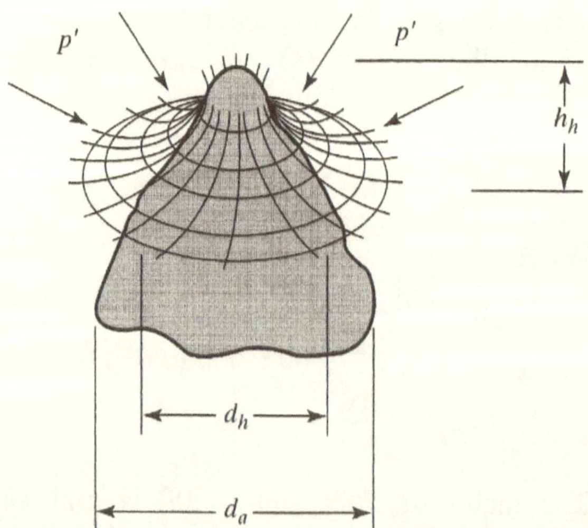

Figure 2.24 Visualization of a stone puncturing a geotextile as pressure is applied from above.

where

F_{reqd} = required vertical puncturing force to be resisted,
d_a = average diameter of the puncturing aggregate or sharp object,
p' = pressure exerted on the geotextile (approximately 100% of tire inflation pressure at the ground surface for thin covering thicknesses),
S_1 = protrusion factor of the puncturing object (see table 2.9),
S_2 = scale factor to adjust the ASTM D4833 puncture test value that uses a 8.0 mm diameter puncture probe to the actual puncturing object (see table 2.9),
S_3 = shape factor to adjust the ASTM D4833 flat puncture probe to the actual shape of the puncturing object (see table 2.9).

TABLE 2.9 RECOMMENDED VALUES FOR INDEPENDENT FACTORS USED IN PUNCTURE ANALYSIS (DIMENSIONLESS)

Puncturing Object	S_1	S_2	S_3
angular & relatively large	0.9	0.8	0.9
angular & relatively small	0.6	0.6	0.7
subrounded & relatively large	0.7	0.6	0.6
subrounded & relatively small	0.4	0.4	0.5
rounded & relatively large	0.5	0.4	0.4
rounded & relatively small	0.2	0.2	0.3

S_1 = protrusion factor
S_2 = scale factor } see Eq. 2.30
S_3 = shape factor

Example 2.10:

What is the factor of safety against puncture of a geotextile from a subrounded 25 mm diameter stone on the ground surface mobilized by a loaded truck with tire inflation pressure of 550 kPa traveling on the surface of the base course? The geotextile has an ultimate puncture strength of 300 N according to ASTM D4833.

Solution: Using the full stress on the geotextile of 550 kPa and intermediate factors from table 2.9 of 0.55, 0.50, and 0.55 for S_1, S_2 and S_3 respectively.

$$F_{reqd} = p' d_a^2 S_1 S_2 S_3$$
$$= (550)(1000)(25 \times 0.001)^2 (0.55)(0.50)(0.55)$$
$$F_{reqd} = 52 \text{ N}$$

Assuming that the cumulative reduction factors are 2.0, the factor of safety is as follows:

$$FS = \frac{F_{allow}}{F_{reqd}}$$
$$= \frac{300/2.0}{52}$$
$$FS = 2.9, \text{ which is acceptable}$$

2.5.5 Impact (Tear) Resistance

As with the puncture requirement just described, the resistance of a geotextile to impact is as much a survivability criterion as it is a separation function. Yet in many instances of separation, the geotextile must resist the impact of various objects. The most obvious one is that of a rock falling on it, but there are also situations in which construction equipment and materials can cause or contribute to impact damage on geotextiles.

The problem addresses the energy mobilized by a free-falling object of known weight and height of drop. Rarely will an object be intentionally impelled onto an exposed geotextile with additional force, so only gravitational energy will be assumed.

To develop a design procedure, we assume a free-falling rock of specific gravity of 2.60, varying in diameter from 25 to 600 mm and falling from heights of 0.5 to 5 m. Using this data the design curves of figure 2.25 are developed. The relationship used is as follows:

$$\begin{aligned} E &= mgh \\ &= (V \times \rho)gh \\ &= [V \times (\rho_w G_s)]gh \\ &= \left(\frac{\pi(d_a/1000)^3}{6}\right)\left(\frac{1000kg}{m^3}\right)(2.6)(9.81)h \end{aligned}$$

$$E = 13.35 \times 10^{-6} d_a^{\ 3} h \tag{2.31}$$

where

E = energy developed (joules),
m = mass of the falling object (kg),
g = acceleration due to gravity (m/sec^2),
h = height of fall (m),
V = volume of the object (m^3),
ρ = density of the object (kg/m^3),
ρ_w = density of water (kg/m^3),
G_s = specific gravity of the object (dimensionless), and
d_a = diameter of the object (mm).

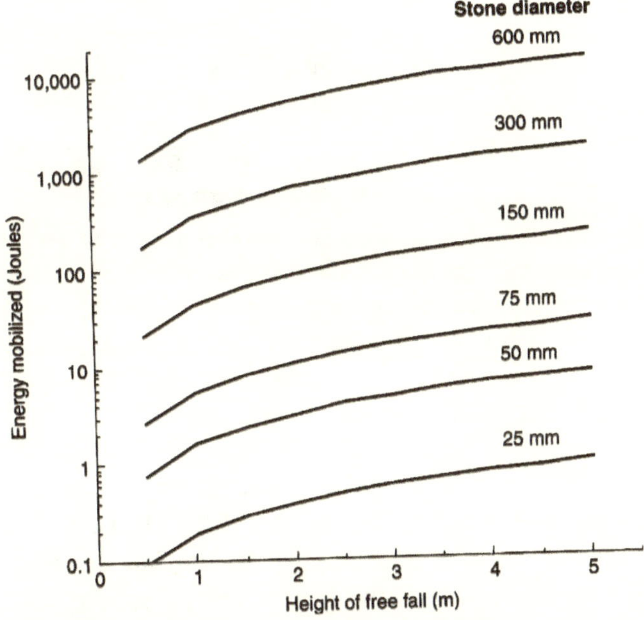

Figure 2.25 Energy mobilization by a free-falling rock on a geotextile with an unyielding support.

Note that these calculated energies are based on the geotextile resting on an unyielding surface that is the worst possible condition. As the soil beneath the geotextile deforms, the geotextile can absorb greater amounts of impacting energy. Since this is usually the case, the modification factors of figure 2.26 are to be used in conjunction with the curves of figure 2.25. Once the required energy is calculated, it should be compared to the allowable impact strength of the geotextile (e.g., the Elmerdorf tear or dynamic cone drop test as discussed in section 2.3.3).

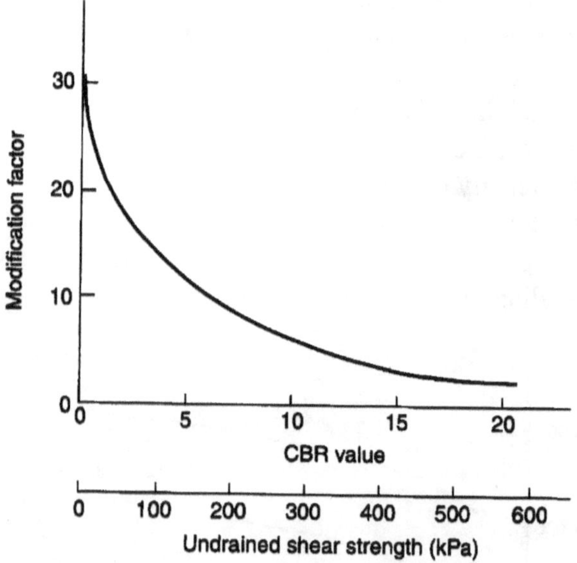

Figure 2.26 Modification factor to be used with energy mobilized by objects falling on geotextiles of varying support resistances characterized by their unsoaked CBR values or undrained shear strength.

Example 2.11:

What energy is mobilized by a free-falling rock of 300 mm size falling 1.5 m onto a geotextile? The geotextile is supported by a poor subsoil having an unsoaked CBR strength of 4. If the geotextile has an allowable impact strength of 36 J, what is the factor of safety?

Solution: Using equation 2.31 one calculates the required impact energy:

$$E_{max} = 13.35\times10^{-6}\left(d_a^3\right)(h)$$
$$= 13.35\times10^{-6}(300)^3(1.5)$$
$$E_{max} = 540J$$

Note that this value is substantiated by the design chart of figure 2.25. Of course, other design charts can be made for different assumptions. This value is now reduced according to the subgrade conditions of figure 2.26.

$$E_{reqd} = 540/13$$
$$= 41.5J$$

This results in a global factor of safety calculation as follows:

$$FS = \frac{E_{allow}}{E_{reqd}}$$
$$= \frac{36}{41.5}$$
$$FS = 0.87, \textit{which is not acceptable.}$$

Thus, holes are likely to be formed when free-falling objects of this size fall directly on the exposed geotextile. Not included in this analysis is the effect of the contact area of the falling object on the geotextile; for a very rounded rock, the effect is much less severe than for a sharp, angular one, which could easily cut through the fabric.

It should be emphasized that the last two methods of puncture and impact design refer not only to roadway separation per se, but to construction survivability of geotextiles in general, recall table 2.2a. In all cases the considerations of this section should be examined, for they are critical in many situations.

2.5.6 Summary

Separation, the *most underrated of all geosynthetic functions*, was addressed in this section. It is underrated because every use of *geosynthetics* carries with it the separation function, yet rarely is separation designed on its own merit. Hopefully, the designs in this section will allow the engineer to determine quantitatively that geotextile is suitable for a specific situation.

Last, and in a sense most important, is the economic justification for the use of geotextiles in the separation function. It lies in the greater use and service lifetime of the system with geotextiles than without. When a geotextile separator is used in roadway cross sections, geotextiles could well double or triple lifetime; however, field data for such quantification is just now becoming available [69, 70].

2.6 DESIGNING FOR ROADWAY REINFORCEMENT

The combined use of soil (good in compression and poor in tension) and a geotextile (good in tension and poor in compression) suggests a number of situations in which geotextiles have made existing designs work better or provided for the development of entirely new applications. These applications were previewed in section 1.3, together with a brief history of the original uses in this particular application area. This section focuses on the design methods using geotextiles for various roadway reinforcement applications.

2.6.1 Unpaved Roads

Overview. The application in this section is for use of geotextiles in unpaved roads, in which soft soil subgrades have sand or stone aggregate placed directly above. No permanent surfacing, such as concrete or asphalt pavement, is immediately placed on the stone. At most, the road is surfaced with quarry crusher run or chip seal for reasonable ridability. There are many thousands of kilometers of unpaved secondary roads, haul roads, access roads, and the like, with no permanent surfacing on them. At a later time, perhaps years after settlement takes place and ruts are backfilled, a permanent surfacing may be placed.

This particular application triggered the high-volume use and acceptance of geotextiles in the 1970s, since calculations can be made for the thickness of stone required without a geotextile, then with a geotextile; the difference being the thickness of stone that is saved. By determining the cost of saved stone versus the cost of the geotextile, the value of using a geotextile is known immediately. The particular design process used in arriving at the respective thicknesses is the focus of this section.

Before beginning, however, it is important to realize that the geotextile must have its tensile modulus or strength mobilized via deformation of the soil subgrade. Although this can be done intentionally by prestressing the fabric, this is usually not the case because of the construction difficulties involved. Instead, the yielding of the soil subgrade by the imposed traffic is the triggering phenomenon, allowing for geotextile deformation and the mobilization of its tensile properties. How much deformation is necessary with regard to the vehicular loading, the particular geotextile, the time it takes for adequate strength mobilization, and so on, are all pressing questions, but the deformation characteristics of the soil subgrade takes precedence. A soft, yielding soil subgrade is needed to mobilize the geotextile's strength—but how soft? In light of the tremendous variety of situations, we must use a broad generality; and in this case, it will be based on the California Bearing Ratio (CBR) of the soil subgrade. The CBR test is used throughout the world and standardized accordingly (e.g., see ASTM D1883 or ISO 12236). The CBR value is a comparison of the subgrade soil's resistance to the force of a 50 mm diameter plunger at a given deformation, with that of a standardized crushed stone base material. It is actually a percentage value, although rarely expressed as such. The test on the soil subgrade can be performed at the in-situ moisture content, or the soil can be saturated for 24 hr. and then tested. These two conditions give rise to unsoaked and soaked CBR values respectively. Typical unsoaked CBR values are given in table 2.10, where a considerable body of empirical correlations is presented. Soaked CBR values are generally lower than unsoaked values, but the difference depends on the soil type. Table 2.11 is given as a guide in this regard; note that it correlates well to the Penn DOT specification of table 2.1.

It should also be noted that *resilient modulus* (obtained from a cyclic triaxial soils test) is beginning to replace CBR as an indicator of soil subgrade strength in pavement design. Charts comparable to those in table 2.10, however, are not yet available.

TABLE 2.10 CORRELATION CHART FOR ESTIMATING UNSOAKED CBR VALUES FROM SOIL STRENGTH OR PROPERTY VALUES

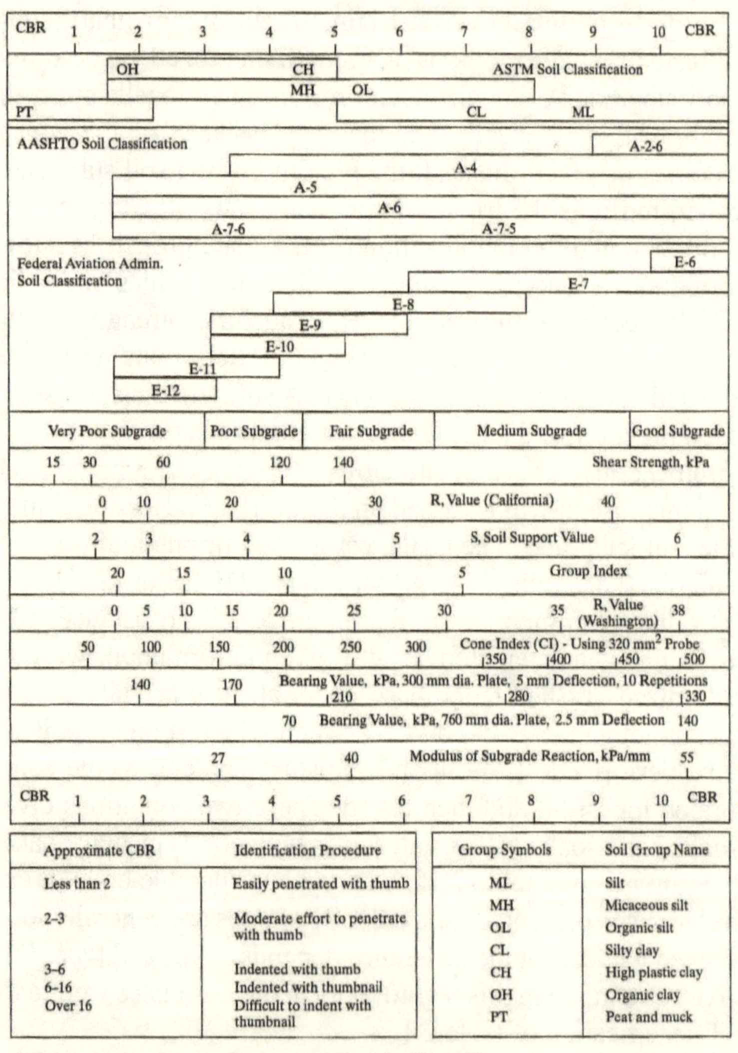

Approximate CBR	Identification Procedure
Less than 2	Easily penetrated with thumb
2–3	Moderate effort to penetrate with thumb
3–6	Indented with thumb
6–16	Indented with thumbnail
Over 16	Difficult to indent with thumbnail

Group Symbols	Soil Group Name
ML	Silt
MH	Micaceous silt
OL	Organic silt
CL	Silty clay
CH	High plastic clay
OH	Organic clay
PT	Peat and muck

Source: After Portland Cement Association and E.I. DuPont literature.

For the purposes of using geotextiles in roadway applications on soil subgrades of different strength characteristics, we will subdivide the functions per table 2.11. Here we can see that with medium to firm soil subgrades [CBR (unsoaked) ≥ 8 and CBR (soaked) ≥ 3] the function is uniquely separation. The design for this condition was described in section 2.5. For poor to very soft soil subgrades—CBR (unsoaked) ≤ 3 and CBR (soaked) ≤ 1—the function is both reinforcement and separation. This is the topic of this section. The intermediate category is only loosely defined; it is generally called *stabilization*. This term represents an interrelated group of functions (separation, reinforcement and filtration) and is essentially a transition category between the two extremes.

Manufacturers' Methods. All the major geotextile manufacturers have an unpaved-road design method for use with their particular geotextiles. They usually show CBR (or other related soil strength values) on the *x*-axis and the required stone thickness (with and without a geotextile) on the *y*-axis. All result in logical behavior, with the geotextile providing greater savings in stone aggregate as the soil subgrade becomes weaker. Since most manufacturers have a range of geotextiles available for reinforcement of unpaved roads, it is also seen that the heavier and stronger geotextiles result in greater stone savings than the lighter and weaker ones. Because each manufacturer's set of curves has its own background (based on theory, laboratory work, field observation, or empirical observation), it is nearly impossible to compare one method with another. Yet the designs have served the industry well and generally with excellent success. Their use is certainly acceptable and, if only one geotextile is available, its manufacturer's method should continue

TABLE 2.11 RECOMMENDED SOIL SUBGRADE CBR VALUES TO DISTINGUISH DIFFERENT GEOTEXTILE FUNCTIONS IN ROADWAY CONSTRUCTION

Geotextile Function(s)	CBR - Value	
	Unsoaked	Soaked
separation only	≥ 8	≥ 3
stabilization* and separation	8 to 3	3 to 1
reinforcement and separation	≤ 3	≤ 1

*a frequently used but poorly defined transition term which always includes separation, some unknown amount of reinforcement, and usually filtration as well.

to be used. If, however, a number of geotextiles are available, a method that views them on the basis of a specific, well-defined property is needed. Such a property could well be the geotextile's tensile modulus, which is the basis of design in the procedure to follow. It should be noted, however, that a number of generic techniques are available, and that Hausmann [71] has assessed and compared them to one another.

Analytic Method. Giroud and Noiray [72] use the geometric model shown in figure 2.27 for a tire wheel load of pressure p_{ec} on a $B \times L$ area, which dissipates through h_o thickness of stone base without geotextile and h thickness of stone base with a geotextile. The geometry indicated results in a stress on the soil subgrade of p_o (without geotextile) and p (with geotextile) as follows:

$$p_o = \frac{P}{2(B + 2h_o \tan \alpha_o)(L + 2h_o \tan \alpha_o)} + \gamma h_o \tag{2.32}$$

$$p = \frac{P}{2(B + 2h \tan \alpha)(L + 2h \tan \alpha)} + \gamma h \tag{2.33}$$

where

P = axle load, and
γ = unit weight of the stone aggregate.

Since the pressure exerted by the axle load through the aggregate and into the soil subgrade is known, the shallow-foundation theory of geotechnical engineering can now be utilized. It is assumed throughout the analysis that the soil is functioning in its undrained condition, and thus, its shear strength is represented completely by the cohesion (i.e., $\tau = c$). In this regard, the tacit assumption is that the soil subgrade consists of saturated fine-grained silt and clay soils. Critical in this design method are the assumptions that without the geotextile the maximum pressure that can be maintained corresponds to the elastic limit of the soil, that is,

$$p_o = \pi c + \gamma h_o \tag{2.34}$$

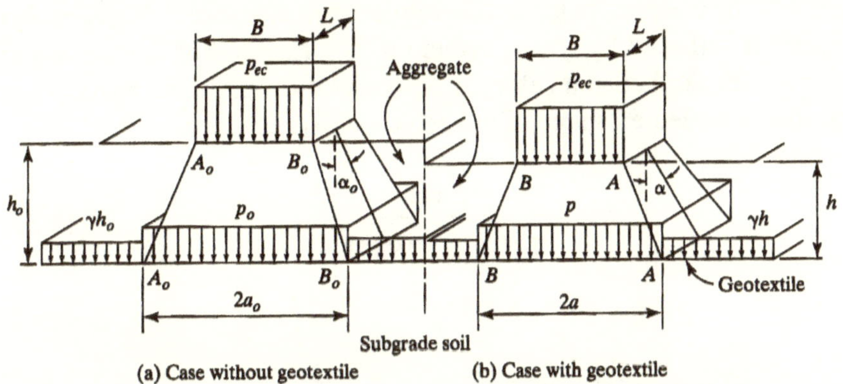

Figure 2.27 Load distribution by aggregate layer. (After Giroud and Noiray [72]): (a) Case without geotextile (b) Case with geotextile

and that with the geotextile the limiting pressure can be increased to the ultimate bearing capacity of the soil, that is,

$$p^* = (\pi + 2)c + \gamma h \tag{2.35}$$

These assumptions reasonably agree with the earlier findings of Barenberg and Bender [73] using small-scale laboratory tests, where on a deformation basis they found that large-scale ruts began at a 3.3c value with no fabric reinforcement, versus a 6.0c value with fabric (where c is the undrained soil shear strength).

Thus, for the case of no geotextile reinforcement, equations (2.32) and (2.34) can be solved, resulting in equation 2.36, which can result in the desired aggregate thickness response curve without the use of a geotextile:

$$c = \frac{P}{2\pi\left(\sqrt{P/p_c} + 2h_o \tan\alpha_o\right)\left(\sqrt{P/2p_c} + 2h_o \tan\alpha_o\right)} \tag{2.36}$$

where

c = soil cohesion,
P = axle load,
p_c = tire inflation pressure,
h_o = aggregate thickness, and
α_o = angle of load distribution ($\cong$ 26 deg.).

For the case where geotextile reinforcement is used, p^* in equation (2.35) is replaced by $(p-p_g)$, where p_g is a function of the tension in the geotextile; hence its elongation is significant. On the basis of the probable deflected shape of the geotextile-soil system,

$$p_g = \frac{E\varepsilon}{a\sqrt{1+(a/2S)^2}} \tag{2.37}$$

where

E = modulus of geotextile,
ε = elongation (strain),
a = geometric property (see figure 2.27), and
S = settlement under the wheel.

Combining equations (2.33), (2.35), and (2.37), and using $p^* = p - p_g$ gives equation 2.38, where h is the unknown aggregate thickness. It can be graphed for various rut-depth thicknesses and various moduli of geotextiles.

$$(\pi + 2)c = \frac{P}{2(\mathrm{B} + 2h\tan\alpha)(\mathrm{L} + 2\mathrm{h}\tan\alpha)} + \frac{\mathrm{E}\varepsilon}{a\sqrt{1+(\mathrm{a}/2\mathrm{S})^2}} \tag{2.38}$$

Using equations 2.36 and 2.38, the design method is essentially complete, since both h_o (thickness without a geotextile) and h (thickness with a geotextile) can be calculated. From these two values $\Delta h = h_o - h$ can be obtained, which represents the savings in aggregate due to the presence of the geotextile. For convenience, however, it can be read directly from figure 2.28. This figure also considers the effects of traffic. In this case, the required thickness h' becomes $h' = h'_0 - \Delta h$, which is obtainable from the curves by subtracting the two ordinate values of h'_o and Δh. Note that the effect of service lifetime takes the form of number of vehicle passages.

Two examples follow: one illustrating the general design procedure [72] and the other showing a specific example with an economic analysis included. The influence of rut depth has been further evaluated by Holtz and Sivakugan [74].

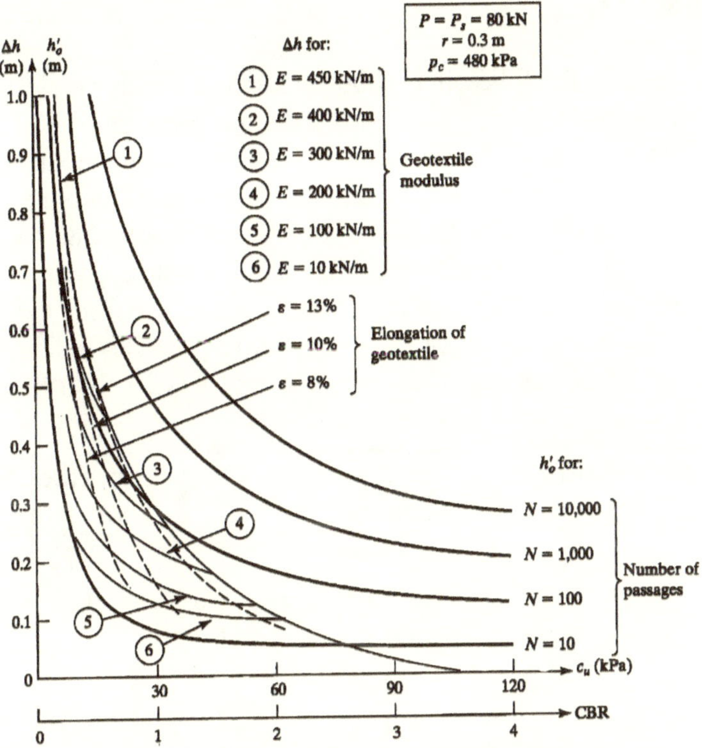

Figure 2.28 Design curves for determining aggregate thickness of unpaved roads and aggregate saved using various geotextiles. (After Giroud and Noiray [72])

Example 2.12:

Given 340 passages of 80 kN single axle-load vehicles with tire inflation pressure = 480 kPa; soft soil CBR = 1; geotextile modulus $E = 90$ kN/m; and an allowable rut depth = 0.3 m. What is the required aggregate thickness of an unpaved road?

Solution: Figure 2.28 gives $h'_0 = 0.35$ m for $N = 340$ and $CBR = 1$ (i.e., thickness when no geotextile is used). It also gives $\Delta h = 0.15$ m for $E = 90$ kN/m and $CBR = 1$ (i.e., the reduction in aggregate thickness when a geotextile-reinforcement layer is used). The difference between the two values, or 0.20 m, is the required aggregate depth when the geotextile is used.

Example 2.13:

Assume the following: 1000 passages of 80 kN single-axle-load vehicles with a tire inflation pressure of 480 kPa on a soaked soil $CBR = 2$, a candidate geotextile modulus $E = 170$ kN/m, and an allowable rut depth = 0.3 m. (a) Plot the response curve from figure 2.28, (b) determine the aggregate savings, and (c) do an economic analysis based on the distance the project is from the stone quarry and geotextile supplier respectively, using the following data (the stone unit weight is 20 kN/m^3):

Distance (km)	Aggregate cost (dollars/kN)	Geotextile cost (dollars/m^2)
< 5	1.90	0.72
5-20	2.03	0.76
20-50	2.50	0.78
50-100	2.90	0.84
100-200	4.00	0.90

Solution: (a) The required complementary curve to figure 2.28 is shown below.

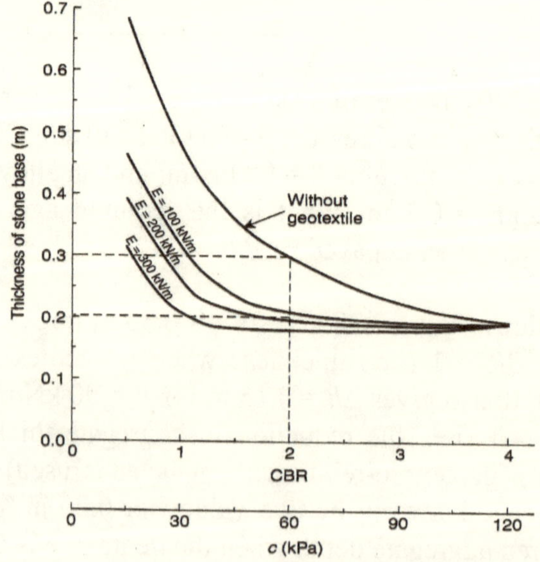

(b) At $CBR = 2.0$
h'_o (without geotextile) = 300 mm
h' (with geotextile) = 205 mm
Δh (savings in stone) = 95 mm
(c) Based on 20 kN/m³, this is a 0.020 kN/m²-mm stone thickness, which results in the table below. It is easily seen that the use of the geotextile as a roadway separator is very economical and becomes more so as the distance from the stone quarry to the project site becomes greater, i.e., it is much more cost effective to transport geotextiles as it is for stone.

Distance (km)	Aggregate total cost (dollars/kN)	Aggregate unit cost (dollars/m²-mm)	Aggregate unit savings (dollars/m²)	Geotextile unit cost (dollars/m²)	Geotextile total savings (dollars/m²)
<5	1.90	0.038	3.61	0.72	0.99
5-20	2.03	0.041	3.90	0.76	1.55
20-50	2.50	0.050	4.75	0.78	2.53
50-100	2.90	0.058	5.51	0.84	3.95
100-200	4.20	0.084	7.98	0.90	6.23

Laboratory Method. If laboratory facilities are available, it is possible to model the situation so as to arrive at a reinforcement ratio provided by the geotextile. The procedure is as follows:

1. Take the lower portion of a standard laboratory CBR mold and fill it with the soil in question at its in situ density and water content.
2. Place crushed stone in the upper portion of the mold.
3. With the load piston on top of the stone, perform a load-versus-deflection test at discrete intervals of piston deflection and record the data.
4. Using a CBR mold that has been modified to hold a geotextile at the interface between the soil subgrade and the crushed stone, repeat the test with the candidate geotextile in position and record the data. (The modification can be made by welding flanges to the upper and lower sections of the CBR mold and clamping or bolting the geotextile between the flanges).

5. Calculate the ratio of the loads at each deflection increment. The data of table 2.12 show this reinforcement ratio for four separate test sets of a geotextile placed on a kaolinite clay at different water contents. Here we see that the reinforcement ratio increases as both the deflection and the water content increase.
6. Assuming that this reinforcement ratio can be used as a multiplier to the in situ CBR of the soil, a number of accepted design procedures can be used to arrive at an aggregate thickness with and without geotextiles.

TABLE 2.12 LABORATORY OBTAINED REINFORCEMENT RATIOS* (WITH AND WITHOUT A GEOTEXTILE) FROM MODIFIED CBR TESTS

	Kaolinite Clay** Soil at Water Content			
Deflection (mm)	32%	35%	38%	41%
3.3	1.0	1.0	1.2	1.4
6.7	1.0	1.1	1.3	1.7
10	1.0	1.2	1.5	2.0
13	1.1	1.3	1.7	2.2
25	1.3	1.5	2.0	2.4
37	1.5	1.8	2.4	3.0
50	1.8	2.2	3.0	3.4

*ratio = (soil with geotextile/soil without geotextile)
**shrinkage limit, w_s = 18%; plastic limit, w_p = 32%; liquid limit, w_l = 41%

Example 2.14:

Using the US Army Corps of Engineers Modified CBR Design Method (WES TR 3-692), calculate the required stone base thickness for an unpaved road carrying 5000 coverages of 45 kN equivalent single-wheel loads using a tire contact area of 300 × 450 mm for (a) stone on a kaolinite clay soil at 41% water content with a *CBR* = 1.0 with no geotextile reinforcement, and (b) the same conditions but with a geotextile whose data are typical of table 2.12 at 25 mm deflection. Then (c) compare the resulting thicknesses.

Solution: The essential formula is the following:

$$h = (3.24 \log C + 2.21)\left(\frac{P}{36.0 \times CBR} - \frac{A}{2030}\right)^{1/2} \tag{2.39}$$

where

h = aggregate thickness (mm),
C = traffic in terms of coverages,
P = equivalent single wheel load (N), and
A = tire contact area (mm^2).

This leads to the general relationship:

$$h = (3.24 \log 5000 + 2.21)\left(\frac{45{,}000}{36.0 \times CBR} - \frac{(300)(450)}{2030}\right)^{1/2}$$

$$h = (14.19)\left(\frac{1250}{CBR} - 66.5\right)^{1/2}$$

(a) For no geotextile reinforcement and CBR = 1.0, the required thickness is

$$h'_o = (14.19)(34.4)$$
$$= 488mm$$

(b) When using a geotextile that results in an equivalent CBR = 2.4 (from table 2.12; 2.4 ×1.0 = 2.4), the thickness is

$$h' = (14.19)(21.3)$$
$$= 302mm.$$

(c) Thus, the savings in stone base (Δh) afforded by using a geotextile is

$$\Delta h = h'_o - h'$$
$$= 488 - 302$$
$$\Delta h = 186mm \ (i.e., \ a \ 38\% \ savings).$$

Sewn Seams. With the soft compressible subgrade soils under consideration in this section on unpaved roads, the matter of geotextile overlap for transferring stress across rolls becomes an issue. This overlap affects both the longitudinal sides and the transverse ends of the geotextile rolls. As expected, the softer the soil, the greater the necessary amount of overlap. Figure 2.29 gives a guide for different types of use on the basis of overlap required. It is easily seen that large overlap distances are required for low-strength soils. Not only is this wasted geotextile but it necessitates the calculation of geotextile-to-geotextile friction (recall the shear tests of section 2.3.3). As a result, field-sewing of the geotextiles is generally preferred (see figure 2.30). Example 2.15 illustrates how field-sewing of geotextiles can be very economical.

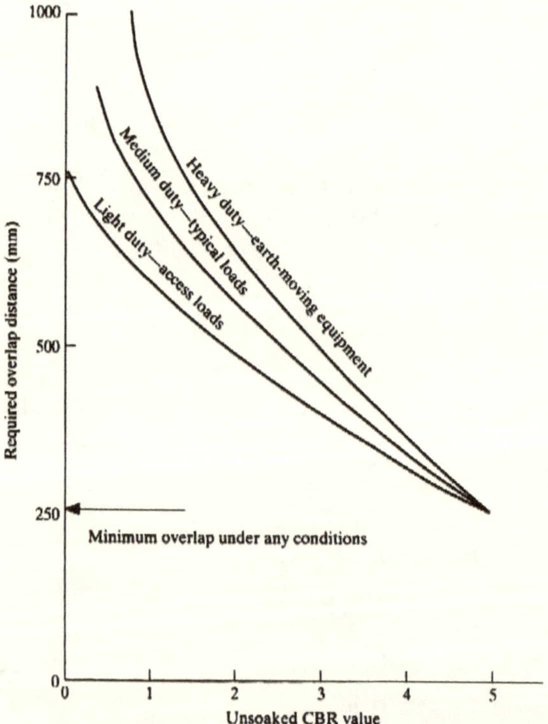

Figure 2.29 Recommended overlap for geotextiles used in unpaved roads as a function of unsoaked soil subgrade CBR value.

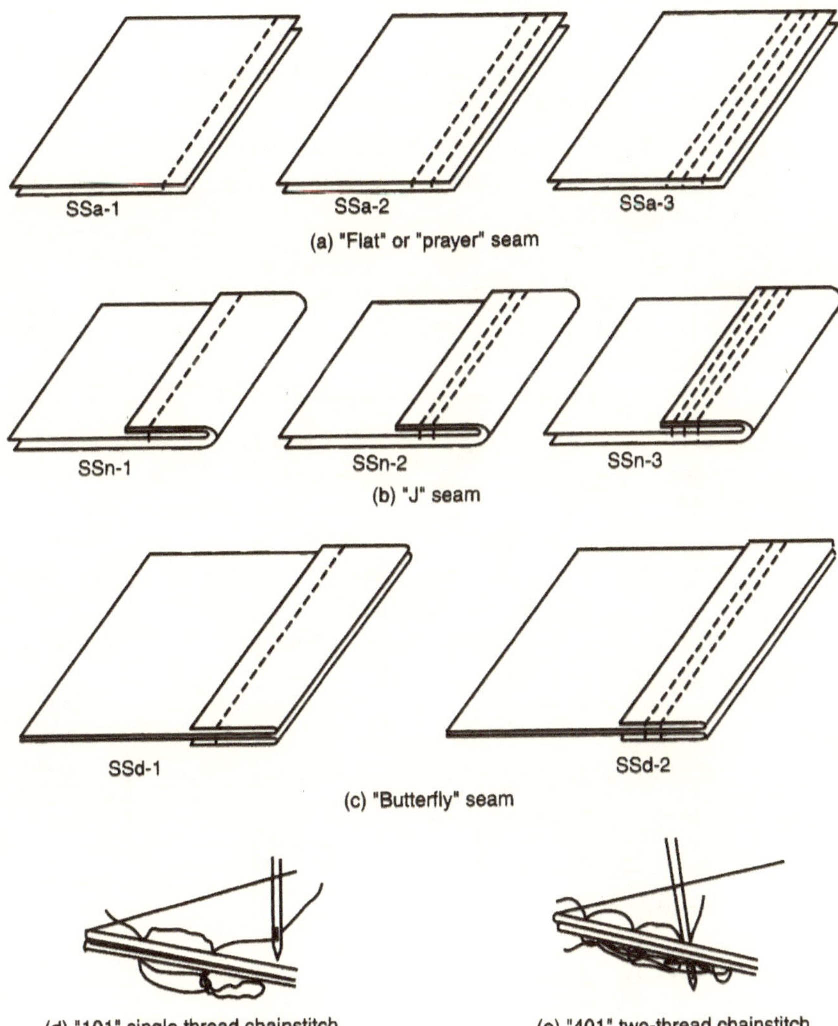

Figure 2.30 Various types of sewn seams for joining geotextiles. (After Diaz [75])

Example 2.15:

Using the overlap guide of figure 2.29, calculate when geotextile sewing becomes more economical than geotextile overlap for a single seam down the center of a 2000 m long access road. The costs are $1.75/m^2 for heavy geotextile, $1.37/ m^2 for medium geotextile, and $1.00/ m^2 for light geotextile. Sewing costs are $400 per day for sewing machine rental and thread and $175 per day for three laborers each, who can easily sew a 2000 m seam in one day.

Solution: The following chart is taken from data in figure 2.29 and the costs provided.

Unsoaked Subgrade CBR	Type of Loads	Overlap Required (m)	Geotextile Required (m^2)	Overlap Geotextile Cost ($/2000 m)		
				Heavy ($1.75/ m^2)	Medium ($1.37/ m^2)	Light ($1.00/m^2)
1.0	Heavy	0.86	1720	3010	2360	1720
	Medium	0.71	1420	2480	1940	1420
	Light	0.58	1160	2030	1590	1160
2.5	Heavy	0.57	1140	1990	1560	1140
	Medium	0.51	1020	1780	1400	1020
	Light	0.44	880	1540	1200	880
4.0	Heavy	0.37	740	1300	1010	740
	Medium	0.35	700	1220	960	700
	Light	0.33	660	1150	900	660

Since the total sewing costs are $925 per day, it is seen that only on the lower-right portion of the chart (on relatively strong subgrades with light geotextiles for light—and medium-duty vehicles) does sewing not pay for itself.

When considering the field-sewing of geotextiles, a number of details must be addressed. They are as follows:

- *Thread type*: The choices being polypropylene, polyester and polyamides (it is necessary to use the same thread type as geotextile fiber type and certainly not a stronger type).
- *Thread tension*: Usually adjusted in the field, this should be sufficiently tight without cutting the geotextile.
- *Stitch density*: Two, three or four stitches per 25 mm are customary.
- *Stitch type*: The choices being prayer, J-type, or butterfly (see figure 2.30), the strongest being the butterfly type.
- *Number of rows*: One, two, or three can be used, but generally two are recommended, see figure 2.30).
- *Type of chain stitch*: The 401 two-thread is recommended.

The sewing of geotextiles has rapidly advanced to the point where all geotextile construction on soft soil sites should consider its use. A geotextile sewing guide is provided in [76]. Tensile seam strengths of 170 kN/m have been attained (recall figure 2.10), and productivity has reached a point where sewing is no longer an obstacle for rapid progress of the work.

2.6.2 Paved Roads

Whenever geotextile use in unpaved roads is discussed, the question of the material's use in paved roads usually follows. To address this properly, we must focus on the general characteristics of the situation. It is most important to recognize that if the road is to be paved with concrete or asphalt immediately (i.e., during initial construction), it cannot be placed on an excessively yielding soil subgrade. If the subgrade yields, the road section will deform and the surfacing will simply crack after a few load repetitions. Many agencies put the *lower limit* of acceptable unsoaked CBR values in the range 10 to 15. As just described, however, the geotextile must deform in order to mobilize its strength, and the *upper limit* of soil subgrade strength for such mobilization as suggested in table 2.11 is an unsoaked CBR of 3 to 8. This contradiction begs the question of how the geotextile is to reinforce if it is not significantly deformed. Advocates of a reinforcement function in paved roads on firm soil subgrades will suggest that the geotextile deformation around the coarse-aggregate base course (when heavily rolled) is sufficient to mobilize the geotextile's strength. Thus, the design can proceed in a manner similar to unpaved roads.

Those who feel this is not the case still desire a geotextile under the stone base, but for reasons other than soil subgrade reinforcement. Here the primary function becomes separation, discussed in sections 2.2.1 and 2.5, or filtration, handled in section 2.2.3 and 2.8. The economic justification is the longer service lifetime with a geotextile separator than without. Recall the discussion in section 2.5.6 and Perkins et al. [77] who provide the current status of research in this regard.

When separation is the primary function in paved road applications, it is important to recognize where the geotextile is located with respect to the pavement cross section and applied loads. In a trial test site with 40 mm of asphalt paving, 150 mm of base course, and 100 mm of large crushed stone, a lightweight geotextile (150 g/m^2) failed under 165,000 repetitions of a standard 80 kN axle load [78]. This premature geotextile failure was evidenced by abrasion of the yarns followed by fines pumping up from the subgrade into the stone base. Although no specific design is available to explain the situation, it does illustrate that a minimum set of geotextile properties is required in most situations. In other words, an adequate survivability criterion is required to ensure reasonable performance in general situations.

To specifically add reinforcement for paved roads on firm subsoils, a geotextile pretensioning system is required. By pretensioning the geotextile, the stone base will be placed in compression (i.e., thereby providing a lateral confinement) and will effectively increase its modulus over the nonreinforced case. Some of these concepts are discussed in section 2.7.4; however, they are extremely difficult to implement.

2.7 DESIGNING FOR SOIL REINFORCEMENT

This section continues the discussion of the use of geotextiles in the primary function of reinforcement. Since this was the topic of the preceding section involving road systems, it could easily have been incorporated into that section. However, this type of soil reinforcement raises a unique set of design issues, whereby the geotextile in horizontal layers and the interspersed soil form a mechanically stabilized earth (MSE) system rather than acting as discrete material elements. The applications here involve wall reinforcement (facing angle ≥ 70° to the horizontal), embankment (slope) reinforcement (facing angle < 70° to the horizontal), and foundation or basal reinforcement (typically horizontal placement).

2.7.1 Geotextile-Reinforced MSE Walls

Background. Conventional gravity and cantilever wall systems made from masonry and concrete resist lateral earth pressure by virtue of their large mass. They act as rigid units and have served the industry well for centuries. However, a new era of retaining walls was introduced in the 1960s by H. Vidal with Reinforced Earth. Here metal strips extending from exposed facing panels back into the soil being reinforced and serve the dual role of anchoring the facing units and being restrained through frictional stresses mobilized between the strips and the backfill soil. The backfill soil both creates the lateral pressure and interacts with the strips to resist it. The walls called mechanically stabilized earth (MSE) walls with geosynthetic reinforcement are very flexible compared to conventional gravity structures. They offer many advantages, including significantly lower cost per square meter of exposed surface. A steady series of variations followed Vidal's steel strips, all of which can be put into the MSE wall category:

- Facing panels with metal strip reinforcement
- Facing panels with metal wire mesh reinforcement
- Solid panels with tieback anchors
- Anchored gabion walls
- Anchored crib walls
- Geotextile-reinforced walls (to be described here)
- Geogrid-reinforced walls (to be described in chapter 3)

In all cases, the reinforced soil mass behind the wall facing consists of *mechanically stabilized earth (MSE)* and the wall system is generically called an MSE wall.

Construction Details. Critical in the proper functioning of a geotextile-reinforced MSE wall is its proper construction, which is done on a carefully planned and sequential basis. Upon preparing an adequate soil foundation, which consists of removing unsuitable material and compacting in-situ or replacement foundation soils, the wall itself is begun. There is no concrete footing with these walls, and the lowest geotextile layer is placed directly on the foundation soil. An iterative construction sequence, developed by the US Forest Service and illustrated in figure 2.31, is followed. This type of wall is referred to as a wraparound wall facing.

1. A wooden form of a height slightly greater than the individual soil layer thickness, called the lift height, is placed on the ground surface or on the previously placed lift after the first layer is completed. This form is nothing more than a series of metal L brackets with a continuous wooden brace board running along the face of the wall.
2. The geotextile is then unrolled and positioned so that approximately 1 m extends over the top of the form and hangs free. If it is sufficiently wide, the geotextile can be unrolled parallel to the wall. In this way the geotextile's cross machine direction is oriented in the maximum stress direction. This will depend on the required design length and geotextile strength, which will be discussed later. If a single roll is not wide enough, two of them can be sewn together. The sewn strength is then a governing factor. Alternatively, the geotextile can be deployed perpendicular to the wall in full-width strips and adjacent roll edges can be overlapped or sewn. In this way the geotextile's machine direction is oriented in the maximum stress direction.
3. Backfill is now placed on the geotextile for 1/2 to 3/4 of its lift height and compacted. This is typically 200 to 400 mm and is done with light weight construction equipment. The choice of backfill soil type is important. If angular gravel, drainage can easily occur but high installation damage to the geotextile must be considered. If fine-grained silts or clays, drainage cannot occur and hydrostatic pressures must be considered by using back and base drains. This leaves sand, which the

author considers the ideal backfill soil for MSE walls that are reinforced by geotextiles or geogrids.

4. A windrow is made 300 to 600 mm from the face of the wall with a road grader or is dug by hand. Care must be exercised not to damage the underlying geotextile.
5. The free end of the geotextile, that is, its "tail," is then folded back over the wooden form into the windrow.
6. The remaining lift thickness of soil is then completed to the planned lift height and suitably compacted.
7. The wooden form is then removed from in front of the wall, and the metal brackets from beneath the lift, and the assembly is reset on top in preparation of the next higher lift. Note that it is usually necessary to have scaffolding in front of the wall when the wall is higher than 1.5 or 2.0 m.

When completed, this sequence provides walls as appear in figure 2.32. The exposed face of the wall must now be covered to prevent the geotextile's weakening due to UV exposure (recall section 2.3.6) and possible vandalism. Bituminous emulsions or other asphalt products have been used for covering the wall face and have the advantage of being flexible, as is the wall itself. Unfortunately, oxidation of the bitumen causes deterioration after a few years, and it must be periodically reapplied. Alternatively, the surface of wraparound geotextile walls have been covered with shotcrete (wet-mixed cement / sand / water paste with air supplied at the nozzle) or gunite (dry cement / sand mix with water and air supplied at the nozzle). A wire mesh anchored between the geotextile layers may be necessary to keep the coating adhered to the vertical face of the wall. Still further, a precast concrete or cast-in-place concrete facing can be used but only after deformation equilibrium of the wraparound wall has occurred.

Design Methods. There are two somewhat different approaches to the design of geotextile walls: that used by Broms [79] and that used by the US Forest Service, Steward et al. [80], and Whitcomb and Bell [81]. The latter method will be followed in this book. This method follows the work that Lee et al. [82] did on reinforced earth walls with metallic strip reinforcement and was originally adapted to geotextile walls by Bell et al. [83]. The design progresses in parts as follows:

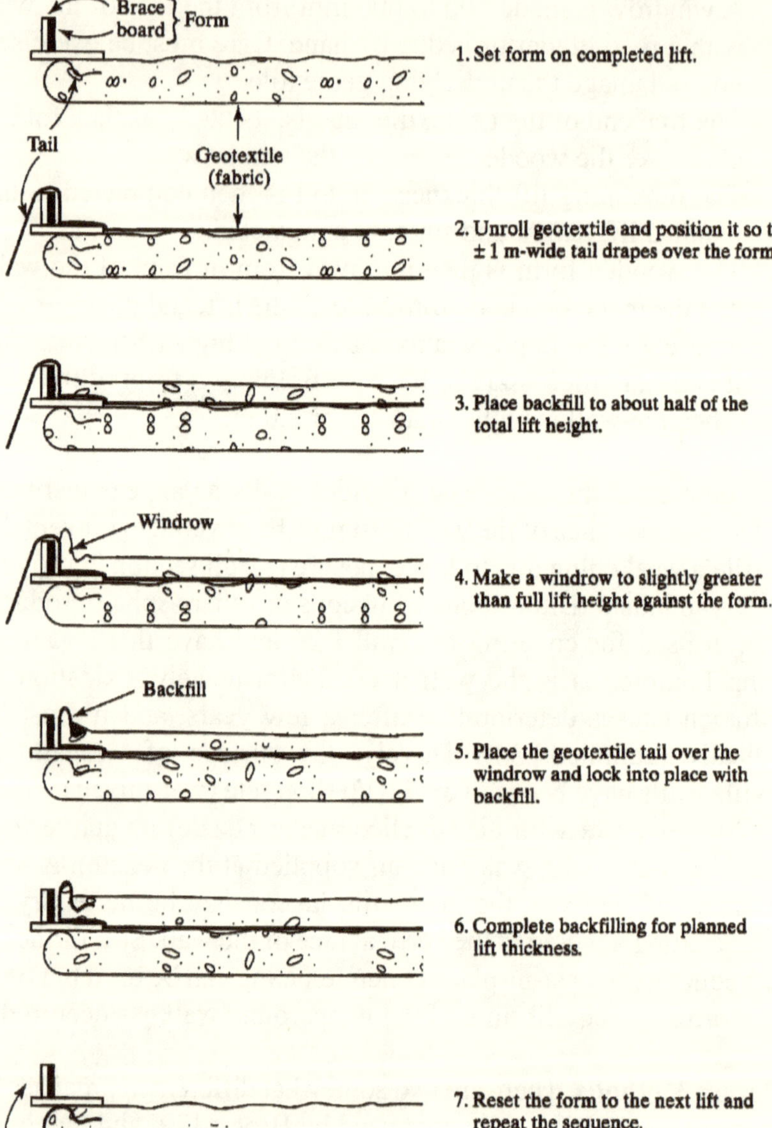

Figure 2.31 Construction sequence for geotextile wrap-around walls suggested by U. S. Forest Service.

- Internal stability is first addressed to determine geotextile spacing, geotextile length, and overlap distance. This establishes the integrity of the MSE mass.
- External stability against overturning, sliding, and foundation failure is investigated and the internal design verified or modified accordingly. This establishes that the MSE mass will remain stationary.
- Miscellaneous considerations, including wall-facing details and external drainage are completed. This addresses other possible stability issues.

Figure 2.32 Geotextile wrap-around walls. (Compliments of Crown Zellerbach Corp.)

To determine the geotextile layer separation distances, earth pressures are assumed to be linearly distributed using Rankine active earth-pressure conditions for the soil backfill and at-rest conditions for the surcharge. A prediction conference at the Canadian Royal Military College, however, showed that the entire design to be presented here is quite conservative, see Jarrett and McGown [84]. Therefore, active earth pressure (K_a) conditions will be used throughout. Note that cohesion is assumed to be zero per discussion by Leshchinsky [85]. A less-conservative approach would be to use a Coulomb analysis for the earth pressure values. This is the approach used in several computer codes and will be discussed later. Boussinesq elastic theory for live loads on the soil backfill is used. As shown in figure 2.33, the following earth pressures result:

$$\sigma_{hs} = K_a \gamma z \tag{2.40}$$

$$\sigma_{hq} = K_a q \tag{2.41}$$

$$\sigma_{hl} = P \frac{x^2 z}{R^5} \tag{2.42}$$

$$\sigma_h = \sigma_{hs} + \sigma_{hq} + \sigma_{hl} \tag{2.43}$$

where

σ_{hs} = lateral pressure due to soil;
K_a = $\tan^2(45 - \varphi/2)$ = coefficient of active earth pressure, where
φ = angle of shearing resistance of backfill soil;
γ = unit weight of backfill soil;
z = depth from ground surface to layer in question;
σ_{hq} = lateral pressure due to surcharge load;
q = $\gamma_q D$ = surcharge load on ground surface, where
γ_q = unit weight of surcharge soil, and
D = depth of surcharge soil;
σ_{hl} = lateral pressure due to live load;
P = concentrated live load on backfill surface;
x = horizontal distance load is away from wall;
R = radial distance from load point on wall where pressure is being calculated
σ_h = total, or cumulative, lateral earth pressure on wall.

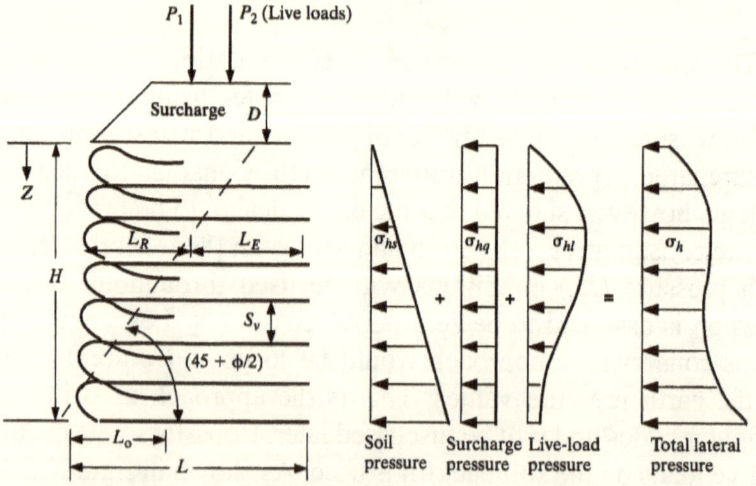

Figure 2.33 Earth pressure concepts and theory for geotextile wall design.

The calculations of σ_{hs} and σ_{hq} are quite straightforward, but σ_{hl} presents problems, particularly for multiwheeled truckloads where superposition of each wheel must be performed. Figure 2.34 greatly aids in such calculations.

By taking a free body at any depth in the total lateral pressure diagram and then summing the forces in the horizontal direction, one obtains the equation for the lift thickness:

$$\sigma_h S_v = \frac{T_{allow}}{FS} \tag{2.44}$$

$$S_v = \frac{T_{allow}}{\sigma_h FS} \tag{2.45}$$

where

S_v = vertical spacing (lift thickness),
T_{allow} = allowable stress in the geotextile (recall equation 2.24 and table 2.8a),
σ_h = total lateral earth pressure at depth considered, and
FS = factor of safety (use 1.3 to 1.5 when using T_{allow} as determined above).

The same free-body approach can be taken for obtaining the length of embedment of the geotextile layers in the anchorage zone, L_e. Note that when these values are obtained they must be added to the nonacting lengths (L_R) of geotextile within the active zone for the total geotextile lengths (L); that is,

$$L = L_e + L_R \tag{2.46}$$

where

$$L_R = (H - z)tan\left(45 - \frac{\phi}{2}\right) \tag{2.47}$$

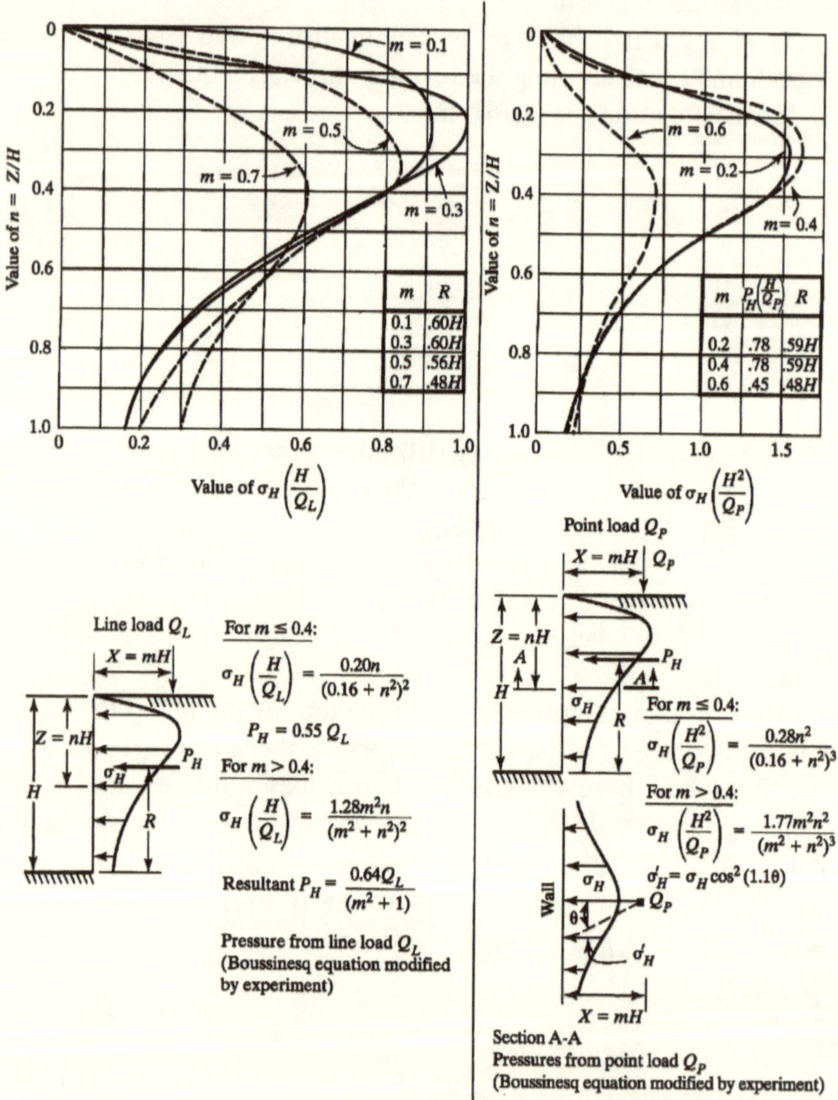

Figure 2.34 Lateral earth pressure due to a surface load. Left side is for line load; right side is for point load. (After NAVFAC [86])

and

$$\begin{aligned} S_v \sigma_h FS &= 2\tau L_e \\ &= 2(c_a + \sigma_v \tan\delta)L_e \\ &= 2(c_a + \gamma Z \tan\delta)L_e \end{aligned}$$

$$L_e = \frac{S_v \sigma_h FS}{2(c_a + \gamma Z \tan\delta)} \tag{2.49}$$

where

τ = shear strength of the soil to the geotextile,
L_e = required embedment length (minimum is 1.0 m),
S_v = vertical spacing (lift thickness),
σ_h = total lateral pressure at depth considered,
FS = factor of safety,
c_a = soil adhesion between soil and geotextile (zero if granular soil is used),
γ = unit weight of backfill soil,
Z = depth from ground surface, and
δ = angle of shearing friction between soil and geotextile.

Finally, the overlap distance L_o is obtained in a manner similar to that above with a few exceptions, namely that the distance Z should be measured to the middle of the layer, and σ_h is not as large as illustrated in figure 2.33. It is reasonably well established that the stress in reinforcement elements is maximum near the failure plane and falls off sharply to either side [87]. As an approximation, $0.5\sigma_h$ will be used, which results in the following equation.

$$L_o = \frac{S_v \sigma_h FS}{4(c_a + \gamma Z \tan\delta)} \tag{2.50}$$

where L_o is the required overlap length (minimum is 1.0 m).

Next, we must consider external stability of the geotextile-reinforced MSE mass, which includes overturning, sliding, and foundation failures. These are illustrated in figure 2.35. These features are common to all wall systems and can be treated exactly the same way as with gravity or crib walls. They are generally site-specific insofar as calculations are concerned. In general, it is recommended that for overturning and foundation bearing capacity the *FS* value ≥ 2.0 and for sliding the *FS* value ≥ 1.5.

The miscellaneous considerations that generally must be addressed are facing details; facing connections (if applicable); seaming methods (if necessary); drainage behind, beneath and in front of wall; erosion above and in front of wall, guard post, light posts, fencing, and other appurtenances with or without deep foundations.

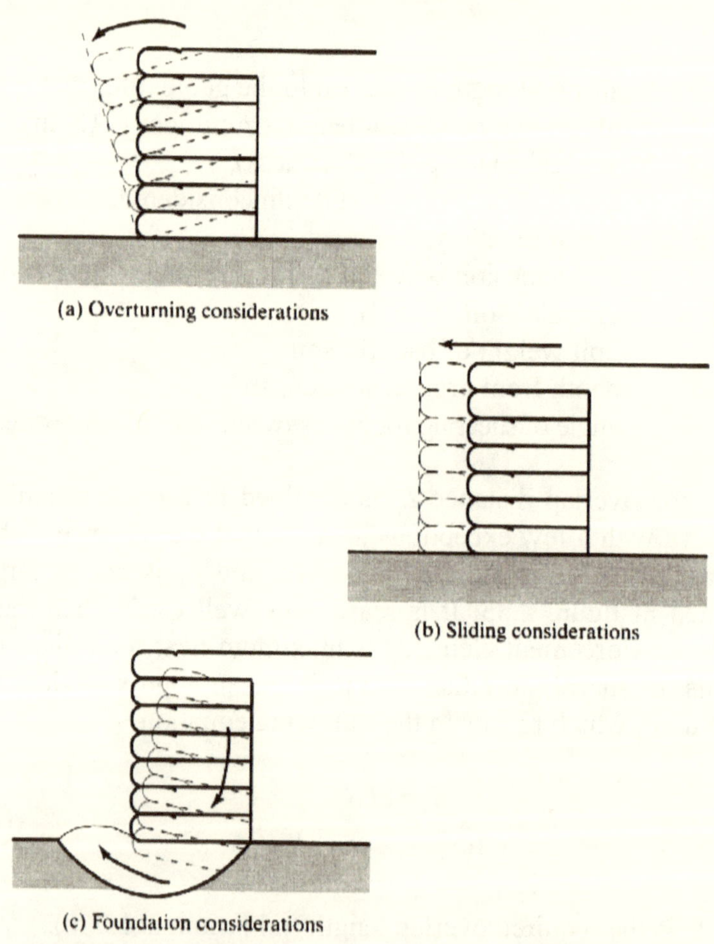

Figure 2.35 External stability considerations for geotextile walls.

Example 2.16:

Design a 6 m high wraparound type of geotextile wall that is to carry a storage area of equivalent dead load of 10 kPa. The wall is to be backfilled with a granular soil (SP) having properties of γ = 18 kN/m^3, φ = 36 deg., and c = 0. A woven slit-film geotextile with warp (machine) direction ultimate wide-width tensile strength of 50 kN/m and friction angle with granular soil of $\delta = 24°$ (recall table 2.4) is intended to be used in its construction. The orientation of the geotextile is perpendicular to the wall face and the edges are to be overlapped or sewn to handle the weft (cross machine) direction. A factor of safety of 1.4 is to be used along with appropriate geotextile reduction factors.

Solution: (a) Determine the horizontal pressure as a function of the depth Z in order to calculate the spacing of the individual layers; i.e., the S_v value.

$$\begin{aligned} K_a &= \tan^2(45-\phi/2) \\ &= \tan^2(45-36/2) \\ &= 0.26 \end{aligned}$$

$$\begin{aligned} \sigma_h &= \sigma_{hs} + \sigma_{hq} \\ &= K_a\gamma Z + K_a q \\ &= (0.26)(18)(Z) + (0.26)(10) \\ \sigma_h &= 4.68Z + 2.60 \end{aligned}$$

The allowable geotextile strength is obtained using the following reduction factors:

$$\begin{aligned} T_{allow} &= T_{ult}\left[\frac{1}{RF_{ID} \times RF_{CR} \times RF_{CBD}}\right] \\ &= 50\left[\frac{1}{1.2 \times 2.5 \times 1.27}\right] \\ &= 50/3.79 \\ &= 13.2\, kN/m \end{aligned}$$

Now using equation 2.46 for varying depths, calculate the geotextile layer spacings.

- At full wall height $Z = 6$ m:

$$S_v = \frac{T_{allow}}{\sigma_h (FS)}$$

$$= \frac{T_{allow}}{[4.68(Z) + 2.60]1.4}$$

$$= \frac{13.2}{[4.68(6.0) + 2.60]1.4}$$

$$S_v = 0.307\ m;\ use\ 0.30\ m$$

- By trial and error, see if the spacing can be opened up to 0.50 m at $Z = 3.3$ m.

$$S_v = \frac{13.2}{[(4.68)(3.3) + 2.60]\,1.4}$$

$$S_v = 0.52 \text{ m; use } 0.50 \text{ m.}$$

- Again by trial and error, see if the spacing can be further opened up to 0.65 m at $Z = 1.3$ m.

$$S_v = \frac{13.2}{[(4.68)(1.3) + 2.60]\,1.4}$$

$$S_v = 1.08\ m;\ use\ 0.65\ m$$

Thus, the layers and their spacings are shown in the figure below. At this point we have a mechanically stabilized earth (MSE) mass, actually a geotextile stabilized earth mass, which is self-contained within itself.

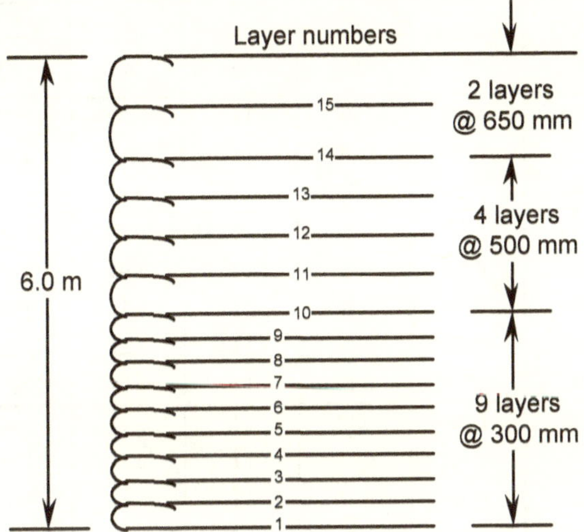

Due to the critical nature of some MSE walls a *serviceability* criterion can be imposed limiting the possible outward deflection of the wall to some acceptable value, e.g., 20 mm. This is done utilizing the wide-width stress versus strain response of the reinforcement at the design (allowable) stress in the reinforcement. This latter value includes both reduction factors and the factor of safety.

(b) Determine the length of the fabric layers (L) using equation 2.49 for L_e with δ = 24 deg. and c = o. Note that L_R uses a Rankine failure plane and is calculated from equation 2.48.

$$L_e = \frac{S_v \sigma_h (FS)}{2(c + \gamma Z \tan\delta)}$$

$$= \frac{S_v (4.68Z + 2.60)1.4}{2(0 + 18Z \tan 24°)}$$

$$L_e = \frac{S_v (6.55Z + 3.64)}{16.0Z}, and$$

$$L_R = (H - Z)\tan\left(45 - \frac{36}{2}\right)$$

$$L_R = (6.0 - Z)(0.509)$$

Layer No.	Depth Z (m)	Spacing S_v (m)	L_e (m)	L_e min. (m)	L_R (m)	L_{calc} (m)	L_{spec} (m)
15	0.65	0.65	0.49	1.0	2.72	3.72	use 4.0
14	1.30	0.65	0.38	1.0	2.39	3.39	
13	1.80	0.50	0.27	1.0	2.14	3.14	
12	2.30	0.50	0.26	1.0	1.88	2.88	use 3.0
11	2.80	0.50	0.25	1.0	1.63	2.63	
10	3.30	0.50	0.24	1.0	1.37	2.37	
9	3.60	0.30	0.14	1.0	1.22	2.22	
8	3.90	0.30	0.14	1.0	1.07	2.07	
7	4.20	0.30	0.14	1.0	0.92	1.92	use 2.0
6	4.50	0.30	0.14	1.0	0.76	1.76	
5	4.80	0.30	0.14	1.0	0.61	1.61	
4	5.10	0.30	0.14	1.0	0.46	1.46	
3	5.40	0.30	0.14	1.0	0.31	1.31	
2	5.70	0.30	0.14	1.0	0.15	1.15	
1	6.00	0.30	0.13	1.0	0.00	1.00	

Note that the calculated L_e values are very small (this is typically the case with all types of geosynthetic reinforced walls) and the minimum value of 1.0 m should be used. When this is added to L_R for the total length, you should round up to a even number of meters. Also, the important consideration of total geotextile width must be addressed. Three situations can be envisioned.

Case 1: If the geotextile rolls are wide enough, they can be deployed parallel to the wall, and the weft or cross machine direction is the important property insofar as its wide-width strength is concerned. Although this is possible for the lower fabric layers, it is not for the uppermost, since, 4.0 + 0.65 + 1.0 = 5.65 m, which is wider than many commercially available geotextiles.

Case 2: Alternatively, two adjacent rolls of fabric can be used parallel to the wall, but a sewn seam, or large overlap, must be used for the uppermost layers. If sewn seams are used, an appropriate reduction factor must be used.

Case 3: The fabric layers can be deployed perpendicular to the wall, thereby utilizing their warp or machine direction wide-width strength in the major principal stress direction. This was the case posed in this example. This requires sewn seams, or overlaps, in the opposite direction. However, in this (the minor principal stress) direction the required forces are significantly lower, e.g., 33 to 50% of the major principal stress direction.

(c) Check the overlap length L_o, to see if it is less than the 1.0 m recommended value using equation 2.50.

$$L_o = \frac{S_v \sigma_h (FS)}{4(c_a + \gamma Z \tan\delta)}$$
$$= \frac{S_v [4.68(Z) + 2.60] 1.4}{4[0 + (18)Z \tan 24°]}$$

This is maximum at the upper layer at $Z = 0.65$ m:

$$\mathrm{L_o} = \frac{0.65[4.68(0.65) + 2.60] 1.4}{4[0 + (18)(0.65) \tan 24°]}$$
$$= 0.25\ m;\ \text{use } 1.0\ m \text{ throughout}$$

The solution at this point in the design, appears as in the following sketch. Note that the P_a-vector is at an inclination δ, thus in so doing this is a modified Rankine analysis.

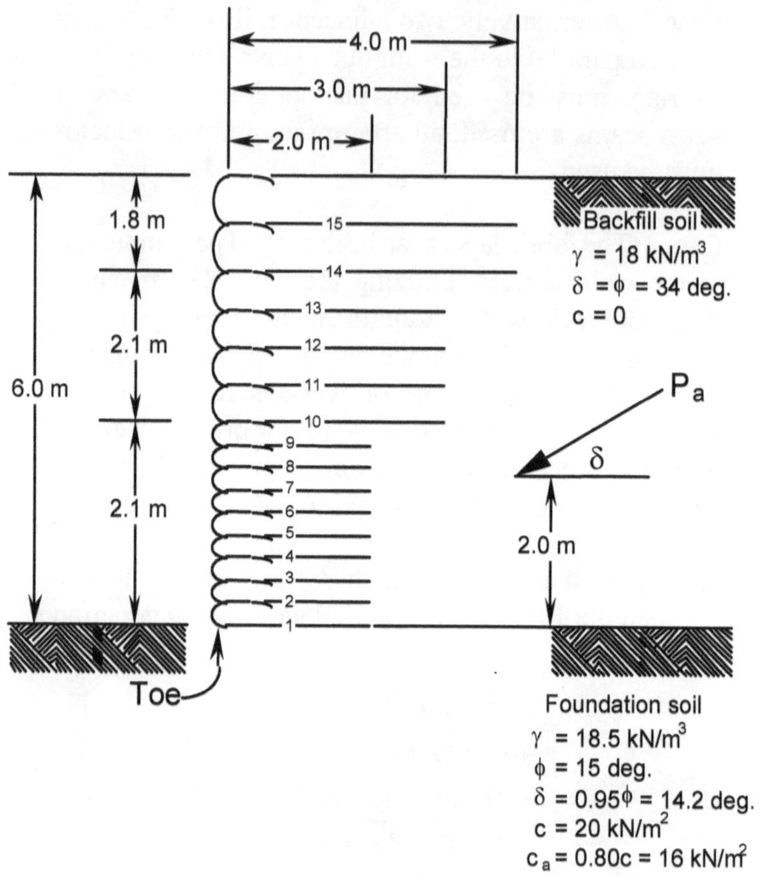

(d) Since the internal stability of the wall has been provided for, focus now shifts to external stability. Standard geotechnical engineering concepts are used to analyze overturning, sliding, and bearing capacity. See the above figure where as follows:

$$K_a = \tan^2(45 - \varphi/2) = \tan^2(45 - 34/2)$$
$$= 0.28$$
$$P_a = 0.5\gamma H^2 K_a$$
$$= 0.5(18)(6)^2(0.28)$$
$$= 90.7 \text{ kN/m}$$
$$P_a \cos 34 = 75.2 \text{ kN/m}$$
$$P_a \sin 34 = 50.7 \text{ kN/m}$$

For overturning, moments are taken about the toe of the wall to generate a FS_{OT} value.

$$FS_{OT} = \sum \frac{resisting\,moments}{driving\,moments}$$

$$= \frac{w_1 x_1 + w_2 x_2 + w_3 x_3 + P_a \sin\delta(4)}{P_a \cos\delta(2)}$$

$$= \frac{(6)(2)(18)(1) + (3.9)(1)(18)(2.5) + (1.8)(1.0)(18)(3.5) + (50.7)(4)}{(75.2)(2.0)}$$

$FS_{OT} = 4.7 > 2.0$; which is acceptable.

This high value of calculated factor of safety is very typical of walls of this type. Even further, overturning is not a likely failure mechanism since one has a very flexible mechanically stabilized earth system that cannot support bending stresses. Thus, many designers do not even include an overturning calculation in the design process.

For sliding, horizontal forces at the bottom of the wall are summed to obtain a FS_s value.

$$FS_s = \sum \frac{resisting\ forces}{driving\ forces}$$

$$= \frac{\left[c_a L + \left(\frac{w_1 + w_2 + w_3 + P_a \sin\delta}{2}\right)\tan\delta\right]2}{P_a \cos\delta}$$

$$= \frac{\left[32 + \left(\frac{216 + 70.2 + 32.4 + 50.7}{2}\right)\tan 14.2\right]2}{75.2}$$

$FS_s = 2.1 > 1.5$; acceptable

For foundation failure, use shallow foundation bearing capacity theory for a FS_{BC} (see, e.g., [88]).

$$
\begin{aligned}
p_{ult} &= cN_c + qN_q + 0.5\gamma BN_\gamma \\
&= (20)(10.98) + 0 + 0.5(18.5)(2)(2.65) \\
&= 219.6 + 49.0 \\
&= 269 kN/m^2 \\
p_{act} &= (18)(6) + (10) \\
&= 118 kN/m^2 \\
FS_{BC} &= \frac{p_{ult}}{p_{act}} \\
&= \frac{269}{118} \\
FS_{BC} &= 2.3 > 2.0; \text{which is acceptable}
\end{aligned}
$$

Both internal and external designs are now complete. The wall uses 15 layers of fabric (the lowest nine at 0.30 m spacing; the middle four at 0.50 m spacing; the upper two at 0.65 m spacing). The fabric lengths are 3.3 m (2 + 0.3 + 1) at the lowest level, 4.5 m (3 + 0.5 + 1) at the intermediate level and 5.65 m (4 + 0.65 + 1) at the upper level.

While this example illustrates the design of a wraparound geotextile retaining wall design, it does not address for the incorporation of live loads as produced by traffic. Example 2.17 will illustrate how this is done, but just to the point of calculating the additional horizontal stress distribution against the wall. Beyond this, the design proceeds as in example 2.16.

Example 2.17:

For the 200 kN dual-tandem-axle truck whose eight wheel dimensions are shown below, calculate the horizontal wall stresses for a 6 m high wall at 1 m increments.

Solution: Using figure 2.34, with $n = Z/H$, $m = X/H$, $H = 6.0$ m, $Q_p = 25$ kN, and $\sigma'_h = \sigma_h \cos^2 (1.1\theta)$, each wheel gives the following tabulated horizontal stresses (in kN/m²) as shown in the table below.

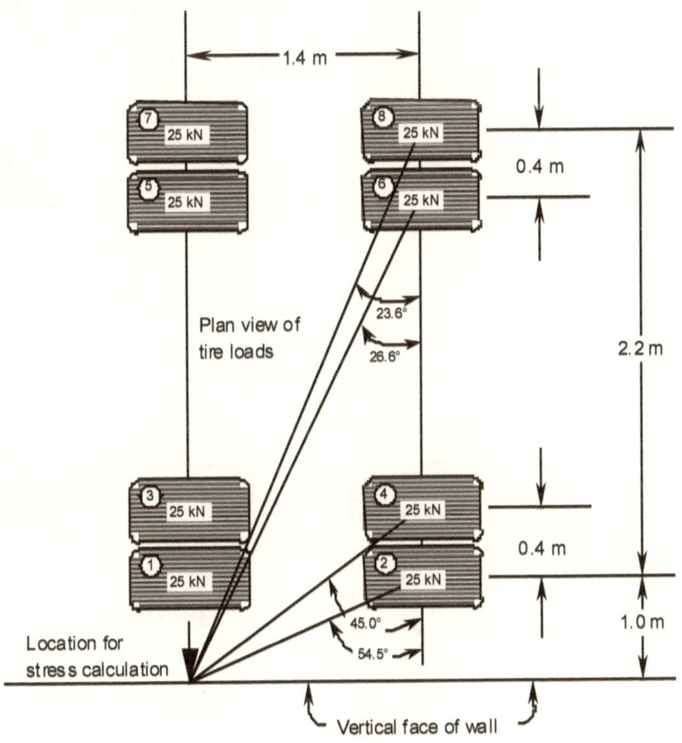

Wheel 1						Wheel 2
z	n=Z/H	X	m=X/H	$\sigma h\ H^2/Q_p$	σ_h	$\sigma_h' = 0.25\ \sigma_h$
0	0.00	1.0	0.17	0.0	0.00	0.00
1	0.17	1.0	0.17	1.2	0.82	0.18
2	0.33	1.0	0.17	1.6	1.08	0.24
3	0.50	1.0	0.17	1.0	0.71	0.16
4	0.67	1.0	0.17	0.6	0.39	0.09
5	0.83	1.0	0.17	0.3	0.22	0.05
6	1.00	1.0	0.17	0.2	0.12	0.03
Wheel 3						Wheel 4
z	n=Z/H	X	m=X/H	$\sigma h\ H^2/Q_p$	σ_h	$\sigma_h' = 0.42\ \sigma_h$
0	0.00	1.4	0.23	0.0	0.00	0.00
1	0.17	1.4	0.23	1.2	0.82	0.34
2	0.33	1.4	0.23	1.6	1.08	0.46
3	0.50	1.4	0.23	1.0	0.71	0.30
4	0.67	1.4	0.23	0.6	0.39	0.16
5	0.83	1.4	0.23	0.3	0.22	0.09
6	1.00	1.4	0.23	0.2	0.12	0.05
Wheel 5						Wheel 6
z	n=Z/H	X	m=X/H	$\sigma h\ H^2/Q_p$	σ_h	$\sigma_h' = 0.76\ \sigma_h$
0	0.00	2.8	0.47	0.0	0.00	0.00
1	0.17	2.8	0.47	0.7	0.50	0.38
2	0.33	2.8	0.47	1.2	0.84	0.64
3	0.50	2.8	0.47	0.9	0.65	0.50
4	0.67	2.8	0.47	0.6	0.41	0.31
5	0.83	2.8	0.47	0.4	0.24	0.19
6	1.00	2.8	0.47	0.2	0.15	0.11
Wheel 7						Wheel 8
z	n=Z/H	X	m=X/H	$\sigma_h\ H^2/Q_p$	σ_h	$\sigma_h' = 0.81\ \sigma_h$
0	0.00	3.2	0.53	0.0	0.00	0.00
1	0.17	3.2	0.53	0.5	0.32	0.26
2	0.33	3.2	0.53	0.9	0.63	0.51
3	0.50	3.2	0.53	0.8	0.57	0.46
4	0.67	3.2	0.53	0.6	0.40	0.33
5	0.83	3.2	0.53	0.4	0.26	0.21
6	1.00	3.2	0.53	0.2	0.16	0.13

$\Sigma\ (\sigma_h + \sigma_h')$	
z	stress kPa
0	0.00
1	3.62
2	5.47
3	4.05
4	2.48
5	1.47
6	0.89

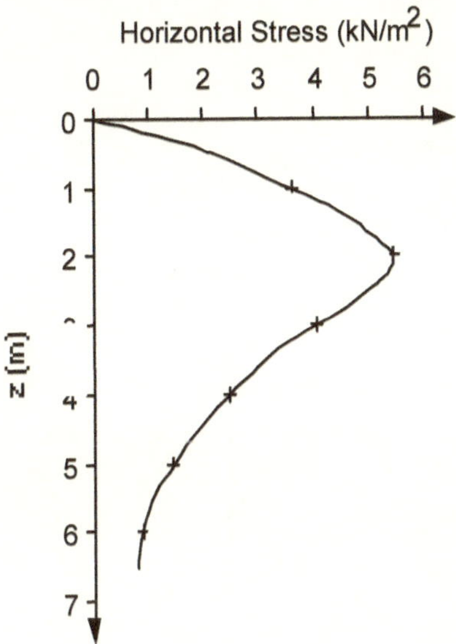

The stress distribution diagram shown above for the live load is now added to that from the soil and surcharge (if any) to obtain the resultant horizontal (or lateral) pressure distribution as illustrated in figure 2.33.

Summary. It is easily seen that the geotextile wall design just completed (with or without live loads) is not trivial, and to do such designs continuously is a very time-consuming task. In the case of a manufacturer of a particular geotextile style, it would be preferred to develop design guides by systematically varying certain parameters in the analysis (for example, the height of wall and the slope angle of the wall face). Innovative design graphs can be generated; an example using Polyfelt's TS styles of geotextiles is shown in figure 2.36. Graphs for different geotextiles could be similarly developed, or the type of loading could be included as a separate variable. The variations are essentially limitless. There are numerous computer codes available from manufacturers for their particular products as well as several generic programs that are commercially available. These will be discussed in chapter 3, on geogrids.

Regarding the performance of geotextile walls, one of the most carefully developed, constructed, and monitored was built in 1982 near Glenwood Springs, Colorado. It is 5.0 m high and 90.0 m long, with

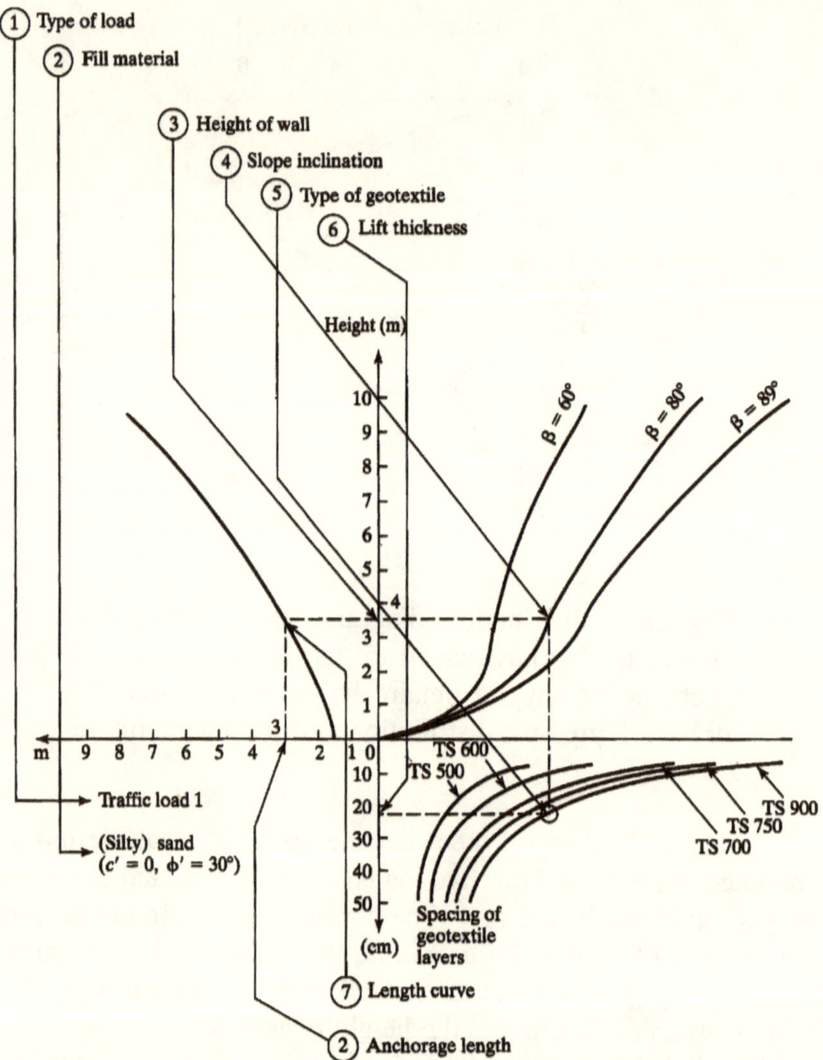

Figure 2.36 Design guide for geotextile walls using Polyfelt geotextiles. (Compliments of TenCate Geosynthetics, Inc.)

the length consisting of ten 9.0 m segments. These segments consist of different nonwoven geotextiles supplied by separate manufacturers. They were sewn together in order to form the 90.0 m length. Two of the segments were purposely underdesigned to provide for controlled failure. When such a failure did not occur, 5.2 m of surcharge soil was added, but still there was no failure. Each segment of wall continues to remain serviceable, with no noticeable creep, even though part of the wall is founded on soft soil, which has settled more than 600 mm [89]. Clearly, geotextile-reinforced walls are intrinsically sound in concept, but this case history suggests that the design method may be quite conservative. Several other noteworthy geotextile-reinforced walls have appeared in the literature. For example, a 12.6 m high wall was constructed to form support for an additional 5.0 m of surcharge soil fill in Seattle, Washington (Allen et al. [90]), and a 12.2 m high vertical wall was used to support a high-bridge approach while its adjacent section was being constructed (Stevens and Souiedan [91]).

When we compare geotextile walls to gravity walls (and, to a lesser extent, other types of flexible walls), there are the following advantages and disadvantages to using geotextiles:

- *Advantages:* A flexible wall system is created, a minimum excavation is needed beneath the face of wall, there are no corrosion problems, the backfill need not be gravel, drainage can occur using certain geotextiles, unskilled labor can be used, no heavy equipment is required, and the cost per square meter of exposed wall is very low.
- *Disadvantages:* The design method appears to be quite conservative, the geotextile interaction in the analysis (perhaps via arching theory) is not currently considered; creep is potentially a problem, thereby requiring a relatively high reduction factor; the wall face must be coated to prevent UV degradation and vandalism; and the coatings (shotcrete, gunite, or asphalt) are not particularly attractive.

The above disadvantage regarding the appearance of wraparound walls is being dealt with ever more effectively. For instance, it is becoming popular to use wall-facing panels made from large-sized timbers, gabions, or decrotive concrete panels. The geotextiles are fixed to or placed between the facing and extend into the backfill soil exactly as described and designed in this section. For walls less than 5 m in height, timber can be used for the facing. The attachment detail is important, as described by Richardson and Behr [92].

Perhaps the greatest economy can be realized by the nature of the backfill soil. Large stone (e.g., AASHTO no. 8 or no. 57) should not be used, since the installation damage is likely to be excessive; rather,

TABLE 2.13 RECOMMENDED SOIL BACKFILL GRADATION FOR GEOTEXTILE AND GEOGRID REINFORCEMENT APPLICATIONS (WALLS AND SLOPES) TO PROVIDE ADEQUATE DRAINAGE AND TO AVOID EXCESSIVE INSTALLATION DAMAGE

Sieve Size (No.)	Particle Size (mm)	Percent Passing
4	4.76	100
10	2.0	90-100
40	0.42	0-60
100	0.15	0-5
200	0.074	0

Source: After Koerner et al. [93]

a sand backfill, with sufficient permeability for drainage, is recommended. Table 2.13 gives a suggested gradation. In many cases, sand is locally available and is generally significantly less expensive than quarried stone or river gravel. Even backfill soils that are finer-grained than sand are possible, but then a geosynthetic drainage system behind or within the backfill must be considered. Use of external or internal drainage systems to accompanying silt or clay backfills will be addressed in chapters 3 and 8.

2.7.2 Geotextile-Reinforced MSE Slopes and Embankments

Background. It should come as no surprise that if vertical walls can be built using geotextiles, steep soil slopes, and embankments can be stabilized by them also. In fact, as the slope angle with the horizontal (β) decreases, a wall transitions into a slope or embankment, albeit one in which the exposed face is not covered with anything except vegetation aided by some type of erosion-control material. In this section, geotextiles will be deployed in horizontal layers with no upturned facing treatment or hard-faced wall surface. When this is the case, the design methodology transitions from lateral earth-pressure theory to slope stability analyses. At what β value, the transition occurs is an interesting issue, and the US Federal Highway Administration sets the value at 70° with the horizontal. Various geotextile deployment schemes for embankments are shown in figure 2.37. The (a) pattern is typical. The uneven spacing pattern of (b) reflects those cases where stresses are higher in the lower regions of the slope than in the top. The short-edged strips shown in (c) and (d), sometimes called *secondary* reinforcement, represent compaction aids, necessary since high compaction at the edge of the slope is difficult to achieve. These short geotextile layers also eliminate shallow sloughing failures between widely spaced reinforcement layers. Note that all these schemes require the embankment to be built at the same time as placement of the geotextile proceeds; that is, they are not in-situ stabilization schemes (these are discussed later in this section).

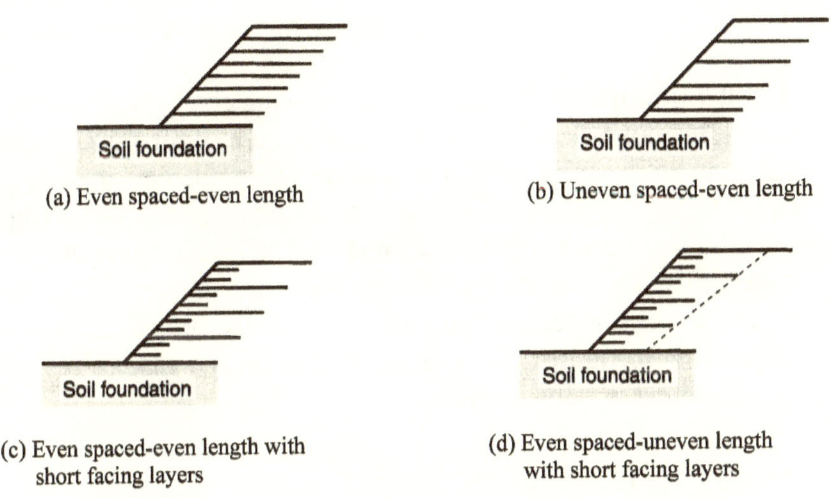

Figure 2.37 Various geotextile deployment schemes for stabilizing steep soil embankments on firm foundations.

Construction Details. Geotextile placement in embankment stabilization situations is relatively simple in that the sheets are usually placed horizontally as directed by the design. When using woven geotextiles (moreso than with nonwovens) it is important to recognize the direction of maximum stress. For two-dimensional plane strain cases, this is typically in the direction of the embankment face.

If the geotextile is sufficiently wide, it can be used parallel to the face of the slope and the weft, or cross machine direction, must carry the major principal stresses; the fabric must be manufactured and specified accordingly. If the geotextile is not sufficiently wide and is still to be used parallel to the face of the slope, sewn seams will be required, which will have to be included with an appropriate reduction factor in arriving at a T_{allow} value.

If the geotextile is oriented perpendicular to the face of the slope, the warp or machine direction will carry the major principal stress. Now the weft or cross-machine direction of these rolls will have to carry the minor principal stress, which is typically 33 to 50% of the major principal stress and can be handled by sewing or by an overlap sufficient to mobilize the required strength by frictional resistance. Although either method can effectively transmit the mobilized tensile stresses in the minor principal direction from one geotextile sheet to the next, the labor cost of sewn seams usually becomes small when many seams are required. Thus, the cost comparison illustrated in section 2.6.1 swings even farther than illustrated in favor of sewn seams.

Limit Equilibrium Design. The usual geotechnical engineering approach to slope-stability problems is to use limit equilibrium concepts on an assumed circular arc failure plane, thereby arriving at an equation for the factor of safety. Alternatively, a two-part wedge (or compound) analysis can be used and will be illustrated in chapter 3 on geogrids. The resulting equations for a circular arc failure for total stresses and effective stresses respectively are given below corresponding to figure 2.38. It is illustrated for the case of several layers of geotextile reinforcement.

$$FS = \frac{\sum_{i=1}^{n}\left(N_i \tan\phi + c\Delta l_i\right)R + \sum_{i=1}^{m} T_i y_i}{\sum_{i=1}^{n}\left(W_i \sin\theta_i\right)R} \tag{2.51}$$

$$FS = \frac{\sum_{i=1}^{n}\left(\overline{N}_i \tan\overline{\phi} + \overline{c}\Delta l_i\right)R + \sum_{i=1}^{m} T_i y_i}{\sum_{i=1}^{n}\left(\overline{W}_i \sin\theta_i\right)R} \tag{2.52}$$

where

FS = (global) factor of safety,
N_i = $W_i \cos\theta_i$,
$W_i, \overline{W}_i$ = total and effective weight of each slice,
θ_i = angle of intersection of horizontal to tangent at center of each slice,
Δl_i = arc length of each slice,
R = radius of failure circle,
$\varphi, \overline{\varphi}$ = total and effective angles of shearing resistance respectively,
$c, \overline{c}$ = total and effective cohesions respectively,

T_i = allowable geotextile tensile strength,
y_i = moment arm for geotextiles (note that in large-deformation situations these moment arms could become equal to R, which is generally a larger value):
n = number of slices;
m = number of geotextile layers;
$\overline{N}_i$ = N_i - u_i Δx_i, in which
u_i = $h_i \gamma_\omega$ = pore-water pressure,
h_i = height of water above base of circle for each slice,
γ_ω = unit of weight of water, and
Δx_i = width of slices.

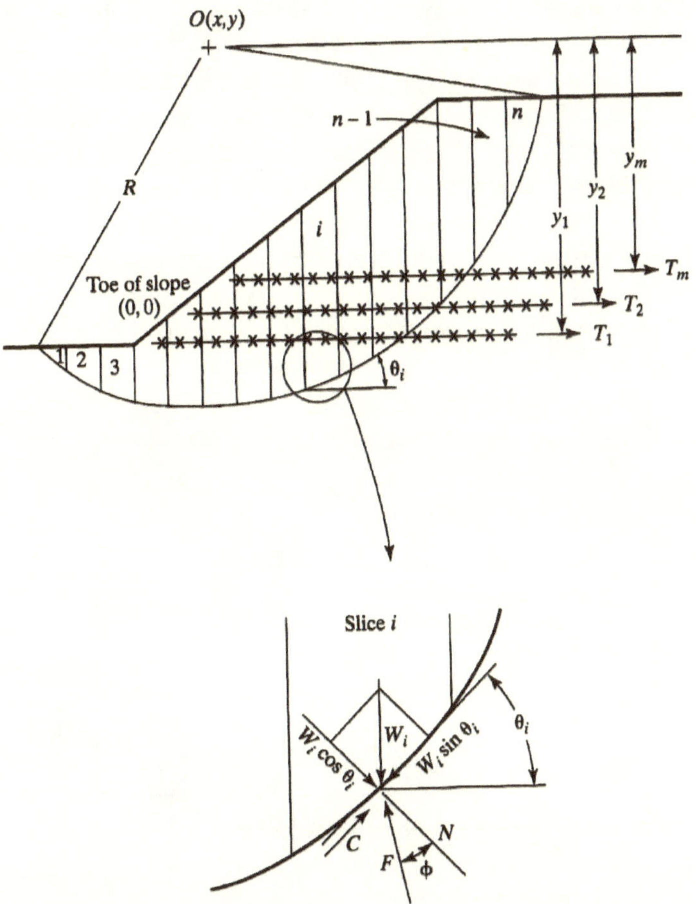

Figure 2.38 Details of circular arc slope stability analysis for (c, ϕ) shear strength soils.

Use of the total stress analysis equation, equation 2.51, is recommended for embankments where water is not involved or when the soil is at less-than-saturation conditions. The effective stress analysis equation, equation 2.52, is for conditions where water and saturated soil are involved—conditions typical of earth dams and delta areas involving fine-grained cohesive soils.

These equations are tedious to solve, and when additional consideration is given to finding the minimum value of factor of safety by varying the radius and coordinates of the origin of the circle, the process becomes unbearable to do by hand. Fortunately, many computer codes exist that either include, or can readily be modified to include, the contribution of the geotextile reinforcement—i.e., the $\Sigma T_i y_i$ term.

The moment arm y has been the topic of discussion in that its initial placement as shown is likely to be distorted as rotational deformation tends to occur. In the limit, this distortion could orient the geotextile along the potential failure arc, changing (and increasing) the moment arm from y to R. Kaniraj [94] has found that this transition can add as much as 45% to the resisting moment of the geotextile. While this is certainly possible, it is quite site-specific and the more conservative value of y is preferred.

For fine-grained cohesive soils whose shear strength can be estimated by undrained conditions, the entire analysis becomes much simpler. (Recall that this is the same assumption as was used in section 2.6.1 on unpaved roads.) Here slices need not be taken, since the soil strength does not depend on the normal force on the shear plane. Figure 2.39 gives the details of this situation, which results in equation 2.53. Examples 2.18 and 2.19 illustrate its use.

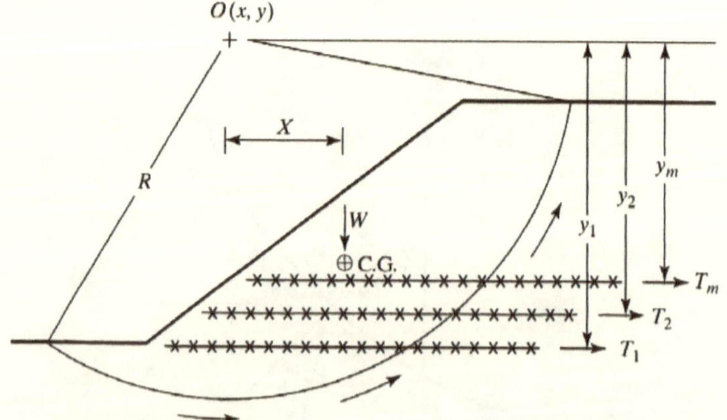

Figure 2.39 Details of circular arc slope stability analysis for soil strength represented by undrained conditions.

$$FS = \frac{cL_{arc}R + \sum_{i=1}^{m} T_i y_i}{WX} \tag{2.53}$$

where

FS = (global) factor of safety,
c = cohesion = 0.5 q_u,
q_u = unconfined compression strength of soil,
L_{arc} = length of the failure arc,
R = radius of the failure circle,
T_i = allowable tensile strength of various geotextile layers,
y_i = moment arm of geotextile layers,
W = weight of failure zone, and
X = moment arm to center of gravity of failure zone.

Example 2.18:

Assume you are dealing with the 10.0 m high, 50 deg. angle embankment shown below, which consists of a silty clay soil (γ = 19 kN/m^3, φ = 0 deg., c = 15 kPa, area = 60 m^2, center of gravity as indicated) on a silty clay foundation (γ = 20 kN/ m^3, φ = 0 deg., c = 18 kPa, area = 55 m^2, center of gravity as indicated). (a) Determine the factor of safety with no geotextile reinforcement. (b) Determine the factor of safety with a geotextile of *allowable* tensile strength 40 kN/m (note that with a cumulative reduction factor of 3.0, this is an ultimate strength geotextile of 120 kN/m) placed along the surface between the foundation soil and the embankment soil. (c) Determine the factor of safety with ten layers of the same geotextile placed at one meter intervals from the foundation interface to the top of the embankment. Assume that sufficient anchorage behind the slip circle shown is available to mobilize full geotextile strength. (d) Discuss how the problem solutions would differ if granular soil were involved?

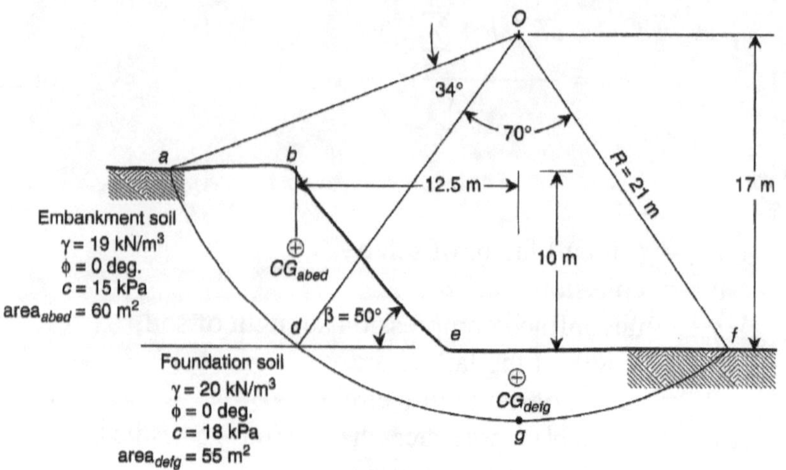

Solution: The following computational data are needed in all parts of the problem:

$$W_{abed} = (60)(19) = 1140 kN/m$$

$$W_{defg} = (55)(20) = 1100 kN/m$$

$$L_{ad} = 2(21)\pi\left(\frac{34}{360}\right) = 12.5m$$

$$L_{df} = 2(21)\pi\left(\frac{70}{360}\right) = 25.7m$$

(a) Slope as shown (with no geotextile reinforcement):

$$FS = \sum \frac{resisting\ moments}{driving\ moments}$$

$$FS = \frac{\left(c_e L_{ad} + c_f L_{df}\right) R}{W_{abed}(12.5) + W_{defg}(0)}$$

$$= \frac{\left[(15)(12.5) + (18)(25.7)\right] 21}{1140(12.5) + 0}$$

$$FS = \frac{13650}{14250} = 0.96; \text{ not acceptable and failure is indicated}$$

(b) Slope with a geotextile along surface *ed* with sufficient anchorage beyond point *d*:

$$FS = \frac{13650 + 40(17)}{14250}$$

$FS = 1.01$; still not acceptable and failure is incipient

(c) Slope with ten layers at one meter intervals from surface *ed* upward, all of which have sufficient anchorage behind the slip surface:

$$FS = \frac{13650 + 40(17 + 16 + 15 + \ldots\ldots + 9 + 8)}{14250}$$

$FS = 1.31$; which is marginally acceptable

(d) Regarding geotextile reinforcement considering *granular* embankment soils—that is, those with a frictional component, the design method involves taking slices and making the modifications described, recall figure 2.38 and equation 2.51 or 2.52. This is certainly possible, but it requires a computer code with search capabilities to find the critical arc radius and coordinates. Note that even in the undrained example presented here, this same type of search is required, but the calculations are much simpler because slices are not necessary.

Example 2:19:

Assume you are dealing with the slope shown in example 2:18. (a) Determine how much embedment (or anchorage length) is required *behind* the potential slip circle in order to mobilize the allowable tensile strength of the geotextile. Assume that the transfer efficiency of the geotextile to the shear strength of the soil is 0.80 and base the calculation on a $FS = 1.5$. (b) Determine the total length of the geotextile, using the maximum distance from the slope face to the failure plane to be 8.0 m. (c) Comment on the effect of the possible placement orientations of the geotextile rolls.

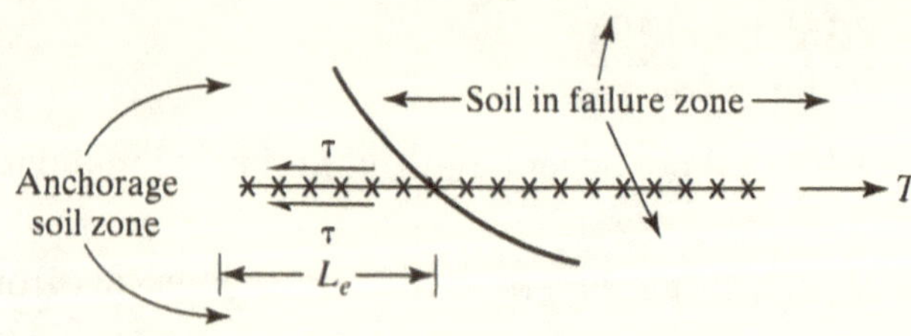

Solution: (a) When the anchorage test was explained in section 2.3.3, it was assumed that the resistance was uniformly distributed over the geotextile's embedment and that the strength was entirely mobilized. This is almost certainly not the case. It appears that the concentration decreases rapidly as the embedment length increases and that separate mobilized and fixed portions of the geotextile exist. For this problem, however, a linear distribution will be assumed over a continuous displaced length, since it results in a conservative length. Taking force summation in the x-direction results in the following equation:

$$2\tau E L_e = T_{allow}(FS)$$
$$2(15)\,(0.8)\,(L_e) = 40\,(1.5)$$
$$L_e = 2.5 \text{ m},$$

which is the required embedment length beyond the potential slip circle for sufficient anchorage of each geotextile layer

(b) The total length of each geotextile layer will be 2.5 + 8.0 = 10.5 m.

(c) Since the typical widths of commercially available geotextiles range from 3.0 to 4.5 m, seams will be necessary in at least one direction. If the geotextile rolls are oriented parallel to the slope face at least two seams will be required and they will be oriented perpendicular to the major stress direction. If this is the case, a reduction factor for seam strength will have to be included to arrive at the value of T_{allow}. Figure 2.10 provides guidance in this regard. However, if the geotextile is oriented perpendicular to the slope face, sewn seams will be required along the edges of each roll of geotextile and will therefore be in the minor principal stress direction. Overlaps could also be considered, but they would probably not be an economic solution for the higher-strength geotextiles used in this problem; for these, sewn seams are the logical choice.

Summary. Geotextile-reinforced slopes and embankments have been shown to be a practical expedient in many situations. When reinforced, slope heights and/or angles can be significantly increased over the nonreinforced situation. Designwise, the process involves modifications to limit equilibrium procedures that are within the realm of geotechnical engineering practice and seem to be a rational approach. There are several uncertainties in the analysis, however, and additional research is warranted in the following areas:

- What reduction factors should be used to adapt an ultimate strength value to an allowable value?
- Should load factors be used on the other side of the equation as in load and reduction factor design (LRFD) methods?
- Should the (global) factor of safety be on the soil and the geotextile and, if so, should they be the same values?
- What moment arm should be used?
- How are shear stresses transmitted from the soil to the geotextile?
- Is there interaction between closely spaced geotextile layers? If so, how does one address this interaction?
- What anchorage is needed, and how is it mobilized?
- How is strain compatibility of the soil and geotextile(s) considered?
- What type of surface slope treatment is necessary to provide erosion protection?
- Would a shift in design using finite element methods give insight (or eliminate) some of the aforementioned items, see Guler and Dosay [95].

2.7.3 Geotextile-Reinforced Foundation Soils

The purpose of this section is to focus on stable embankment soils or other structures placed above unstable and/or weak foundation soils. This is in contrast to the previous section, in which the embankment itself consisted of either unsuitable or weak soils or the situation required very steep and/or high slopes.

Background. Fine-grained saturated soils exist near most river estuaries and harbor areas around the world. Unfortunately, these are often the areas where industrialization is the most intense. Buildings, factories, freight yards, stockpiles, storage tanks, access roads, roadways, highways, railroads, and other appurtenances of industry are all incompatible with such weak foundation soil conditions. The traditional foundation options given to the owners of such facilities have been to drive deep foundations through the unsuitable soils, thereby avoiding them altogether, excavate and replace the soil with suitable soil, stabilize the soil with injected additives, or surcharge and wait until consolidation strength occurs. All these methods have a degree of applicability, but all suffer from being either expensive, time-consuming, or both.

An alternative method, which is the focus of this section, is to deploy a high-strength geotextile over the site, place a sand drainage layer/working blanket above it, install vertical wick drains (to be described in chapter 8) to the bottom of the soft foundation layer, and then complete the surcharge fill up to the equivalent design load. The sand blanket and surface layers can be placed by conventional earth-moving equipment or by dredging. As will be illustrated later, the entire process can be accomplished under water if the need arises.

Clearly, the geotextile acts as a reinforcement material, since the shear strength of the foundation soils are often less than 10 kPa, which would hardly support the weight of an individual. The vertical wick drains (also called prefabricated vertical drains or strip drains) are geosynthetic composites and are used to drain the excess pore water from the foundation soil as it is mobilized by the surcharge fill. The surcharge fill usually consists of locally available soils.

There are two somewhat different variations for the configurations of these projects. One is a large *areal fill* in which the length and width of the site are approximately equal. In such cases there is no clearly defined principal stress direction, and the strength of the geotextile must be equally balanced in all directions. This, of course, is required of the seams in both directions as well as for the material itself. Actual situations in this category are often industrial—or building-site development projects. The second variation is one in which the length of the fill is much larger than the width, called a *linear fill*. In these cases, the major principal stress direction can be identified and the geosynthetic reinforcement can be aligned accordingly. Seams can often be avoided or placed in the minor principal stress directions. Situations in this category are roadway embankments and containment dikes. Both areal and linear fill situations will be illustrated by case histories.

Construction Details. The US Army Corps of Engineers has been the leading force behind the use of high-strength woven geotextiles to reinforce very soft foundation soils [96,97]. Quite often the task has been to construct permanent linear dikes for the containment of dredged soil or flood protection. The failed dikes in New Orleans via Hurricane Katrina are a case in point [98]. As such, the necessity for high-strength seams can be somewhat avoided by placing the warp (strong) direction transverse to the dike's alignment, in the direction of the major principal stress. This allows for the weft (weaker) direction to be seamed and placed in alignment with the minor principal stress. See Sprague and Koutsourais [99] for a compilation of projects of this type. The foundation soil strengths for these projects were generally very low, from 1 to 8 kPa. The geotextiles used had wide-width tensile strengths 80 to 700 kN/m. They were all relatively heavy woven fabrics made from polyester or polypropylene fibers. The embankments placed above the geotextiles varied in height from 2 to 7 m. Postconstruction consolidation settlements varied from 1 to 5 m. To my knowledge all have been successful with only one known problem: at one site a propagating tensile failure occurred in the geotextile between instrumentation holes as surcharge was near full height. This illustrates the importance of reducing geotextile strength to allow for planned holes such as those made in the fabric for installation of instrumentation devices or wick drains [100].

The Wilmington Harbor South Disposal Area project [101] is an example of a *linear fill*. The project consisted of the construction of a U-shaped dike from land out into a major river to provide storage capacity for dredged material from maintenance dredging.

The foundation soils, which are under as much as 5 m of water, consist of weak, highly compressible silts and clays that form the tidal flats and shallows. Unconfined compression shear strengths range from nil to 10 kPa for depths averaging 27 m, where firm sands and gravels are eventually encountered.

The poor foundation conditions just described, the limited quantity of granular borrow soil available, and environmental considerations led to the adoption of a wide-bermed embankment to enclose this disposal area. The concept behind the design involves "floating" the dike on the soft foundation soil with the use of a high-strength geotextile for tensile reinforcement. The chosen geotextile deployed on this project was a woven polyester fabric and was specifically designed for this application. The geotextile specifications are given in table 2.14. After placement of the high-strength geotextile, the dike was constructed of dredged granular soil placed in two stages. The first stage averaged approximately 3 m deep by 180 m in width and formed the wide berm of the dike section. This first stage construction consisted of five separate hydraulic fills [101]. When constructing the embankment, the outer two fills were placed concurrently, so as to contain the foundation soil and prevent its lateral extrusion. This is a critical aspect of the construction since the lateral containment that is provided forces the central soil to subsequently consolidate (and not laterally extrude) thereby providing long-term stability. This also placed the central section of the geotextile in tension and provided added support for the subsequent three fills. The subsequent three fills applied load whenever the geotextile rose above the level of the water due to high pressures from the underlying foundation soils. This first stage reached an average top-of-fill elevation of 2.1 m that was approximately 1.0 m above mean high water.

Prefabricated vertical wick drains were installed through the granular fill and geotextile to a depth of 12 m. The drains were in a triangular pattern of 3 m. Upon the completion of primary consolidation settlement (measured via piezometers, settlement anchors, and inclinometers), the second stage embankment was placed on top of the first stage up to an elevation of +4.6 m. The outboard slope has a riprap erosion control system protecting it, and the inboard slope has a low-permeability soil liner.

TABLE 2.14 COMPARISON OF SPECIFICATIONS FOR TWO HIGH STRENGTH GEOTEXTILE FOUNDATION STABILIZATION PROJECTS

Geotextile Specification Properties	Linear Fill (Containment Dike; Wilmington Harbor; U. S. Army Corps of Engineers)	Areal Fill (Industrial Development; Seagrit, Maryland; Maryland Port Admin.)
Polymer type	PET	PP & PET
Tensile strength (kN/m)	260	180
Modulus (kN/m)	3300	500
Elongation (%)	10 - 35	15-35
Stiffness (mg-cm)	–	30,000
Friction angle (deg.)	30	30
Seam strength		
Warp (kN/m)	none	105
Weft (kN/m)	140	105
Seam type	J	J
Seam thread type	PET	PA, PET

An example of an *areal fill* on dredged soil using a high-strength geotextile was at Seagirt, Maryland, for the Maryland Port Administration [102]. This 45 ha site contains 6 to 11 m of dredged soil at water contents of 50% to 150%. The goal was to prepare the site for ground surface loads of approximately 30 kPa within the extraordinarily short time of six months. This goal was achieved by deploying a 1000 g/m^2 woven geotextile having wide-width strength of approximately 210 kN/m. The required seam strengths was 105 kN/m, which was the limiting design constraint. See table 2.14 for a comparison of the geotextile specification of this project with the previously described linear fill project. In this areal fill project, sewn seams were particularly critical since the direction of the maximum principal stress was not known. Thus, seam strength dominated the design (recall figure 2.10 for seam efficiencies of high-strength geotextiles that must be included in equation 2.24 as an additional reduction factor). A single 0.75 m thick lift of granular soil, serving as the drainage blanket/working platform, was placed and wick drains at 1.5 m centers were installed to the bottom of the in-situ soil. Note that the allowable geotextile strength must make accommodation for these holes and must be included as still another reduction factor in equation 2.24. The final operation was to place an 2.5 m surcharge fill over the entire site. Settlement plates and piezometers were the main control instruments from which the surcharge fill placement rate and dwell time were controlled. The project was very successful in that consolidation occurred to the anticipated degree within the desired six-month time frame. The entire area is currently paved and used for heavy truck storage, loading, and unloading.

Design Methods. In considering an embankment placed on very soft foundation soil and supported by a geotextile, a number of design elements, all of which are potential failure scenarios, are present. Figure 2.40 illustrates these various possibilities. In sequentially going from one design scenario to the next, the overall geotextile/ embankment design gradually becomes more defined.

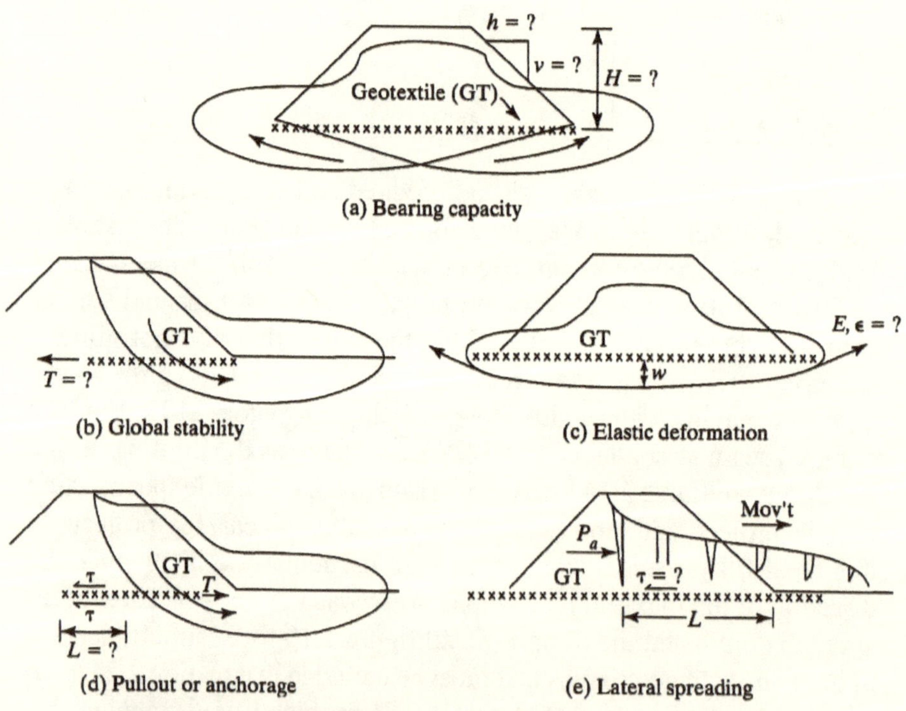

Figure 2.40 Geotextile design models for use in soft stabilization. (After Koerner, et al. [100])

(a) bearing capacity	→	for overall embankment geometry
(b) global stability	→	for strength design in major principal stress direction
	→	for strength design in minor principal stress direction
(c) elastic deformation	→	for required modulus and maximum strain in major principal stress direction
	→	for required modulus and maximum strain in minor principal stress direction
(d) pullout or anchorage	→	for required anchorage length behind slip plane(s)
(e) lateral spreading	→	for frictional resistance against embankment sliding

Bearing Capacity. Regarding bearing capacity as shown in figure 2.40a, the limiting embankment height that can be placed on a given foundation soil is essentially independent of the geotextile. If a mass failure occurs beyond the limits of the reinforced zone, the geotextile will be carried along en masse. Thus, conventional geotechnical engineering theory can be used directly:

$$q_{allow} = cN_c/FS \tag{2.54}$$

where

$q_{allow} = \gamma H_{allow}$ = allowable soil-bearing capacity, in which
γ = unit weight of embankment soil, and
H_{allow} = allowable height of embankment;
c = cohesion of the foundation soil;
N_c = bearing capacity factor (= 3.5 to 5.7); and
FS = factor of safety.

Calculations based on equation 2.54 are surely worst-case situations, for soil strength invariably increases with depth (see Humphrey [103]).

Global Stability. Figure 2.40b shows the type of global stability model that results in the required strength of the geotextile. It is precisely the same as formulated in equation 2.53, since the soil foundation strength can generally be estimated by its undrained shear strength and a single reinforcement layer is usually being placed. The strength of the embankment soil above the geotextile is another matter. Depending on the soil's thickness, this shear strength is often taken as being zero—the assumption being that if tension cracks occur in the embankment soil due to lateral deformation, the shear strength can easily be lost. Shown in figure 2.41 a, b, c are the required geosynthetic strength values for cases of surcharge weight only, surcharge weight plus placement bulldozer, and surcharge weight plus wick drain installation crane, respectively. For weak foundation soils and typical slope angles in the 20° to 40° range, the required tensile strength of the geotextile is seen to be quite high.

Example 2.20:

What are the required, allowable and ultimate wide-width geotextile strengths for a 4.0 m high reinforced embankment whose face is on a slope of 3(*H*)-to-1(*V*) using a factor of safety of 1.3 and reduction factors from table 2.8a. The undrained shear strength of the foundation soil is 2 kPa.

Solution: Using a slope angle of 18.4 deg. and figure 2.41a results in

$$T_{reqd} = 66 \text{ kN/m}$$

from which

$$FS = \frac{T_{allow}}{T_{reqd}}$$

$$1.3 = \frac{T_{allow}}{66}$$

$$T_{allow} = 86 kN/m$$

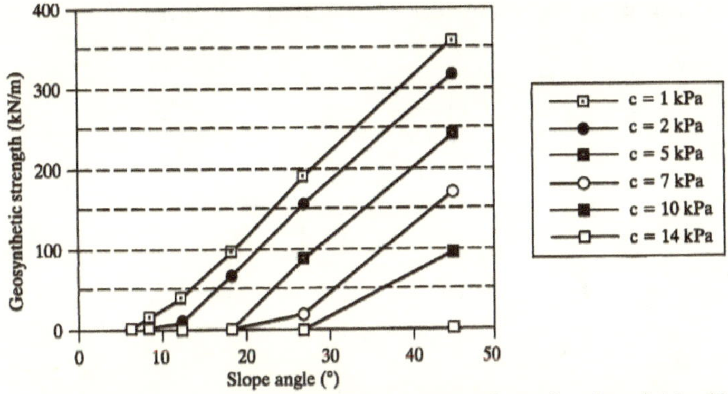

(a) Required geosynthetic strength based on FS = 1.3. Chart reflects soil surcharge height of 4 m.

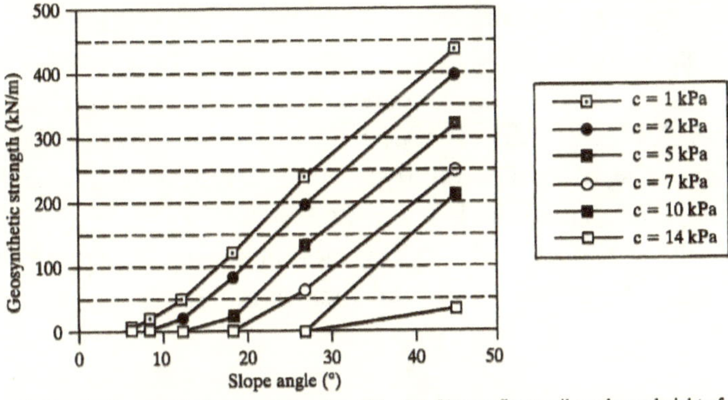

(b) Required geosynthetic strength based on FS = 1.3. Chart reflects soil surcharge height of 4 m plus 13 kPa dozer on embankment.

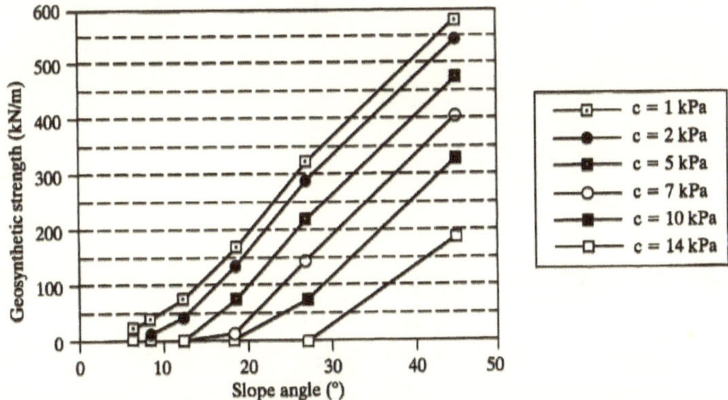

(c) Required geosynthetic strength based on FS = 1.3. Chart reflects soil surcharge height of 4 m plus 42 kPa wick drain installation equipment on embankment.

Figure 2.41 Studies for embankment global stability using geotextile reinforcement.

Using equation 2.24 and table 2.8a gives the necessary ultimate wide-width laboratory test strength.

$$T_{allow} = T_{ult}\left[\frac{1}{RF_{ID} \times RF_{CR} \times RF_{CBD}}\right]$$

$$86 = T_{ult}\left[\frac{1}{1.2 \times 2.0 \times 1.2}\right]$$

$$T_{ult} = 86(2.88)$$

$$T_{ult} = 250 kN/m$$

Note that the above is for the geotextile itself. If seams are involved in the direction of the principal stress, see figure 2.10 for typical efficiencies. Here it is seen that maximum efficiency for a 250 kN/m geotextile seam is approximately 0.75. This is comparable to a reduction factor of 1.33. If holes are involved via the insertion of wick drains, the loss of strength is approximately linear with hole dimension [100]. Thus, wick drains at 3 m centers making geotextile holes of 300 mm size will require another reduction factor of 10/9 = 1.11. Using these two additional factors makes the formulation as follows:

$$T_{allow} = T_{ult}\left[\frac{1}{RF_{ID} \times RF_{CR} \times RF_{CBD} \times RF_{Seams} \times RF_{Holes}}\right]$$

$$86 = T_{ult}\left[\frac{1}{1.2 \times 2.0 \times 1.2 \times 1.33 \times 1.11}\right]$$

$$T_{ult} = 86(4.25)$$

$$= 365 kN/m$$

Elastic Deformation. The amount of elastic deformation allowed by the geotextile will govern the deformation of the embankment, as shown in figure 2.40c. Obviously, too great an amount will cause unwanted embankment deformation and loss of underlying foundation soil. Thus, relatively high-modulus values of the geosynthetic are desirable. Unfortunately, "relatively high" is a poorly defined term. The US Army Corps of Engineers desired value of maximum strain at the required stress is approximately 10%, thus

$$E = T_{reqd} / \varepsilon_f \tag{2.55}$$

$$E = T_{reqd} / 0.10$$

$$E_{reqd} = 10\, T_{reqd} \tag{2.56}$$

However, to obtain this E_{reqd} value requires a significantly stronger geosynthetic than T_{reqd} without this condition. The modulus requirement will dominate over the strength requirement if this condition is imposed. Note that these comments are based on the geosynthetic itself and do not consider seamed areas. The latter situation is difficult to consider, however, deformation monitoring of seams been attempted, Guglielmetti et al. [104].

Pullout or Anchorage. With the mobilization of all, or part, of the geotextile reinforcement's strength comes an equal and opposite requirement that the soil behind the slip zone resist pullout. As shown in figure 2.40d, the situation is one in which an anchorage problem can be envisioned. Extending the work of section 2.7.2, an equation can be formulated as follows:

$$T_{act} = 2\tau L$$

$$= 2(c_a + \sigma_v \tan\delta) L$$

and

$$L_{reqd} = \frac{T_{act}}{2(c_a + \sigma_v \tan\delta)} \tag{2.57}$$

or

$$L_{reqd} = \frac{T_{act}}{2E(c_a + \sigma_v \tan\phi)} \tag{2.58}$$

where

L_{reqd} = required anchorage length behind the slip plane;
T_{act} = actual stress in the geosynthetic;
c = cohesion of the soil;
c_a = adhesion of the soil to the geosynthetic;
φ = friction angle of the soil;

δ = friction angle of the soil to the geosynthetic;
σ_v = average vertical stress = γH, in which
γ = unit weight of embankment soil, and
H = average height of embankment above geosynthetic; and
E = anchorage, or pullout, efficiency of geosynthetic-to-soil.

For geotextiles, E=0.8 to 1.2; for geogrids, E=1.3 to 1.5. If insufficient anchorage distance is available to mobilize the required strength of the geosynthetic, physical methods of attachment—rolling of the material around stone or attachment to timber cribbing—is necessary.

Lateral Spreading. On occasion, tension cracks have been observed on the surface of embankments as shown schematically in figure 2.40e. The situation of lateral spreading can be analyzed using the following equation for active earth pressure and considering granular soil fills to be above the geosynthetic.

$$P_a = \tau L$$
$$P_a = \left(\sigma_{v_{ave}} \tan\delta\right) L$$
$$0.5\gamma H^2 K_a = \left(0.5\gamma H \tan\delta\right) L$$
$$\tan\delta_{reqd} = \left(HK_a\right)/L \qquad (2.59)$$

where

δ_{reqd} = required friction angle of geosynthetic-to-soil;
H = embankment height;
K_a = coefficient of active earth pressure = $\tan^2(45 - \varphi/2)$; in which
φ = friction angle of embankment soil
L = length of zone involved in spreading; and
$\tan\delta_{reqd}$ = $E(\tan\varphi)$, in which
E = shear, or frictional, efficiency of geosynthetic-to-soil; for geotextiles, E = 0.6 to 0.8; for geogrids, E = 1.0 to 1.5.

Geotextile Implications. The previous designs focus entirely on quantitative analyses and procedures that lead to the ultimate selection of the reinforcement geosynthetic. However, there are other considerations, many of which are qualitative, that must be brought into focus. These include the specific gravity and rigidity (or stiffness) of the geosynthetic, and the size and weight of rolls.

If the site under consideration is at, or under, the surface of water, buoyancy is not a desirable feature. Thus, the geosynthetic should not float, and its specific gravity should be greater than one. Rigidity, or stiffness, of the geosynthetic is desirable for providing some type of working platform for deployment. The ASTM stiffness test described in section 2.3.2 can be used for specifications. The minimum value is related to the CBR of the foundation soil (recall table 2.3). This is an important feature and an area where additional investigation is warranted. The size and weight of geosynthetic rolls must be considered by everyone involved in the process particularly the contractor. It is obviously a site-specific situation, but one that is paramount in the success of the project. Designs that cannot be reasonably constructed *should not* be designed to begin with.

Summary. It is the author's perception that soft soil stabilization using geosynthetics has been implemented from two different perspectives. The first is by the geotechnical engineer, who has modified traditional design methods to accommodate the inclusion of reinforcement; the second is by the manufacturer, who has provided a means of accomplishing an end. In this case, the means is a high-strength geosynthetic, either a geotextile or geogrid.

For designs and installations that are both economical and safe, the perspectives of engineer and manufacturer must be brought together.

Beginning with the design elements illustrated in figure 2.40 the stress-strain characteristics of the geosynthetic are progressively defined. The initial technical design is then modified by other site-specific considerations, the applicable yarns, and fabrication of the geotextile or geogrid. The next, and very significant, question is whether the result is a balanced design. For example, where is the critical aspect of the design? If it is the field-sewn seams, then everything (warp, weft, modulus, elongation, etc.) should be formulated from this point. Last, there is the question of whether the resulting geosynthetic is constructible in light of the actual site situation. Here considerations of workability and survivability are very important.

In the final analysis, one should be able to arrive at a geosynthetic design that is both optimally safe and economical. It is indeed a very worthwhile pursuit, for finally the profession has a technique whereby we can almost walk on water.

2.7.4 Geotextiles for Basal Reinforcement between Deep Foundations

A variation on the above theme is to use a high-strength geotextile (or geogrid) to span between pile caps, stone columns or related deep foundation systems thus providing *basal reinforcement* between discrete supports. The reinforcement is considered as a tensioned membrane under an imposed soil loading that may, or may not, have arching considered. The original application of support between deep foundations is covered in the British Standard Code of Practice [105] that presents the following equations for load (W_T) and the required geosynthetic strength (T_{reqd}) when H > 1.4 (s-a).

$$W_T = \frac{1.4 s f_{fs} \gamma (s-a)}{s^2 - a^2}\left[s^2 - a^2 (p'_c / \sigma'_v)\right] \tag{2.60a}$$

$$T_{reqd} = \frac{W_T (s-a)}{2a}\sqrt{1 + \frac{1}{6\varepsilon}} \tag{2.60b}$$

where

W_T = distributed load carried by the reinforcement,
s = center-to-center spacing of piles,
f_{fs} = partial load factor for soil unit weight [= 1.3 (ultimate), = 1.0 (serviceability)],
γ = unit weight of embankment fill,
a = size (or diameter) of piles or pile caps,
p'_c = vertical stress on piles or pile caps,
σ'_v = factored average stress at base of embankment and = $(f_{fs}\gamma H + f_q q)$
q = surcharge intensity on top of the embankment,
H = height of embankment,
f_q = partial load factor for external applied loads [= 1.3 (ultimate), = 1.0 serviceability)]
T_{reqd} = required tensile strength of the geosynthetic reinforcement,
ε = strain in the reinforcement (varies from 5% to 10%).

TABLE 2.15 COMPARISON OF SEVEN BASE REINFORCEMENT METHODS; AFTER FILZ AND SMITH [106]

Method	Stress Reduction Radio (SRR)					
	d/s = 0.25		d/s = 0.33		d/s = 0.50	
	H/s = 1.5	H/s = 4	H/s = 1.5	H/s = 4	H/s = 1.5	H/s = 4
British BS8006 (1995)	0.92	0.34	0.62	0.23	0.09	0.02
Adapted Terzaghi, $K_T = 1$	0.60	0.32	0.50	0.23	0.34	0.13
Adapted Terzaghi, $K_T = 0.5$	0.77	0.52	0.69	0.42	0.54	0.26
Kempfert et al. (2004)	0.55	0.46	0.43	0.34	0.23	0.15
Hewlett and Randolph (1988)	0.52	0.48	0.43	0.31	0.30	0.13
Adapted Guido (1987)	0.12	0.04	0.10	0.04	0.08	0.03
Carlsson (1987)	0.47	0.18	0.42	0.16	0.31	0.12

where: SRR = stress reduction ratio
$= \sigma_{v(\text{on GS})}/\sigma_{z(\text{soil above})}$
$= \sigma_{v(\text{on GS})}/(\gamma H+q)$

d = pile width
s = center-to-center pile spacing
H = embankment height

In calculation of the above equations it has been observed that a very high-strength geotextile used by itself will generally be required. Subsequent research has shown that many other approaches are available most of which allude to the above being generally conservative. Filz and Smith (106) have evaluated seven of these methods finding that the stress reduction ratio (stress on the geosynthetic compared to the full static and surcharge load above it) varies enormously. As seen in table 2.15, this value not only varies within a given pile spacing value, but changes order in going between different design methods. Such variations are obviously quite unsettling.

These limit equilibrium methods are being replaced by finite element and finite difference methods, e.g., Han and Akins [107], which more accurately analyze the incremental stress in the reinforcement. Invariably shown is that the geosynthetic tensile stresses over the edges of the pile caps are enormous. While the pile caps could be rounded, a more flexible and compliant deep foundation system is of great advantage. The emerging area of geogrid encapsulated stone columns and geotextile encapsulated sand columns are ideally suited in this regard [108]. The combined topics, i.e., basal reinforcement and geosynthetic encased stone or sand columns, are reviewed by Koerner and Wong [109]. This is certainly a fruitful topic of research covering a number of possible geotextile (and geogrid) applications.

2.7.5 Geotextiles for Improved Bearing Capacity

With the recognition that high-strength geotextiles can reinforce flexible walls, slopes, and foundations, it follows that soils beneath rigid walls, footings, piers, etc., which have poor bearing

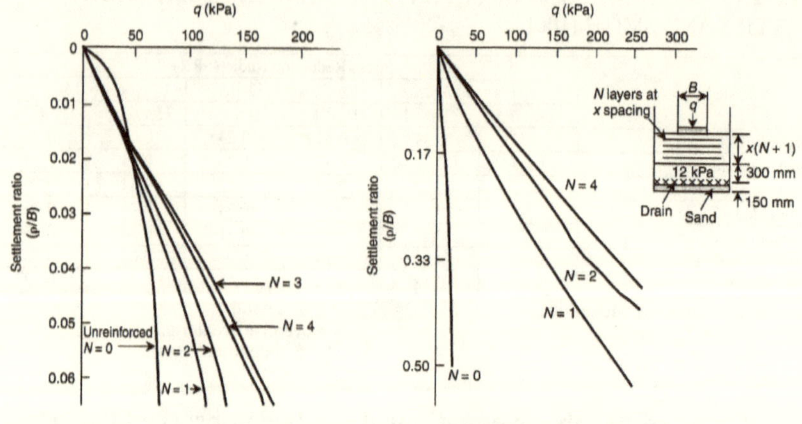

(a) Nonwoven heat-bonded geotextile at 140 mm spacings on loose sand (D = 50%) using 560 mm square footing. (After Guido, et al. [110])

(b) Woven silt-film geotextile at 40 mm spacing on soft saturate clay silt using 150 mm round footing, (Geosynthetic Research Institute)

Figure 2.42 Laboratory developed curves showing improvement in bearing capacity of soils using geotextiles; ρ is the footing settlement and β is the footing width.

capacity should also be a target for improved performance using geosynthetics.

Laboratory studies by Guido et al. [110], using layers of geotextiles on a loose sand, produce the results of figure 2.42a. Here it is seen that multiple (up to three) layers produced beneficial results, but only after a measurable settlement had occurred. This is to be expected, since the geotextile must deform before its reinforcing benefit is realized. Their tests used a nonwoven heat bonded geotextile and varied a number of parameters, including distance to the upper geotextile, the spacing between layers, and the distance the geotextile extends beyond the edge of the footing.

Laboratory studies at the Geosynthetic Research Institute on soft compressible fine-grained soils at saturations above their plastic limit has produced the curves shown in figure 2.42b. Here, using a woven slit film geotextile, similar behavioral trends are noticed; some improvement in bearing pressure is noted throughout, but only at large deformations is the improvement noteworthy. It can be seen that for both of the studies portrayed in figure 2.42, a method of prestressing the geotextile would be an advantage so as to eliminate the required deformation before significant improvement is noted. How to do this in a cost-effective manner, however, is not known. In lieu of prestressing the geotextile, design must consider improved bearing capacity only after relatively large settlement. Also, the laboratory-generated curves shown must be utilized with considerable caution, since scale effects are essentially unknown. In proceeding with a design, four modes of failure must be considered. They are shown schematically in figure 2.43 and are explained below. Also note that geogrids can be used as well as geotextiles.

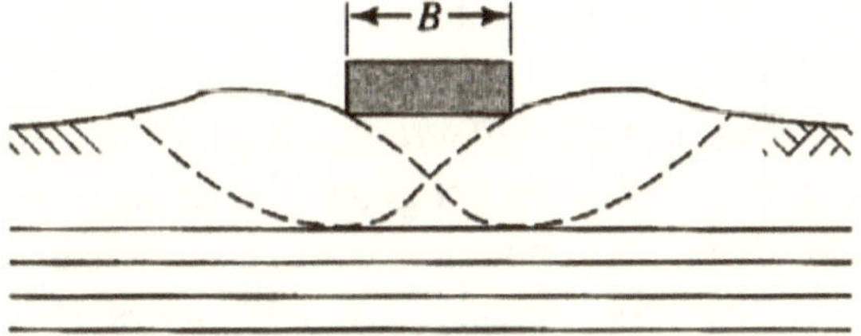

(a) Bearing capacity failure above upper geotextile layer

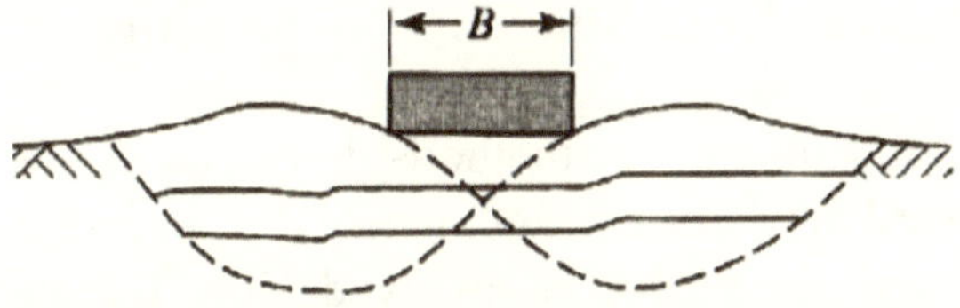

(b) Anchorage pullout of geotextiles due to insufficient embedment length

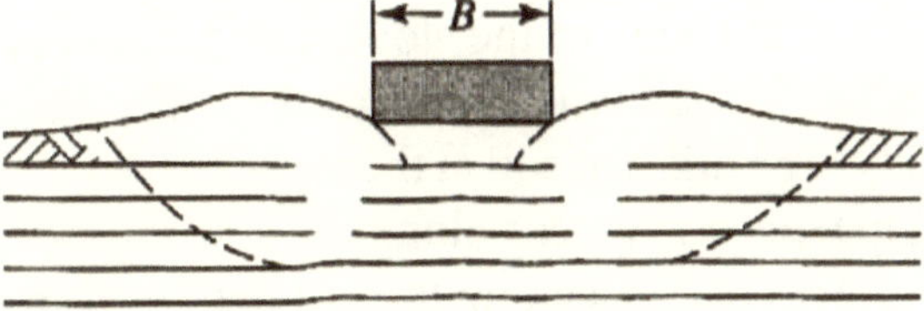

(c) Tensile failure (breaking) of geotextiles

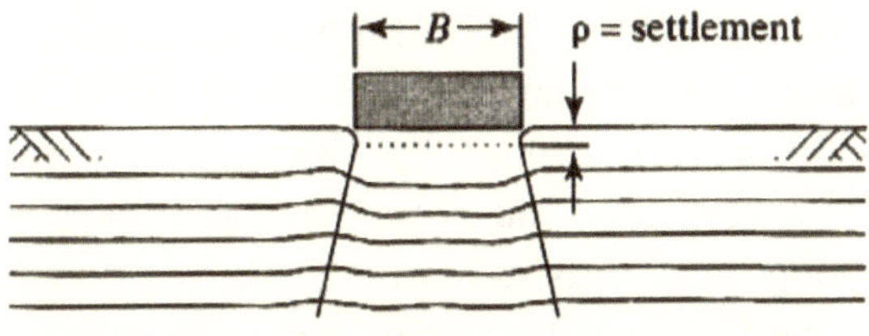

(d) Excessive long-term deformations (creep)

Figure 2.43 Possible modes of failure of geosynthetic reinforced shallow foundations.

- *Bearing capacity failure of the soil above the upper geotextile.* This is probably avoidable if the upper geotextile is within 300 mm of the ground surface.
- *Insufficient embedment length.* This causes anchorage pullout, which is avoidable if the geotextile extends far enough beyond the potential failure zone to mobilize the required resisting anchorage force. This is described in sections 2.7.2 and 2.7.3.
- *Tensile failure of the geotextile(s) due to overstressing.* This is the main design element and uses information such as that presented in figure 2.41.
- *Excessive long-term (creep) settlement.* This is due to the sustained surface loads and subsequent geotextile stress relaxation, which can be avoided if low-enough allowable stresses in the geotextile are used. Allowable tensile strength values should be very conservatively chosen (see equation 2.24 and table 2.8a).

2.7.6 Geotextiles for In-Situ Slope Stabilization

Background. Thus far, the methods described in this section are all oriented toward new construction, where the geotextiles are placed simultaneously with the earthwork involved. However, there are many situations in which existing soil slopes and embankments are at or near to their failure state. Oftentimes, homes or other structures are precariously close to the edge of the slope. Landslides cause losses of billions of dollars and, sometimes, lives as well. The usual indicators of slope instability are bulges of soil at the toe of the slope, springs near the toe of the slope, tension cracks along the soil slope or at the crest, and vegetative growth that is leaning toward the downstream toe. The basic problem in such cases is insufficient shear strength of the soil with respect to the slope angle and/or height. This often suggests low relative density in granular soils and high water content in cohesive soils. Both of these situations could be positively influenced by some type of in-situ stabilization system. The addition of a surface deployed geotextile that is "nailed" into the slope could provide such a system. In *soil nailing* long steel rods are driven into the ground and then shotcreted or gunited to form temporary retaining walls, see Shen et al. [111]. Instead of using a rigid impermeable facing such as shotcrete or gunite, however, the designer could consider using a geotextile strengthened locally at the points where it is nailed to the soil slope.

Such a system, developed by the author [112, 113], is shown in figure 2.44. Here the anchored geotextiles (called *anchored spider netting* since the surface appears quilted and tucked into the soil on regular patterns) are used in the compaction and/or consolidation of the in-situ soils. With both the geotextile netting and the steel-rod nails in tension, the set of free-body diagrams of the system is shown in figure 2.45. The system should be reanchored periodically, since loss of pore volume (either air or water) will result in a relaxed tensile stress of the geotextile. During this time, of course, the soil itself is gaining in shear strength either by increased densification and/or by consolidation. This improvement in shear strength is precisely what is necessary to reinstate the slope's stability. Once the soil properties are sufficiently improved, the geotextile no longer serves a useful strengthening purpose. However, it continues to serve as an erosion control material, thus UV inhibitors must be included in the geotextile during its manufacture.

Design Method. While the role of the surface-tensioned geosynthetic is clear (along with its compressive influence on the enclosed soil mass), the role of the nails protruding into and beyond the potential failure plane is not. Their contribution toward stabilizing the slope is difficult to assess. Using concepts taken from soil nailing, there are a number of potential benefits of the rods. Some of these are as follows: friction along the surface of the rods, adhesion along the surface of the rods, suction at the rod ends if saturated conditions exist, bearing at any protrusions along the surface of the rods, the bending resistance of the rods due to the downward movement of the soil mass, and the torsional resistance of the rods due to any out-of-plane bending forces. These are very difficult phenomena to include in a theoretical analysis and will be lumped together into one single soil-modification parameter in the following analysis.

For the effect of the tensioned net on the soil's surface, the analysis is much more understandable and analytically tractable. The purpose of the tensioned net is clearly to place the encapsulated soil mass in compression. This will add normal forces to the slope, which will increase its stability by causing the soil to densify. For granular soils this densification is rapid and takes the form of compaction where pore air or pore water is expelled. For saturated cohesive soils this densification is slower and takes the form of consolidation where the

Figure 2.44 Cross section and photographs of in-situ slope stabilization using anchored geotextiles. (After Koerner [112,113])

pore water is expelled, but over a considerably longer time than when dealing with granular soil masses. In both cases, the need for a porous geosynthetic at the soil's surface is obvious. Use of an impervious geomembrane is not applicable for this method.

The time-dependent densification process just described will ultimately cause an increase in soil shear strength by means of an increase in friction and/or cohesion, thereby increasing the slope's stability. This densification (hence volume reduction) process will require redriving of the anchors to greater depths on a periodic basis. Eventually the increased shear strength parameters will allow for the slope to support itself without the need for the anchored spider netting at all.

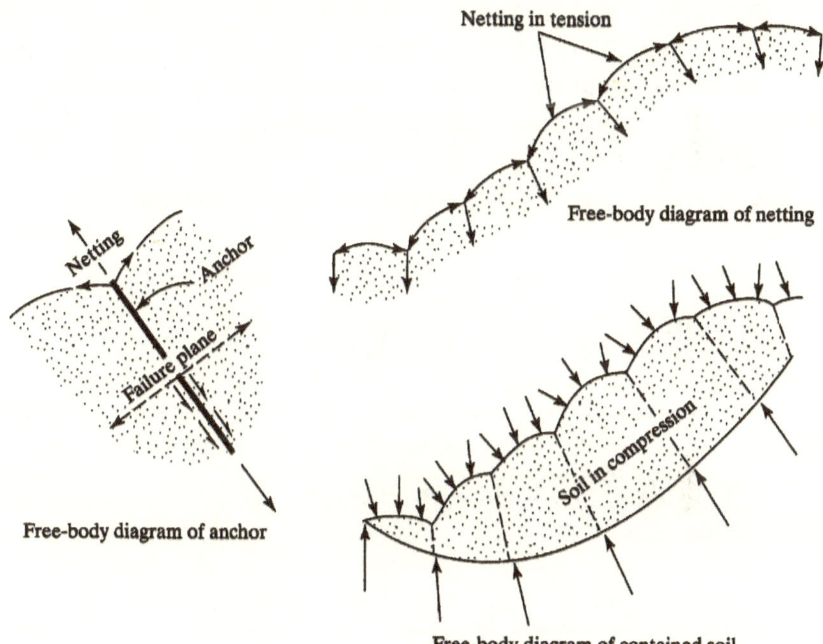

Figure 2.45 Free-body diagrams of various components of anchored geotextiles. (After Koerner [112, 113])

In the analysis to follow, the modified Bishop method, based on effective stresses is used [114]. Effective stress analyses are necessary since the slope soils are often wet and invariably have a frictional component requiring use of the method of slices. Both moment and vertical force equilibrium are satisfied in the following analysis after Koerner and Robbins [115], resulting in the following equations.

$$\bar{S}_i = (\bar{c}l_i + \bar{\sigma}l_i \tan\bar{\phi}) / FS \tag{2.61}$$
$$\bar{\sigma}l_i = (W_i - \mu_i l_i \cos\theta_i - \bar{S}_i \sin\theta_i) \sec\theta_i \tag{2.62}$$

Solved simultaneously, equations (2.61) and (2.62) result in the following equation for the factor of safety:

$$FS = \sum_{i=1}^{n} \frac{\bar{c}l_i + (W_i - \mu_i l_i \cos\theta) \tan\bar{\phi} \sec\theta_i}{(W_i \sin\theta_i)\left(1 + \dfrac{\tan\bar{\phi}\tan\theta_i}{FS}\right)} \tag{2.63}$$

Note in equation (2.63) that the factor of safety is not an explicit function and an iterative solution is necessary. Added to the complexity of the equation is the necessity of summing each of the individual slices and finding the minimum factor of safety. Thus, a computer solution is required.

With the addition of anchored spider netting on the surface of the slope as shown in figure 2.44, a number of features are added. Moment and force equilibrium now yield the following equations:

$$\bar{S}_i = \left[(1+f)\left(\bar{c}_m l_i + \bar{\sigma}l_i \tan\bar{\phi}_m\right)\right] / FS \tag{2.64}$$
$$\bar{\sigma}l_i = \left(W_i + F_i \cos\beta_i - \mu l_i \cos\theta_i - \bar{S}_i \sin\theta_i\right) \sec\theta_i \tag{2.65}$$

which when solved simultaneously result in equation (2.66) for the desired factor of safety.

$$FS = (1+f)\sum_{i=1}^{n} \frac{\bar{c}_m l_i + (W_i + F_i \cos\beta_i - \mu_i l_i \cos\theta_i) \tan\bar{\phi}_m \sec\theta_i}{\left[W_i \sin\theta_i - (F_i d_i / R)\right]\left[1 + \dfrac{(1+f)\tan\bar{\phi}_m \tan\theta_i}{FS}\right]} \tag{2.66}$$

where

- f = factor to account for soil anchors (nails),
- $\bar{c}$ = effective cohesion,
- $\bar{\varphi}$ = effective angle of shearing resistance,
- W_i = slice weight,
- l_i = arc length of slice,

μ_i = pore water pressure in the slice,
θ_i = angle that the midpoint of the slice makes with the horizontal, and
n = number of slices, which is arbitrary.

By now comparing equation (2.63) to equation (2.66), the influence of the anchored spider netting can be seen. These features (all of which positively influence the factor of safety) are as follows:

$\overline{c}_m$ = modified effective cohesion (where $\overline{c}_m \geq \overline{c}$),
$\overline{\varphi}_m$ = modified angle of shear resistance (where $\overline{\varphi}_m \geq \overline{\varphi}$),
$(1+f)$ = contribution of the anchors (nails) penetrating the failure plane stability,
$(F_i d_i/R)$ = moment due to the pressure of the stressed net at the ground surface, and
$(F_i \cos \beta_i)$ = contribution of the stressed net at the bottom of the slice (where the equilibrium equations are taken) to stability.

In order to investigate the numeric influence of these added terms to the slope's factor of safety, a computer-based sensitivity analysis has been performed [115]. The analysis uses a uniform slope of height 7.6 m, a slope angle of 55 deg., a soil unit weight 16.8 kN/m^3, a cohesion of 9.5 kPa, and a friction angle of 20 deg. The factor of safety of the slope using equation (2.63) without anchored spider netting is 0.97.

Now using anchored spider netting on the slope and only the influence of the rods in equation (2.66), with "f" varying from 0 to 25%, the factor of safety increases as shown in figure 2.46a. Using only the influence of the surface loading term, $\Sigma(F_i/d_i)/R$, in equation (2.66), with the netting pressure σ varying from 0 to 2700 Pa, the factor of safety increases as shown in figure 2.46b. Using only the influence of an increased normal force at the base of the slice, $F_i \cos \beta_i$, in equation (2.66), with σ varying from 0 to 2700 Pa, the factor of safety increases as shown in figure 2.46c. Finally, taking all the short-term gains collectively in equation (2.66) gives figure 2.46d. Here the increase in factor of safety of a slope due to a properly designed and constructed anchored spider net is quite obvious and, indeed, very

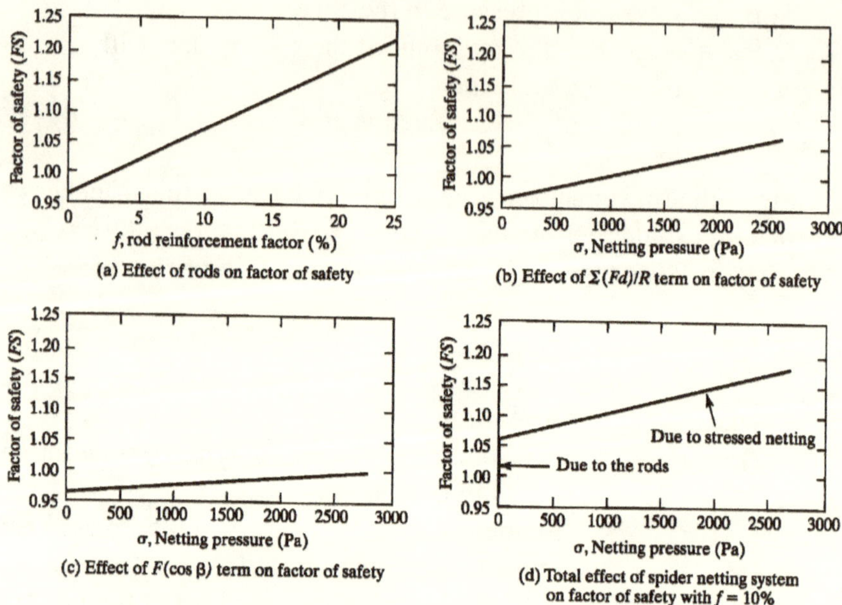

Figure 2.46 Parametric study of factors influencing soil slope stability when using anchored spider netting. (After Koerner and Robins [115])

beneficial. Note that the analysis could also be extended to account for long-term gains in the shear strength parameters ($\bar{c}_m$) and ($\bar{\varphi}_m$). However, their influence is typical of strength gains in any slope stability analysis. They are indeed significant and well understood by geotechnical engineers. The theoretical aspects of the anchored netting concept has been extended and modeled in large-scale laboratory tests, see Ghiassian et al. [116]. In a companion paper, Ghiassian et al. [117] have investigated the seepage considerations in the soil slope beneath the geosynthetic using a number of surface deformation patterns.

2.8 DESIGNING FOR FILTRATION

When liquid flows *across the plane* of the geotextile, the geotextile acts and is designed as a filter. Unfortunately, most literature; such as many highway specifications and numerous manufacturers brochures, identifies this topic incorrectly as drainage. Drainage, a distinct and separate topic, will be treated later in section 2.9.

2.8.1 Overview of Applications

There are a huge number of applications in which geotextiles are used adjacent to soil for the purpose of allowing liquid to pass through them while retaining the soil on the upstream side. Generally, the situations represent liquid moving in one direction only, but in some cases, reversing flow is necessary—for example, in tidal areas. Furthermore, the situations that will be discussed here all involve flow conditions designed on a worst-case scenario basis. This should come as no surprise, since it is the same type of conservatism that is used in all engineering design. Time-dependent random or dynamic flow situations will only be considered peripherally, because too little information is currently available for handling such situations. The factor of safety can always be increased to account for such considerations. Hopefully, the designs to follow cover most of the commonly encountered situations. The specific designs to be treated are the following: geotextile filters behind retaining walls, geotextile filters wrapped around underdrains, geotextile filters used beneath erosion control systems, and geotextiles used as silt fences.

2.8.2 General Behavior

The general behavior of a filter simultaneously demands adequate permeability and proper soil retention. Permeability is required to allow the liquid to pass the filter so as not to build up excess hydrostatic pore pressure. At the same time, it is necessary to retain the majority of the upstream soil to avoid soil piping. By soil piping, we mean the gradual loss of upstream soil fines, which results in higher flow rates, which in turn causes more soil loss, and so on. Thus, we are asking for the design of an open geotextile structure that allows liquid to pass and, at the same time, a closed geotextile structure that retains the soil on the upstream side of the geotextile. While these are indeed contradictory demands, such a geotextile filter design is possible because the amount of liquid flow through the soil is related primarily to particle size. For example, a commonly used empirical relationship between permeability coefficient and soil particle size is the following:

$$k = Cd_{10}^2 \tag{2.67}$$

where

k = permeability coefficient (hydraulic conductivity) of the soil,
C = site-specific constant,
d_{10} = effective soil particle size (i.e., size at which 10% of the soil is finer).

Thus, large particle-size soils can generate high-flow conditions (requiring geotextiles with relatively large voids), while small particle size soils are associated with low flow conditions (requiring geotextiles with relatively small voids). This results in a completely possible design situation, which is addressed in this section.

Before beginning with actual designs it must be repeated that long-term soil-to-geotextile compatibility is also a necessary requirement. This third aspect of the design was treated in section 2.2.3, and the particular details will not be repeated here.

2.8.3 Geotextiles Behind Retaining Walls

Behind conventional reinforced concrete retaining walls there must be a vertical drainage layer, typically consisting of granular soil that serves as a flow path allowing water from the backfill soil to escape into an underdrain system (figure 2.47a), or through weep holes (figure 2.47b). Without this drainage layer, hydrostatic pressures will build up and together with the horizontal soil pressure, could easily cause failure. Such hydrostatic pressures, if not dissipated by adequate drainage, can double the pressure against the wall. Also, the drainage sand must stay free-draining for the lifetime of the wall. If it excessively clogs with migrating soil from the retained backfill within this time span, the sand becomes as useless as if it was not there at all. Thus, it must be protected by a soil filter (which is expensive and difficult to place in a vertical orientation) or by a geotextile filter.

An identical situation, as far as the geotextile is concerned, is in the construction of flexible wall systems that are free-draining in themselves but would, without a soil filter or geotextile filter, allow the backfill soil to move into and through the open spaces. Such walls are illustrated in figures 2.47c and *d* in which the gabion style consists of wire baskets filled with 100 mm and larger stones. To

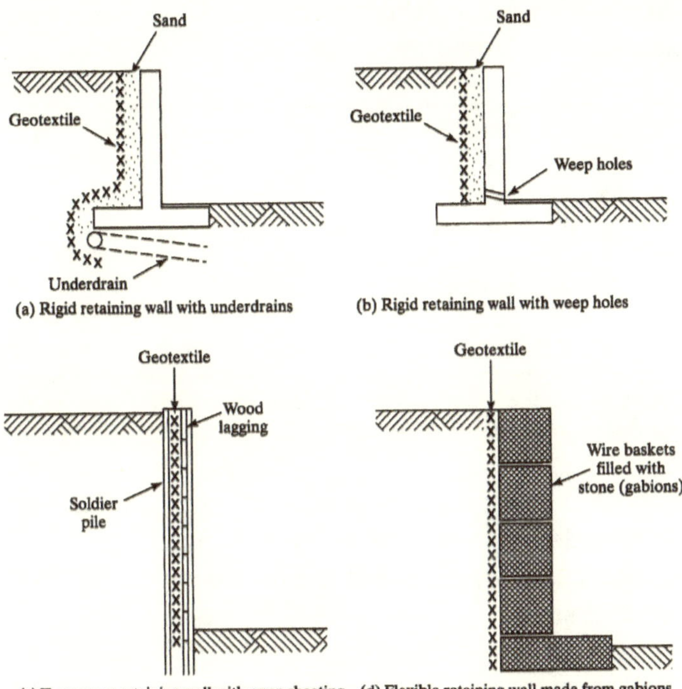

Figure 2.47 Various types of retaining walls in which geotextiles have been used as filters.

backfill against such walls with no filter media (soil or geotextile) would be sheer idiocy. Since the soil filter option is difficult to place in a vertical or near-vertical manner and may even require a series of graded filter soils, the geotextile filter becomes very attractive. Geotextile filter design is illustrated as follows:

Example 2.21:

Given a 3.5 m high gabion wall consisting of three 1 × 1× 3 m long baskets sitting on 0.5 × 2 × 3 m long mattresses as shown below, the backfill soil is a medium-dense silty sand of d_{10} = 0.03 mm, CU = 2.5, k = 0.0075 m/s, and D_R = 70%. Check the adequacy of three candidate geotextiles whose laboratory test properties are given below. Use a cumulative reduction factor in equation 2.25b of 15.0, in order to adjust the ultimate laboratory-obtained permittivity value to an allowable field-oriented value.

Geotextile		Permittivity	AOS*
No.	Type	(sec^{-1})	(mm)
1	nonwoven needle-punched	2.0	0.30
2	woven monofilament	1.2	0.42
3	nonwoven heat-bonded	0.4	0.21

*Note that if the AOS is given in sieve size number, it must be converted to 0_{95} in mm using table 2.5.

Solution: The design is in two stages, with the first being a determination of the flow factor of safety of the geotextile; the second being a consideration of opening size.

(a) The first is done by calculating the required permittivity, ψ, which is $\psi = k/t$

- Calculate the actual flow rate using a flow net as shown.

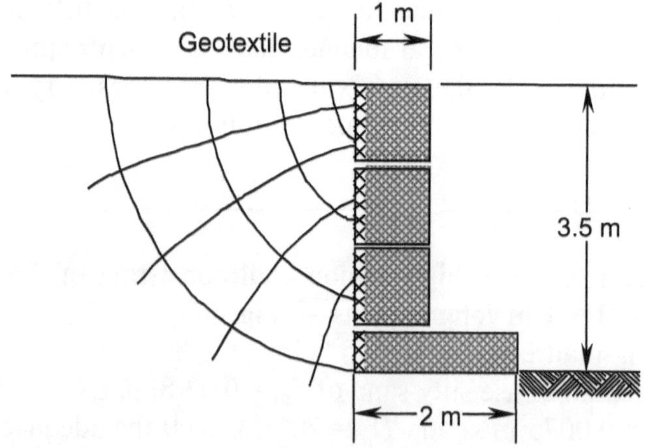

$$q = kh\left(\frac{F}{N}\right)$$
$$= (0.0075)(3.5)\left(\frac{4}{5}\right)$$
$$q = 0.021 m^2 / \sec$$

- Calculate the required permittivity.

$$q = kiA$$
$$q = k\frac{\Delta h}{t}A$$
$$\frac{k}{t} = \frac{q}{(\Delta h)(A)}$$
$$\Psi_{reqd} = \frac{0.021}{(3.5)(3.5\times 1)}$$
$$= 1.71\times 10^{-3}\sec^{-1}$$

- Check against the allowable permittivity of the candidate geotextiles.
  Geotextile 1 (nonwoven needle-punched):

$$\Psi_{ult} = 2.0 sec^{-1}$$
$$\Psi_{allow} = \Psi_{ult}\left(\frac{1}{RF_{SCB}\times RF_{CR}\times RF_{IN}\times RF_{CC}\times RF_{BC}}\right)$$
$$= \frac{2.0}{15.0}$$
$$= 0.13 sec^{-1}$$
$$FS = \Psi_{allow} / \Psi_{reqd}$$
$$= \frac{0.13}{0.00171}$$

$FS = 76$ acceptable; the geotextile has a high factor of safety

Geotextile 2 (woven monofilament):

$$\Psi_{ult} = 1.2 sec^{-1}$$

$$\Psi_{allow} = \frac{1.2}{15.0}$$

$$= 0.080 \text{ sec}^{-1}$$

$$FS = \Psi_{allow} / \Psi_{reqd}$$

$$= \frac{0.080}{0.00171}$$

$$FS = 47 \text{ acceptable; this geotextile is also permeable enough.}$$

Geotextile 3 (nonwoven heat-bonded):

$$\Psi_{ult} = 0.40 sec^{-1}$$

$$\Psi_{allow} = \frac{0.40}{15.0}$$

$$= 0.027 sec^{-1}$$

$$FS = \Psi_{allow} / \Psi_{reqd}$$

$$= \frac{0.027}{0.00171}$$

$$FS = 16, \text{ acceptable, this geotextile is also adequate.}$$

The above shows that many commercially available geotextiles can easily handle the required flow.

(b) The second part of the design relates to the geotextile's opening size, so as to prevent excessive soil loss. The three candidate geotextiles have AOS values of 0.30, 0.42, and 0.21 mm respectively.

- The appropriate criterion for opening size must first be selected. Since this is a noncritical situation, Carroll's criterion (recall section 2.2.3) will be used. This calls for the following:

$$0_{95} < 2.5 \, d_{85}$$

since $d_{10} = 0.03$ mm and $CU = 2.5$, an approximate value of $d_{85} = 0.15$ mm. (Note that the d_{85} value can, and should, be obtained directly by sieving the upstream soil.)

$$
\begin{aligned}
O_{95} &< 2.5(0.15) \\
&< 0.375 \text{ mm}
\end{aligned}
$$

- Check against the candidate geotextiles' AOS values:

Geotextile 1: $AOS = 0.30$ mm < 0.375 mm, acceptable with FS = 1.25
Geotextile 2: $AOS = 0.42$ mm > 0.375 mm, not acceptable with FS = 0.89
Geotextile 3: $AOS = 0.21$ mm < 0.375 mm, acceptable with FS = 1.79

Thus, Geotextile 2 is too open and will experience excessive soil loss based on this soil-retention criterion. (If such a woven monofilament style were used, it would have to have a tighter AOS value, which is possible since the permittivity factor of safety was so high, i.e., 47). Candidate Geotextiles 1 and 3 are both acceptable. The technical decision as to which of these geotextiles to use will be based on the site-specific concern as to which mechanism (permittivity or soil retention) is more important. The nontechnical, but important, final decision is based on cost and availability.

2.8.4 Geotextiles Around Underdrains

Geotextiles have been found to make excellent replacements for uniform or graded soil filters around perforated pipe underdrains. Highways, airfields, and railroads are major application areas. There are numerous cross sections possible, as shown in figure 2.48. Figures 2.48a, b show the geotextile acting as a filter to protect the stone surrounding the perforated pipe or to protect the pipe directly. In the latter case, the pipe is sometimes wrapped with a geotextile stocking in the manufacturing plant and is shipped as a complete unit to the job site. *This design should only be used with slotted corrugated drainage pipe, and even then, design should consider the reduced flow area*

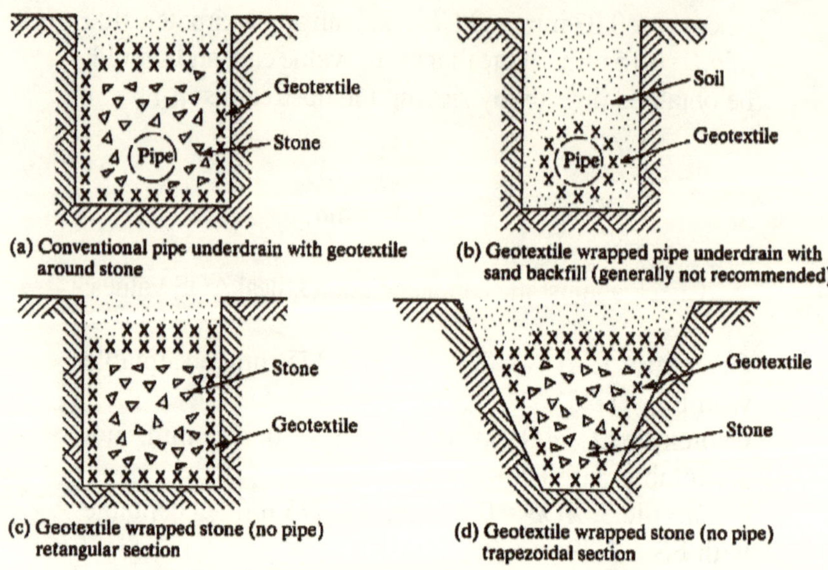

Figure 2.48 Typical cross sections of underdrain systems with and without perforated pipes.

through the geotextile. These sketches also raise the question: "Why have the pipe at all?" Indeed, the transmissivity of an open-graded stone is adequate to handle many flow situations, as long as its long-term protection against fine soil contamination is ensured. The point is precisely why the geotextile is involved. By wrapping the stone as shown in figures 2.48c and d, no pipe at all is required. This is the design for a French drain system, but unlike those in the past, one that will remain free from fine soil contamination if the geotextile is properly designed.

Example 2.22: ______________________________

Design the geotextile filter surrounding an open-graded stone aggregate that, in turn, surrounds a perforated pipe underdrain, as shown in the sketch below. Flow will enter through the stone base from the upper part of the underdrain, while soil infiltration will come from the surrounding native soil. This soil is a dense sandy silt (ML) with the relevant properties of 15% nonplastic fines with $C_c = 2.0$, $C_U = 5.0$, $I_D = 80\%$,

$d_{50} = 0.035$ mm, and $k = 1 \times 10^{-5}$ m/s. The geotextile being considered is nonwoven needle-punched with laboratory tested values of $\psi = 1.5$ sec^{-1} and AOS = 0.212 mm.

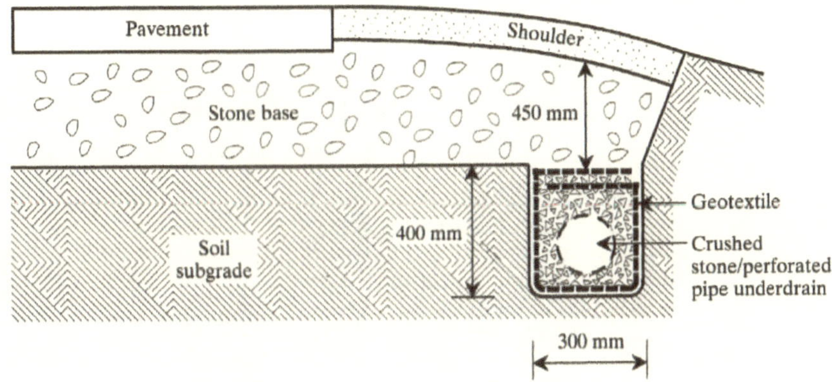

Solution: The solution is again in two parts; one for adequate flow capability, the other for adequate soil-retention capability.

(a) The first part of the design must deal with the flow aspects.

- Estimate the maximum flow coming to the geotextile. This will be through the 450 mm stone base above the underdrain. Cedegren [118] has numerous design charts, from which we have selected a relatively high value of 15 m^3/day-m due to the current tendency to use open graded base courses.
- Calculate the required permittivity.

$$q = kiA = k\frac{\Delta h}{t}A$$

$$\frac{k}{t} = \frac{q}{\Delta hA}$$

$$\psi = \frac{15}{(0.45)(0.30 \times 1)}$$

$$\psi = 111\,\text{day}^{-1} = 1.3 \times 10^{-3}\,\text{sec}^{-1}$$

- Check this required permittivity against the allowable permittivity of the geotextile. Using reduction factor data from table 2.8b in equation 2.25a gives the allowable permittivity (the values of reduction factors were estimated):

$$\psi_{allow} = \psi_{ult}\left[\frac{1}{RF_{SCB} \times RF_{CR} \times RF_{IN} \times RF_{CC} \times RF_{BC}}\right]$$

$$= 1.5\left[\frac{1}{7.0 \times 1.2 \times 1.1 \times 1.4 \times 2.0}\right]$$

$$= 1.5\left[\frac{1}{25.9}\right]$$

$$= 0.058\,\text{sec}^{-1}$$

$$\text{FS}_{flow} = \psi_{allow} / \psi_{reqd}$$

$$= \frac{0.058}{0.0013}$$

$$\text{FS}_{flow} = 45;\ \text{which is acceptable.}$$

(b) The second part of the design has to do with soil retention of all the soils surrounding the geotextile-enclosed drainage stone. Since the stone base course above the geotextile is of no real concern, the finer soil subgrade adjacent to the geotextile and beneath it becomes the focus of attention. Also, to be considered is that siltation of the perforated pipe within the enclosure is a possibility; therefore, the drainage stone must not become contaminated. This situation is considered to be critical.

The criterion we will use for soil retention will be taken from figure 2.4a (recall section 2.2.3). From this table for the site-specific soil we select the following equation:

$$0_{95_{reqd}} < \frac{18}{C_U} d_{50}$$

$$< \frac{18}{5.0}(0.035)$$

$$0_{95_{reqd}} < 0.126 mm$$

Since the candidate geotextile has an opening size $0_{95_{act}} = 0.212\text{mm}$, this geotextile is not acceptable (i.e., the FS = 0.59). The openings are too large and soil will not be properly retained. Another candidate geotextile with a tighter pore structure must be selected.

The design procedure for the alternative geotextile follows along these same lines. Note that it is entirely possible to find a suitable geotextile since the permittivity factor of safety is quite high (i.e., FS = 45) in the analysis. By using a tighter geotextile, the value of 45 will be reduced, but will come into conformance with the soil-retention criterion. This type of tuning the geotextile's properties should be done on the side where the greatest safety is needed, i.e., either permittivity or soil retention.

2.8.5 Geotextiles Beneath Erosion Control Structures

Geotextiles have been used beneath erosion control structures since the late 1950s. Section 1.3.3 mentions some of the original applications. In these applications, both rock riprap and precast concrete blocks or mattresses were placed on the geotextile, and the geotextile was referred to as a *filter fabric*. In other designs, depending primarily on the care exercised by the contractor in placing riprap, a sand cushion may be needed to protect the geotextile from impact damage during installation or abrasion damage during its lifetime (e.g., due to wave action agitating the rock riprap). If precast concrete blocks or mattresses are being used, a geotextile is often used, not so much as a cushion (since these blocks are placed by hand or carefully lowered into position), but as a filter to dissipate pore water, since a major part of the geotextile will be directly covered by the blocks. This feature will be illustrated in example 2.23. Figure 2.49a shows a photograph of a geotextile filter in place with riprap armor above it and figure 2.49b shows the newer concept of a prefabricated concrete mattress. Figure 2.49c shows precast concrete blocks that cover essentially all the underlying geotextile. As seen in figure 2.49d, this resulted in a failure by not allowing for dissipation of pore water from the underlying soil subgrade.

(a) Partial coverage by riprap

(b) Partial coverage by articulated concrete block mattress (Compliments of CETCO Contracting, Inc.)

(c) Complete coverage by paving blocks

(d) Failure occurring from complete coverage

Figure 2.49 Geotextiles being used as filters beneath erosion-control structures.

Example 2.23:

Evaluate the filtration adequacy of a candidate geotextile for placement beneath a rock riprap erosion control system in a coastal inlet area with 1 m tides (i.e., reversing flow conditions with moderate water currents) as shown in the following sketch. The candidate geotextile laboratory properties are $\psi = 0.5$ sec^{-1} and $AOS = 0.21$ mm. The in-situ soil is a beach sand (SP) with $C_u = 3.5$, $d_{50} = 0.10$ mm, $d_{90} = 0.40$ mm and porosity = 0.40.

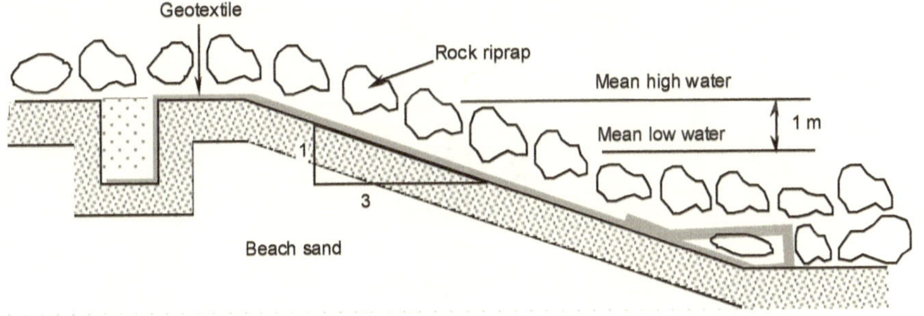

Solution: As with all filtration designs this is a two-part problem, one for adequate flow and the other for soil retention.

1. For adequate flow, the procedure is as follows.
   - Estimate the maximum flow rate due to the 1 m tidal lag. If we assume a water profile as follows:

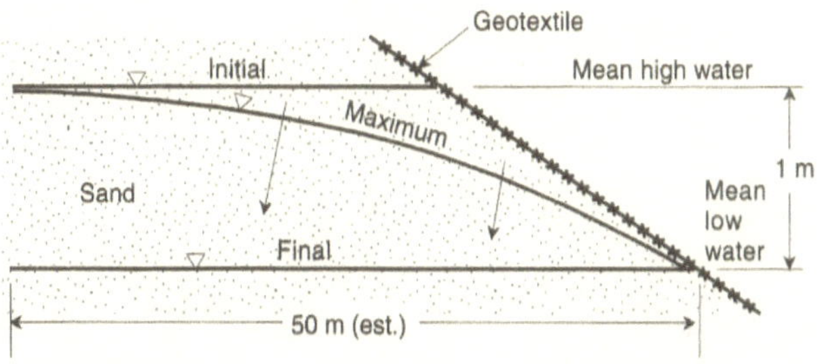

With the tide receding at a maximum rate during an initial 2 hr. period as shown, then

$$q_{\max} = \frac{50 \times 1 \times 1}{2} \times \frac{0.4}{2}$$
$$= 5.0\ m^2 / hr$$
$$q_{\max} = 1.39 \times 10^{-3}\ m^2 / \text{sec}$$

- Calculate the required permittivity.

$$q = kiA = k\frac{\Delta h}{t}A$$
$$\frac{k}{t} = \frac{q}{\Delta h A}$$
$$= \frac{0.00139}{(1.0)(3.16)}$$
$$\Psi_{reqd} = 0.00044 sec^{-1}$$

- Since the candidate geotextile has a laboratory obtained ultimate permittivity of 0.5 sec^{-1}, it must be modified with reduction factors for site-specific conditions.
- The allowable permittivity is found from equation 2.25a and values in table 2.8b, where the reduction factor for blinding is used as its maximum value of 10.0 since the rock riprap will cover a large portion of the geotextile's surface area.

$$\Psi_{allow} = \Psi_{ult}\left(\frac{1}{RF_{SCB} \times RF_{CR} \times RF_{IN} \times RF_{CC} \times RF_{BC}}\right)$$
$$= 0.50\left[\frac{1}{10.0 \times 1.2 \times 1.2 \times 2.5 \times 3.0}\right]$$
$$= 0.50\left[\frac{1}{108}\right]$$
$$= 0.0046 sec^{-1}$$

- The factor of safety is then

$$FS = \frac{\Psi_{allow}}{\Psi_{reqd}} = \frac{0.0046}{0.00044}$$

$$FS = 10, acceptable$$

2. The geotextile is now evaluated with respect to its adequacy to retain the soil beneath it.

- Since these ocean-control structures are destroyed when the contained soil passes through the geotextile voids (resulting in subsidence and loss of stability of the riprap) and the flow regime is pulsating and cyclic, we will use figure 2.4b, which results in the following criterion.

$$d_{50} < 0_{95} < d_{90}$$
or
$$d_{50} < \text{AOS} < d_{90}$$

- Check this against the AOS of the candidate geotextile that is 0.21 mm.

$$0.10 \text{ mm} < 0.21 \text{ mm} < 0.40 \text{ mm; OK}$$

Since 0.21 mm is within the proper bounds, excessive soil loss will not occur and the geotextile is proper as far as soil retention is concerned. The filtration design is essentially complete with the candidate geotextile's flow and retention properties being adequate.

- Another part of the design, however, is to see that the geotextile has adequate strength properties to withstand the impact of falling rip-rap and/or puncture from equipment moving on the surface of the rip-rap. These strength-related design issues were described in section 2.5.

2.8.6 Geotextile Silt Fences

Silt fences consist of above ground fabrics attached vertically onto posts to prevent sediment-carrying sheet runoff from entering into downstream creeks, rivers, or sewer systems. Since all construction activities must have an associated sedimentation and erosion control plan, this concept is used regularly and has replaced bales of straw, hay, and other makeshift methods. The bottom of the silt fence is embedded in a small anchor trench; the posts, to which the geotextile is attached, are usually at 1.5 to 3.0 m spacings. Sometimes a geogrid backup is required behind the geotextile to provide additional support; in this case, the geogrid is attached to the posts first, then the geotextile, both facing upslope. Since the geotextile is exposed to sunlight, it must be UV-stabilized. See figure 2.50 for photographs of typical installations showing (a) as placed, (b) working as designed and intended, and (c) improperly constructed without an adequate anchor trench and being undermined.

The physical model utilized in the original development was developed by Bell and Hicks [119] and has sediment-carrying runoff water depositing soil particles within and on the surface of the geotextile as it first acts as a true filter, then eventually clogs with soil particles in its lower part so that water can no longer pass freely through it. The impoundment created by this action causes coarse sediment to settle behind the clogged portion of the silt fence and the water carrying finer particles to try to reach higher levels on the silt fence above the clogged region. With time, an equilibrium situation is established. This situation is shown in figure 2.51.

Richardson and Middlebrooks [120] provide a design method based on the required site-specific storage volume to be contained by the silt fence. The method assumes that the ground surface is smooth and bare, with sheet erosion (versus rill or gully erosion) being the predominant erosion mechanism. The recommended procedure, albeit somewhat modified here, follows a sequential series of calculations.

(a) Cascading silt fences after installation

(b) Properly functioning silt fence

(c) Silt fence that been undercut by erosion

Figure 2.50 Geotextile silt fence and examples of field performance.

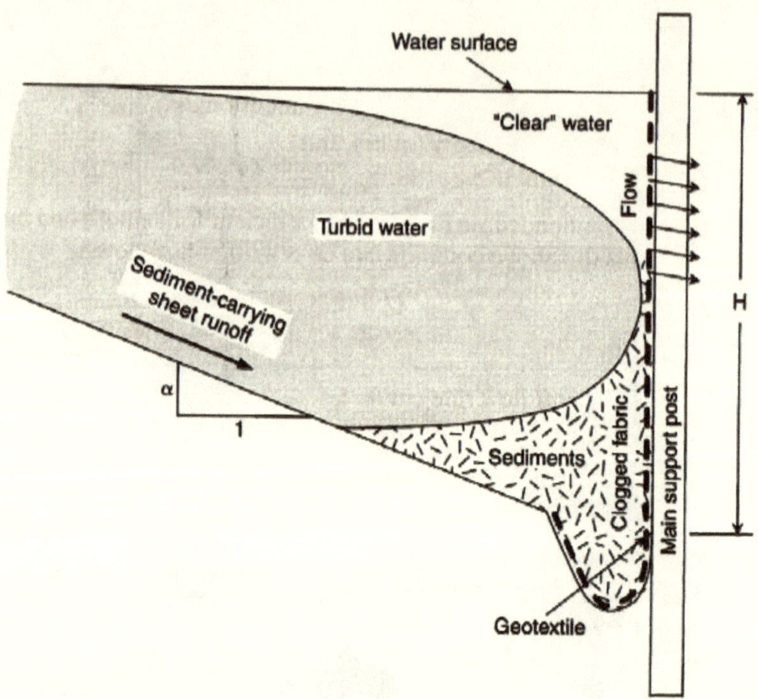

Figure 2.51 Cross section of geotextile silt fence and suggested manner in which system functions.

1. The maximum slope length that can be contained by a single silt fence is obtained from the following equation for a given slope angle under the above-assumed conditions.

$$L_{max} = 36.2e^{-11.1\alpha} \tag{2.68}$$

where

L_{max} = slope length (m) and
α = slope inclination as measured by its steepness, i.e., vertical rise-to horizontal length (dimensionless).

For greater slope lengths than indicated from equation (2.68), a set of cascading silt fences must be used, each of which is individually designed.

2. The runoff flow rate (water plus sediment) is calculated using the site-specific value of rainfall intensity. This is somewhat subjective and may be controlled by local regulations. Recommended is to use the hourly rainfall based on a ten-year recurrence interval.

$$Q = C\,I\,A \tag{2.69a}$$

where

Q = runoff flow rate (m^3/hr),
C = surface runoff coefficient (dimensionless),
I = rainfall intensity (m/hr),
A = area (m^2).

The recommended surface runoff coefficient for smooth and bare soil surfaces is 0.5. Thus, the above equation can be rewritten as follows:

$$Q = 5 \times 10^{-4}\,(I)\,(A) \tag{2.69b}$$

where

Q = runoff flow rate (m^3/hr),
I = rainfall intensity (mm/hr),
A = area (m^2).

3. The sediment flow rate can be estimated to ensure that it does not exceed the total runoff flow rate, which can only occur after repeated storms. Stated differently, the sediment accumulates after every storm and remains behind the silt fence, whereas the clear water flows through the upper portion of the silt fence. This calculation should follow the Universal Soil Loss Equation (described in chapter 8). Typically, silt fences are designed to contain sediments for at least three major storm events. After that, the sediment must be removed and/or the silt fence replaced.

4. The height of the silt fence is then determined using the value of Q obtained previously on the basis of single-event storm intensity. A factor is then applied that represents the number of recurring storm events (i.e., a $FS = 3$ represents three similar storms). Thus,

$$V = Qt = H\left(\frac{H}{\alpha}\right) \tag{2.70}$$

$$H = [(Q)(t)(\alpha)]^{1/2}$$

where

V = total runoff volume (m^3),
Q = runoff flow rate (m^3/hour),
t = storm duration (hr) [assumed to be 1 hr. based on the value of I selected in the calculation of Q],
H = silt fence height to contain a single storm (*m*), and
α = slope inclination (vertical-to-horizontal ratio).

5. The spacing of the silt fence posts is then arbitrarily selected and is integrated into the design per the next two steps.
6. The geotextile is selected on the basis of its ultimate wide-width tensile strength in the weakest direction. Figure 2.52a can be used for this determination. A reduction factor can be included. Note that the opening size of the geotextile is not a governing criterion since excessive clogging via turbidity is the issue, and with individual soil particle sediment, this will occur. Thus, woven slit-film geotextiles predominate in this particular application; they are rapidly clogged, yet possess high-tensile strength.
7. Lastly, the type of fence post is selected. This can be done using the guide given in figure 2.52b.

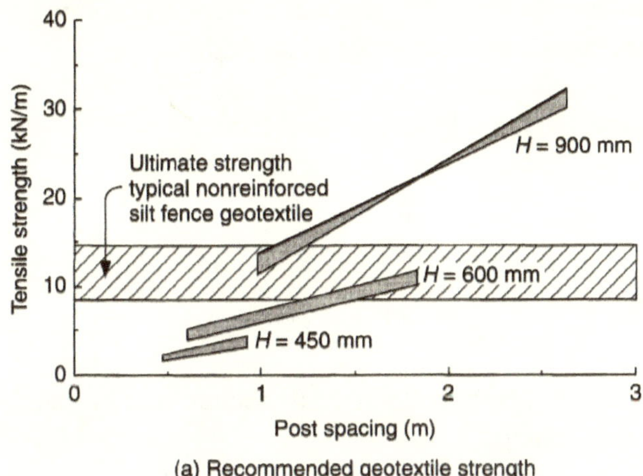

(a) Recommended geotextile strength

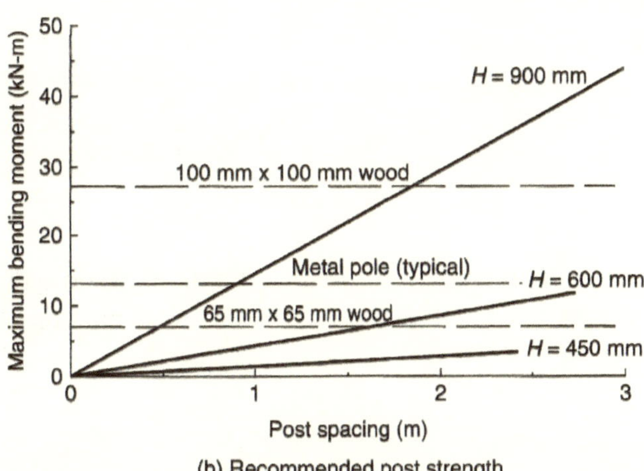

(b) Recommended post strength

Figure 2.52 Design recommendations for silt fence geotextile and post strengths. (After Richardson and Middlebrooks [120])

Example 2.24:

Design a silt fence for a 60 m long relatively smooth surface construction site where topsoil has been stripped and the average slope inclination is 5%. The ten-year recurring single storm intensity is 100 mm/hour.

Solution: Using the procedure just described, the problem is solved in successive steps.

1. Maximum slope length per silt fence:

$$L_{max} = 36.2e^{-11.1\alpha}$$
$$= 36.2\, e^{-11.1(0.05)}$$
$$L_{max} = 21m; \, use \; 20m$$

Therefore, three cascading silt fences will be needed for this construction site.

2. Calculate the runoff flow rate

$$Q = 5\times10^{-4}(I)(A)$$
$$= 5\times10^{-4}(100)(20\times1)$$
$$Q = 1.0m^3 \,/\, hour$$

3. Either calculate the sediment flow rate or assume the number of storm events that the silt fence must contain. Here, we select the following:

number = 3 events

This establishes the factor that will be used on the calculated height of the silt fence.

4. Calculate the height of the silt fence.

$$H = [(Q)(t)(\alpha)]^{1/2}$$
$$= [(1.0)(1.0)(0.05)]^{1/2}$$
$$= 0.22m$$

For three storms:

$$\begin{aligned} 3 \times H &= 3 \times 0.22 \\ &= 0.66m \\ H &= 660mm \end{aligned}$$

5. The spacing of the silt fence posts is assumed to be as follows:

$$S = 1.5 \text{ m}$$

6. The required strength of the geotextile is taken from figure 2.52a

$$T_{reqd} = 10\text{kN / m},$$

which is adequate without geogrid backup support. This can be increased for damage due to stapling/ attachment to the fence post,

$$\begin{aligned} T_{ult} &= FS(T_{reqd}) \\ &= 1.2(10) \\ T_{ult} &= 12kN / m \end{aligned}$$

7. Finally, the type of post is selected from figure 2.52b: Use 65 mm × 65 mm wooden posts.

2.8.7 Summary

Presented in this section were a series of designs in which the geotextile is serving in the filtration function. In such cases the liquid (often water) is moving across the plane of the geotextile. Thus, the designs allow for a required flow capability that is expressed in terms of permittivity (ψ), a term that includes both the permeability coefficient and the geotextile's thickness. However, more than just required flow capability is necessary; upstream soil retention is also necessary. Since these two demands are contradictory (flow requiring large geotextile voids and adequate soil retention requiring small geotextile voids), problems illustrating the criticality of each demand were presented.

The fine soils used behind flexible walls (section 2.8.3) and adjacent to underdrains (section 2.8.4) showed that adequate permittivity is easily achieved, yet small opening-size values are required. This means that the fibers or yarns had to be close to one another, resulting in a relatively tight or dense geotextile structure.

Conversely, the illustration of geotextiles beneath erosion-control structures (section 2.8.5) usually occurs with free-draining sands. Thus, the permittivity of the geotextile is challenged by a lower factor of safety, whereas the opening-size requirement is easily met. Note that if the erosion-control system lies directly on the geotextile and covers a high percentage of it (as with precast-concrete blocks directly on the geotextile), the permittivity design always becomes critical. In this case, the percentage of covered geotextile (sometimes as high as 70%) is included via a high-reduction factor, i.e., RF_{SCB} = 10. A simple proportional factor is appropriate, since flow capability should be a linear function of area remaining open to flow.

In all of these problems, the long-term compatibility of the soil to the geotextile should be considered. In most cases the designs illustrated should suffice; however, when the situation is critical, the flow compatability tests described in section 2.3.5 should be considered.

The geotextile silt fence of section 2.8.6 was also considered. This design simply begs for excessive clogging of the geotextile, which must happen so that a sedimentation reservoir can form behind it. The design then becomes a structural design of the strength of the geotextile, and its supporting post spacings and type. It was discussed in this section because the geotextile above the clogged area is indeed still serving as a true filter.

In closing this section on geotextile filter design, it should be mentioned that field failures have indeed occurred. In a review paper, Koerner and Koerner [121] present eighty-two such failures. The large majority were caused by atypical upstream soils or atypical permeants. Table 2.16 mentions the situations giving rise to the specific failures and adds to a few cases of poor design and poor installation. Taken together all these situations (except for the reversing flow situations) can be readily understood and should have been avoided.

TABLE 2.16 SUMMARY OF GEOTEXTILE FIELD FILTER FAILURES AND APPROXIMATE NUMBER OF OCCURRENCES, KOERNER AND KOERNER [121]

Design Related Failures (21%)

- use of woven slit film fabrics (3)
- excessive surface blockage of geotextile (6)
- geotextile wrapped solid-wall drainage pipe (4)
- reversing flow situations (4)

Atypical Upstream Soil Failures (24%)

- cohesionless fine-grained soils (8)
- cohesionless gap-graded sandy silts (3)
- dispersive clay soils (2)
- ochre-forming soils (7)

Atypical Permeant Failures (37%)

- oily water and sludges (5)
- waters with high turbidity (7)
- highly alkaline water (4)
- landfill leachates (8)
- waste water and agricultural runoff (6)

Installation Related Failures (18%)

- lack of intimate contact (14)
- impermeabilization by excessive gluing (1)

2.9 DESIGNING FOR DRAINAGE

In this book, drainage refers to planar flow within the structure of the geosynthetic. Geotextiles acting as drains will be considered in this section; geonets and drainage geocomposites will be discussed in chapters 4 and 8, respectively.

2.9.1 Overview of Applications

Although closely related to the filtration function just described, geotextile drainage occurs in the in-plane direction rather than cross-plane. The transitive verb *to drain*, according to *Webster's New Collegiate Dictionary*, means "to draw off (liquid) gradually . . .," and it is to precisely this action, as it is performed by geotextiles, that the present section is devoted. Note that all geotextiles are capable of draining liquid in their in-plane directions, but to widely varying degrees. In order of *increasing* in-plane drainage capability, the various manufactured styles are ranked as follows:

- Woven, slit-film
- Woven, monofilament
- Nonwoven, heat-bonded
- Nonwoven, resin-bonded—increasing with increasing weight and decreasing amount of resin
- Nonwoven, needle-punched—increasing with increasing weight (recall figure 2.16)
- Hybrid drainage systems—particularly geonets and drainage geocomposites (see chapters 4 and 8 respectively)

Although all geotextiles possess some in-plane drainage capability, the first three types of geotextiles listed above have too little to take advantage of. The nonwoven resin-bonded and needle-punched groups, however, can be made sufficiently thick to have meaningful quantities of liquid flow within them. These geotextiles, particularly the needle-punched variety, are often 5 mm thick and can be made much thicker in a cost-effective manner. Relatively thick nonwoven needle-punched geotextiles will be the focus of the designs included in this section.

For the flow capability of the geotextiles, we will be considering two general categories: gravity flow (section 2.9.3) and pressure flow (section 2.9.4). This distinction will become obvious in the sections to follow. Some selected applications can be noted in each category:

- Gravity drainage; chimney drains and drainage galleries in earth and earth/rock dams, pore water dissipaters behind retaining walls, flow interceptors (as in fin drains), and placed beneath geomembrane-lined reservoirs for water drainage or gas conveyance.
- Pressure drainage; as vertical drains for rapid soil consolidation, within the soil backfill of reinforced earth walls, within earth embankments and dams, and beneath surcharge fills.

2.9.2 General Behavior

Except for the direction of flow (in-plane rather than cross-plane), the design similarities of this drainage section with the preceding filtration section will be obvious. There are again three aspects of design: adequate flow capacity, proper soil retention, and long-term soil-to-geotextile flow equilibrium. Since soil retention and long-term soil-to-geotextile flow equilibrium are discussed in the previous section, these aspects will not be fully treated in this section; only some brief comments will be included. The reader is referred to sections 2.3.5 and 2.8 for full details.

Regarding the hydraulic design parameter of major concern, we will focus on the transmissivity of the geotextile θ in equation (2.10) (repeated here), where

$$\theta = k_p t \tag{2.10}$$

where

θ = transmissivity,
k_p = in-plane geotextile permeability, and
t = geotextile thickness.

The relationship will appear as $\theta = kt$, the k being understood to be in-plane permeability (rather than the filtration-related cross-plane permeability). Transmissivity will be used in conjunction with Darcy's formula under the assumption that laminar flow exists within the geotextile. This is generally valid, but when very thick geotextiles are used together with high hydraulic gradients, the assumption of a laminar flow regime becomes questionable. (Indeed, with the geonets and drainage geocomposites discussed in chapters 4 and 8, flow is generally turbulent and Darcy's formula should not be used. For geonets and drainage geocomposites a related design based on flow rates will be used.)

2.9.3 Gravity Drainage Design

For gravity drainage problems involving liquid flow in geotextiles, the driving force is merely the slope at which the geotextile is placed. Using the geometry of the particular situation under consideration, a required transmissivity can be calculated using Darcy's formula. This value is then compared to the allowable transmissivity of the candidate geotextile for calculation of a factor of safety. Considering the severity of the situation, these values of factor of safety should be quite high, depending on how the allowable transmissivity is obtained [recall equation (2.25) and table 2.8b].

Note that the allowable geotextile transmissivity is the value at the particular normal stress that is acting on it. This is usually calculated on the basis of the effective stress of the soil placed above, if the geotextile is horizontal. If the geotextile is vertical, the normal stress is the vertical effective stress times the appropriate coefficient of earth pressure. This can usually be taken as the at-rest value, and the relationship $K_o = 1 - \sin \varphi$ is often used. If the friction angle of the soil (φ) is not known, K_o can be taken approximately equal to 0.5. It should be recalled from the testing section involving hydraulic properties (section 2.3.4) that the transmissivity decreases substantially with applied normal stress on the geotextile. Figure 2.16 illustrated this behavior for various mass per unit area nonwoven needle-punched geotextiles. However, a near-constant value of transmissivity is reached above approximately 85 kPa. Examples 2.25 and 2.26 illustrate these concepts.

Example 2.25:

> A 10 m high zoned earth dam is to be used as an irrigation reservoir with a cross section as shown below. A geotextile is being considered behind the clay corewall as a chimney drain and drainage gallery. The geotextile under consideration is a 2000 g/m^2 nonwoven needle-punched geotextile with $\theta_{ult} = 15 \times 10^{-4}$ m^2/min. Use cumulative reduction factors of 3 to convert this to θ_{allow}. What factor of safety does this geotextile have for the amount of flow seeping through the clay corewall, which is a clayey silt of permeability 1×10^{-7} m/s?

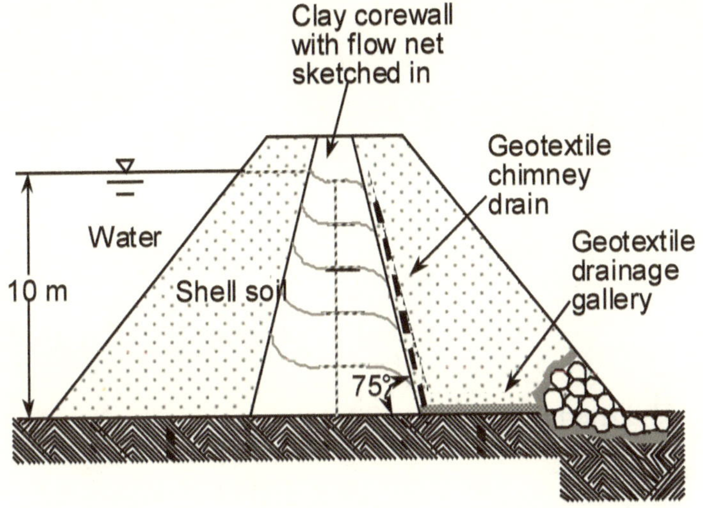

Solution: In stages, the solution is as follows:

(a) Calculate the maximum seepage coming through the clay corewall that the geotextile must carry. The use of a flow net (as shown in the sketch) gives

$$q = kh\left(\frac{F}{N}\right)$$

$$= \left(1\times10^{-7}\right)\left(10\right)\left(\frac{5}{2}\right)$$

$$= 2.50\times10^{-6}\, m^2/\text{sec}$$

$$q = 1.50\times10^{-4}\, m^2/\text{min}$$

(b) Calculate the gradient of flow in the geotextile.

$$i = \sin 75°$$
$$i = 0.97$$

(c) Calculate the required transmissivity, θ_{reqd} using Darcy's formula.

$$
\begin{aligned}
q &= kiA \\
&= ki(t \times w) \\
q &= (kt)(i \times w) \\
kt &= \frac{q}{(i)(w)} \\
\theta_{reqd} &= \frac{1.50 \times 10^{-4}}{(0.97)(1.00)} \\
\theta_{reqd} &= 1.55 \times 10^{-4}\, m^2/\text{min}
\end{aligned}
$$

(d) Determine the factor of safety.

$$
\begin{aligned}
FS &= \frac{\theta_{allow}}{\theta_{reqd}} = \frac{\theta_{ult}/\Pi RF}{\theta_{reqd}} \\
&= \frac{(15 \times 10^{-4})/3.0}{1.55 \times 10^{-4}} \\
FS &= 3.2
\end{aligned}
$$

Due to the critical nature of this application, this *FS* value is low and a minimum value of 5 to 10 is recommended. Two options present themselves: one is to use multiple layers of geotextile (to increase θ_{allow}) in the lower part of the chimney drain and in the drainage gallery (the upper part of the chimney drain could still use one layer); the other is to use the desired *FS* value and then back-calculate the necessary geotextile's transmissivity. This latter suggestion using a $FS = 5.0$ is illustrated as follows:

$$
\begin{aligned}
\theta_{allow} &= \theta_{reqd} \times FS \\
&= (1.55 \times 10^{-4}) \times 5.0 \\
\theta_{allow} &= 7.75 \times 10^{-4}\, m^2/min
\end{aligned}
$$

This, in turn, requires a geotextile to have an ultimate (or as-manufactured) transmissivity considerably in excess of θ_{allow}. If the cumulative reduction factor is 3.0, then

$$\theta_{ult} = \left(7.75 \times 10^{-4}\right) \times 3.0$$

$$\theta_{ult} = 23.2 \times 10^{-4} m^2 / \text{min}$$

As seen in figure 2.16, this is possible only by selecting an extremely heavy nonwoven needle-punched geotextile above 2000 g/m^2. Alternatively, geonets or geocomposites can be considered (chapters 4 and 8 respectively).

(e) We must now do a soil-retention analysis to see that soil particles do not embed in the geotextile and decrease its transmissivity. The analysis is the same as in section 2.8.

(f) Finally, long-term soil-to-geotextile compatibility must be addressed. Here within an earth dam is where long-term flow tests, gradient ratio tests or hydraulic conductivity ratio tests have applicability. See section 2.3.5 for details.

Example 2.26:

Calculate the factor of safety of a 500 g/m^2 nonwoven needle-punched geotextile required to drain water from behind an eight meter high concrete cantilever retaining wall if it has an allowable transmissivity of θ_{allow} = 0.15 $\times 10^{-3}$ m^2/min at is maximum design pressure. The soil backfill is a silty sand (ML-SW) with $k = 5 \times 10^{-5}$ m/s.

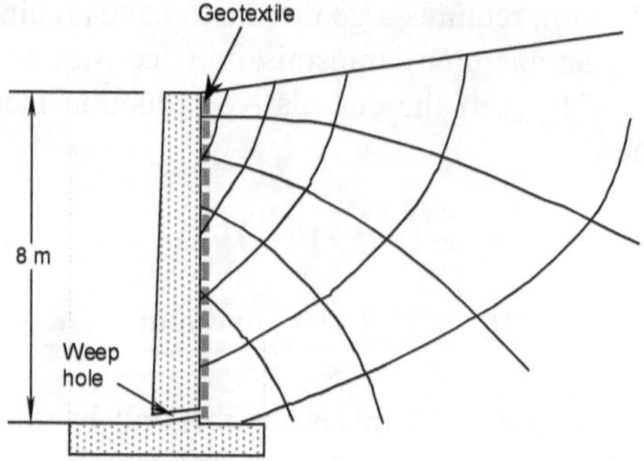

Solution: As before, we proceed in parts:

(a) Calculate the maximum flow rate coming to the geotextile. From the flow net above, we have

$$q = kh\left(\frac{F}{N}\right)$$

$$= \left(5.0\times10^{-5}\right)(60)(8)\left(\frac{5}{5}\right)$$

$$q = 0.024 m^2/\text{min}$$

(b) Determine the flow gradient within the geotextile.

$$i = \sin 90°$$

$$= 1.0$$

(c) Calculate the required transmissivity.

$$q = kiA$$

$$= ki(t\times w)$$

$$q = (kt)(i\times w)$$

$$kt = \frac{q}{(i)(w)}$$

$$= \frac{0.024}{(1.0)(1.0)}$$

$$\theta_{reqd} = 0.024 m^2/min$$

(d) Compare this value to the geotextile's allowable transmissivity to obtain a factor of safety.

$$FS = \frac{\theta_{allow}}{\theta_{reqd}}$$

$$= \frac{0.00015}{0.024}$$

$FS = 0.0062$ which is not nearly acceptable!

It is easily seen is that this application is not suited for geotextiles. It is, however, a perfect situation for geonets or drainage geocomposites that have much greater in-plane flow capacity. Geonets are the subject of chapter 4 and geocomposites are the subject of chapter 8 where we will repeat this exact problem and note the substantial increases in the factor of safety.

2.9.4 Pressure Drainage Design

Geotextile transmissivity is the key parameter in both gravity and pressure drainage; in this sense, the two topics are quite similar. The difference is that for pressure drainage water will flow from locations of higher pressure to locations of lower pressure regardless of the geotextile's orientation. Thus, flow direction depends on each specific situation; some of these situations have been identified. The following equation is formulated for a geotextile placed beneath a surcharge fill on a fine-grained compressible foundation soil, after Giroud [68].

$$\theta_{reqd} = k_p t = \frac{B^2 k_s}{(c_v T)^{1/2}} \tag{2.71}$$

where

θ_{reqd} = geotextile transmissivity,
k_p = in-plane permeability coefficient of the geotextile,
t = thickness of the geotextile,
B = width of the surcharge layer,
k_s = permeability coefficient of the foundation soil,
c_v = coefficient of vertical consolidation of the foundation soil, and
T = time for the surcharge fill to be placed.

Example 2.27 illustrates use of the formula.

Example 2.27:

Consider a variable-width surcharge fill placed in 10 days (14,400 min.) on foundation soil of 1×10^{-9} m/s permeability and 4.6×10^{-6} m²/min. coefficient of consolidation, as shown below. Determine (a) the required geotextile transmissivity as a function of base width of the surcharge fill and graph the result, and (b) using a ultimate geotextile transmissivity of 0.75×10^{-3} m²/min and cumulative reduction factors of 5.0, find the maximum width of surcharge that can be used under these conditions?

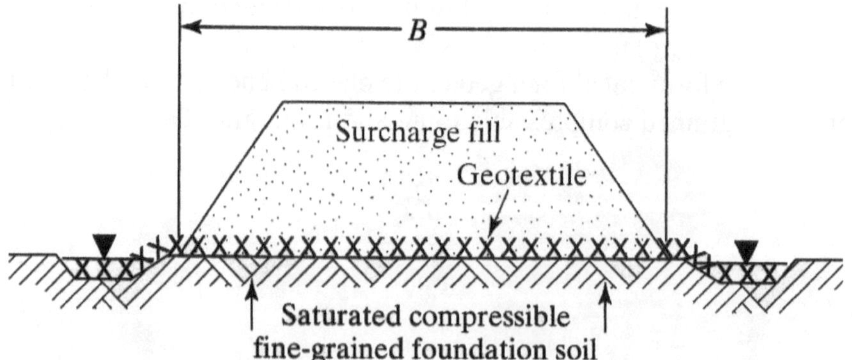

Solution:

(a) Determine the graph of B versus θ

$$\theta_{\text{reqd}} = \frac{B^2 k_s}{(c_v T)^{1/2}}$$

$$= \frac{(1\times10^{-9})(60)B^2}{\left[(4.6\times10^{-6})(0.0144\times10^{6})\right]^{1/2}}$$

$= 2.33\times10^{-7}\ B^2$, *where B is in units of meters.*

When different values of B are calculated the following curve results:

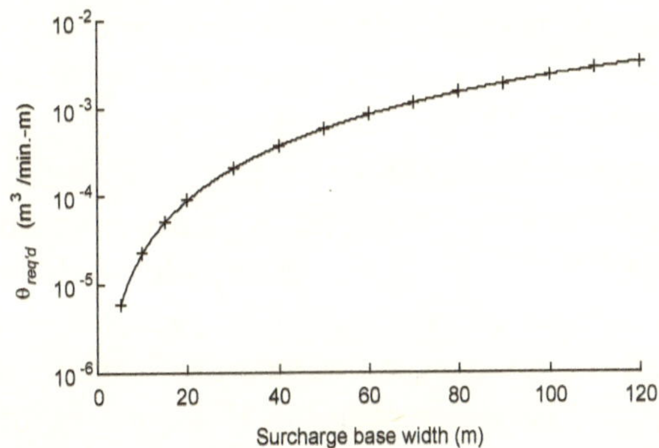

(b) Using the above graph,

$$\theta_{ult} = 0.00075 m^2 / min$$

$$\theta_{allow} = \theta_{ult} / \Pi RF$$

$$= 0.00075 / 5.0$$

$$= 0.00015 m^2 / min$$

and from the above curve the maximum value is as follows:

$$B_{\max} = 30 \text{ m}$$

2.9.5 Capillary Migration Breaks

The upward movement of water in fine grained soils has been known to present problems in two distinctly different areas of the world. One is in cold regions, where temperature gradients below freezing cause moisture to rise above the stationary water table. This rise occurs in the capillary zone and can eventually result in frozen layers (ice lenses), which will expand, lifting any structure placed above it. The phenomenon is called *frost heave* and is well-documented in the geotechnical literature.

Remedies for frost heave usually involve a capillary break or cut-off placed horizontally at a depth beneath the lowest elevation of the freezing isotherm. Sands and gravels have been used, but geotextiles offer an attractive and cost-effective alternative. Use of a thick, nonwoven needle-punched geotextile is easily placed and can be readily graded to drain the rising water away from the area of immediate concern. Several laboratory studies and case histories of geotextiles used in this manner are available [121, 122].

It should be mentioned that geotextiles are generally hydrophobic (that is, they repel water), so there is no wicking action across the plane of the geotextile. This is important to note because there is a popular misconception that geotextiles can wick water, as does the wick of a candle with wax. This is not so since both polypropylene and polyester repel water. Until the geotextile voids are saturated, there is only minor intimate contact of the water with the polymer fibers. This is just the opposite of water in soil, where soil particles are hydrophilic, and the water nests itself around points of contact of adjacent soil particles. Once geotextiles are completely saturated, however, they can be used to siphon water in a continuous manner, thereby maintaining flow within them [123].

In a completely different part of the world, namely, deserts and arid regions, there is a similar problem involving capillary rise. Here, as groundwater rises it brings dissolved salts with it [124]. As this salt-laden water comes near the ground surface, it kills all vegetation, which takes up the salt water through the root system. Equally severe is the salt attacking building foundations of both stone and concrete, making these usually adequate structural materials very friable.

As with the frost cutoff, this application area of salt-migration cutoff can effectively use a geotextile exhibiting proper in-plane drainage characteristics. The design procedure uses the transmissivity parameter, as in the case of gravity drainage. Once again laminar flow is assumed, so that Darcy's formula is employed in the problem solution. Example 2.28, which follows, actually suggests gravity and pressure situations but is solved completely by a gravity approach.

Example 2.28:

A refrigerated storage building for frozen foods is to be founded on the site illustrated below. A capillary break beneath the building's foundation having a geotextile is being considered as a solution. Will a 700 g/m^2 nonwoven needle-punched geotextile having an allowable transmissivity $\theta = 0.0007$ m^2/min at 25 kPa normal stress be adequate if the required $FS = 3.0$?

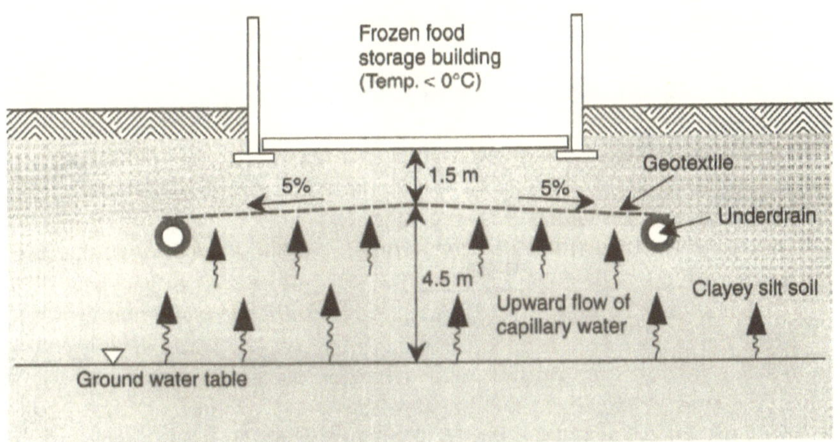

Solution: The problem is similar to the chimney drain example of section 2.9.3 with some obvious exceptions.

(a) Determine the flow rate of upward water migration to the geotextile, which is a function of the soil's permeability and the thermal gradient drawing the water upward. Guides are given in reference 88, where a conservative value is selected for this example.

$$q = 2.7 \times 10^{-5} \text{ m}^2/\text{min}$$

(b) Calculate the gradient of flow in the geotextile.

5% slope = 0.05 gradient

(c) Calculate the required transmissivity, θ_{reqd}.

$$
\begin{aligned}
q &= kiA \\
&= ki(t \times w) \\
q &= (kt)(i \times w) \\
kt &= \frac{q}{i \times w} \\
\theta_{reqd} &= \frac{0.000027}{(0.05)(1.0)} \\
&= 0.00054 \text{ m}^2 / \text{min}
\end{aligned}
$$

(d) Determine the factor of safety

$$
\begin{aligned}
FS &= \frac{\theta_{\text{allow}}}{\theta_{\text{reqd}}} \\
&= \frac{0.00070}{0.00054} \\
FS &= 1.3 < 3.0 \ \textit{not acceptable}:
\end{aligned}
$$

Therefore, at least a triple layer of geotextile is needed or a geonet or drainage geocomposite, as will be discussed in chapters 4 and 8.

2.9.6 Summary

Geotextiles used in the drainage function are quite attractive for use as long as the required planar flow rates are relatively low. Certainly the drainage of fine-grained soil masses (like silts and clays) is possible, along with some examples shown in this section. Nonwoven needle-punched geotextiles are best suited for this function. High mass per unit areas, or multiple layers, can be used to obtain sequentially higher flow rates. Although not shown herein, their flow rates can be further augmented by being constructed with high-denier fibers. Significantly higher flow capacity, however, requires a different approach. That will be seen to be nicely fulfilled by geonets and drainage geocomposites.

Some of the examples of this section used an allowable transmissivity for their calculations. Thus, no reduction on an ultimate transmissivity value by means of equation 2.25 and table 2.8b was illustrated. Other problems used relatively low reduction factors to obtain an allowable transmissivity from the laboratory test value. This is because the current transmissivity test procedures (ASTM D4716 and ISO 12958) are often configured as performance tests. Site-specific conditions and loads can be readily replicated, and the resulting value will then be an allowable value (or near allowable) rather than an ultimate one. If not, the procedures set forth in reducing an ultimate to an allowable value by using reduction factors must be used.

2.10 DESIGNING FOR MULTIPLE FUNCTIONS

2.10.1 Logic for Chapter

In sections 2.5 to 2.9, the primary function of the geotextile was readily apparent. This was to orient the reader's attention toward the design-by-function concept, and also because most applications lend themselves to a readily definable primary function design. Nevertheless, there are other applications in which the geotextile must be designed for multiple functions. In these cases a single dominant (primary) function cannot always be identified. Thus, there are primary, secondary, tertiary, and perhaps even quaternary functions that may vary in a particular application. Furthermore, these functions might vary from site to site. Such multiple-function applications are the focus in this section. They should not be taken lightly or be considered of lesser importance than those discussed previously. Some of the major uses of geotextiles are included in this section on multiple-function applications.

2.10.2 Reflection Crack Prevention in Pavement Overlays

Overview. The resurfacing of existing pavements that have excessive cracks in them represents an ongoing and expensive task for all federal, state, local, and private organizations that own and maintain roads. Such resurfacing is usually done with bituminous overlays ranging in thickness from 25 to 100 mm. Particularly exasperating to the road owners (and to the users and their automobiles as well) is when the cracks in the original pavement reflect up through the new overlay earlier than anticipated. To combat this, thicker overlays than desirable are used, but at the cost of added expense, lower curb heights, and excessive weight and thickness on the subgrade system. Due in part to the magnitude of this problem and the potential market that it represents, the use of geotextiles to remedy the situation has been attempted in a number of ways. In some instances strips of geotextile have been placed over the cracks, spanning them by 150 to 600 mm on each side, and the overlay placed above. Polyester, polypropylene, and fiberglass geotextiles as well as geogrids (see chapter 3) have all been used in this regard. Alternatively, however, a major use has been with full-width geotextile sheets that have been waterproofed with asphalt cement or asphalt emulsion, over the entire pavement surface, and then overlaid with the final bituminous surfacing. The goal of such a process is to either decrease the thickness of the overlay while keeping a lifetime equivalent not using a geotextile, or to increase the lifetime of the overlay while using the same thickness as without the use of the geotextile. It should be noted that this technique is used only for existing bituminous pavements, not for Portland-Cement concrete pavements. The significantly sharper edges of concrete would generally puncture and tear the lightweight geotextiles customarily used for this application.

As with other topics in this particular section, a clear-cut primary function is difficult to identify. It probably involves either the reinforcement from one side of the existing crack to the other (via the geotextile's tensile strength), or moisture-proofing against water moving through the pavement and into the subgrade via the impregnation of the geotextile with asphalt cement or asphalt emulsion. Since either function is possible, both design concepts will be explored after the construction and related details are addressed.

Construction. The use of full-pavement-width geotextiles in reflective crack prevention for bituminous pavement overlays is seemingly quite simple and straightforward. The general process follows; however, each step has its own peculiarities and subtleties.

1. Any existing failures in foundation soils must be repaired before resurfacing is done. A geotextile of 160 to 320 g/m^2 weight cannot be expected to hold up highway vehicles traveling only a few centimeters above it when the foundation soil beneath the stone base is unacceptable. In some instances these primary repairs must be very extensive.

2. Cracks in the existing bituminous pavement might be filled (see figure 2.53a). Cracks up to approximately 6 mm thick are filled with hot-liquid crack filler, while larger cracks are filled with asphalt, hot mix, or cold patch.
3. An asphaltic sealant is then uniformly sprayed over the existing pavement (see figure 2.53b). The amount ranges from 0.2 to 2.3 l/m^2, depending on the porosity of the existing pavement and the absorbency of the geotextile. Recommended sealants are asphaltic cement (AC-5 or AC-20), cationic asphalt emulsions (CRS-2 or CRS-1h), and anionic asphalt emulsions (RS-2 or RS-1).

(a) Filling cracks in existing bituminous pavement

(b) Spraying asphalt-based sealant over existing pavement

(c) Geotextile being placed by mechanical equipment

(d) Hot mix bituminous overlay being placed

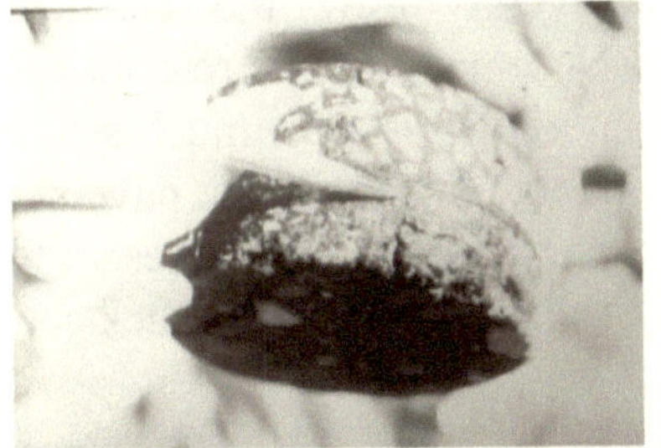

(e) Asphalt pavement core showing the crack-arresting feature offered by the geotextile with the new overlay placed above

Figure 2.53 Construction procedures and equipment for using geotextiles in reflective crack prevention in bituminous overlays. (Compliments of Amoco (now Propex) Fabrics and Fibers Co.)

When using the asphalt emulsions, care is required to ensure that they cure adequately before the geotextile is placed. Curing takes from 30 min. to 4 hr., depending on the temperature and humidity. Cutback asphalts cannot be used with polypropylene geotextiles, since the solvent in them reacts with the polymer at high temperatures.

4. The geotextile is then placed on the sealant by hand or with mechanical equipment (see figure 2.53c). Excessive wrinkles or folds in the geotextile must be cut open and laid flat. Stiff brooms are used to obtain a good bond with the sealant and to smooth out the surface as well. Joint overlaps of 25 to 75 mm are generally used. Additional sealant should be applied at these joints. If sealant comes through the geotextile, sand can be spread over it to absorb the excess.
5. The hot mix overlay is placed directly on the geotextile as soon as possible (see figure 2.53d). The temperature of the mix should be about 150°C with a maximum of 165°C. Care must be taken to avoid movement or damage to the geotextile from turning of the paver or truck movement, since these vehicles are riding directly on the waterproofed geotextile.
6. The resulting system should appear as the core shown in figure 2.53e.

A few points in the procedure need additional comment. The amount of asphalt sealant used is very important. Too little sealant leaves the geotextile unsaturated (thus permeable to infiltrating water), and too much sealant leaves an excess above or below the geotextile (thus forming a potential slip layer when the overlay is placed). Phenomenologically, the hot overlay draws the sealant up into the geotextile, just saturating it. (Note that for this reason, cold patch cannot be used as the overlay material.) Just how much sealant should be used depends on both the quality of the existing pavement and the type of geotextile being used. Button et al. [125] present the following equation for the quantity of sealant (also called a tack coat) to be used:

$$Q_d = 0.36 + Q_s \pm Q_c \tag{2.72}$$

where

Q_d = design sealant quantity (l/m^2),
Q_s = saturation content of the geotextile being used (l/m^2), and
Q_c = correction based on sealant demand of the existing pavement surface (l/m^2).

Concerning the value of Q_s, most manufacturers have specific geotextiles for this application (recall table 2.2g) and are familiar with the required amount. In the absence of this information we can take a flat pan with the candidate geotextile placed in it and experimentally determine the required amount. The geotextile is first saturated in asphaltic cement at 120°C for 1 min. It is allowed to cool and then is pressed with a hot iron between two absorbent papers to remove the excess asphalt. The value of Q_s is measured accordingly. A related procedure can be done in the field with a completely flat piece of sheet metal beneath the geotextile. The quantity Q_s depends mainly on the geotextile's thickness and to a lesser extent on other manufacturing details. Concerning the value of Q_c, see table 2.17, where it is seen that the older more oxidized pavement surfaces require greater amounts of sealant.

TABLE 2.17 SEALANT DEMAND OF EXISTING BITUMINOUS PAVEMENT SURFACES

Surface Condition	Q_c (l/m^2)
Flushed	-0.09–0.09
Smooth, nonporous	0.09–0.23
Slightly porous, slightly oxidized	0.23–0.36
Slightly porous, oxidized	0.36–0.50
Badly pocked, porous, oxidized	0.50–0.59

Source: After Button, et al. [125]

Much has been written concerning geotextile selection. On the basis of the majority of field projects completed to date, lightweight nonwoven needle-punched geotextiles prevail. Such low initial modulus geotextiles make us wonder about the reinforcement possibilities made for this application. They do indeed saturate well with sealant, so that waterproofing may be the dominant function.

Since the decision as to primary function is by no means clear cut, two separate design methods will be presented. The first is based on reinforcement as the primary function, and the second is based on waterproofing as the primary function.

Reinforcement Based Design. The key to the reinforcement-based design of geotextiles in reflective crack prevention in bituminous pavement overlays is the fabric's effectiveness as determined by laboratory testing or by experience. Quantitatively, it is defined as follows:

$$FEF = \frac{N_r}{N_n} \tag{2.73}$$

where

FEF = fabric effectiveness factor,
N_r = number of load cycles to cause failure in the geotextile-reinforced case, and
N_n = number of load cycles to cause failure in the nonreinforced case.

Values of *FEF* vary widely when based on laboratory tests (as they usually are), the range being from 2.1 to 15.9 (see Murray [126]). Upon having this value, however, design can be approached by a number of procedures. Majidzadeh et al. [127], use a mechanistic design procedure influenced by both rutting (distortion) and fatigue (cracking). Another approach, however, is merely to modify existing asphalt overlay design methods. In this regard the design traffic number (*DTN*), on which overlay designs are based, will be modified as follows:

$$DTN_r = \frac{DTN_n}{FEF} \tag{2.74}$$

where

DTN_r = design traffic number in the fabric-reinforced case,
DTN_n = design traffic number in the nonreinforced case, and
FEF = fabric effectiveness factor.

Using the Asphalt Institute's overlay design procedure [128], example 2.29 illustrates the procedure. It is based on an arbitrarily selected *FEF* value of 3.0 and uses figure 2.54 as the basic design guide. The procedure is as follows:

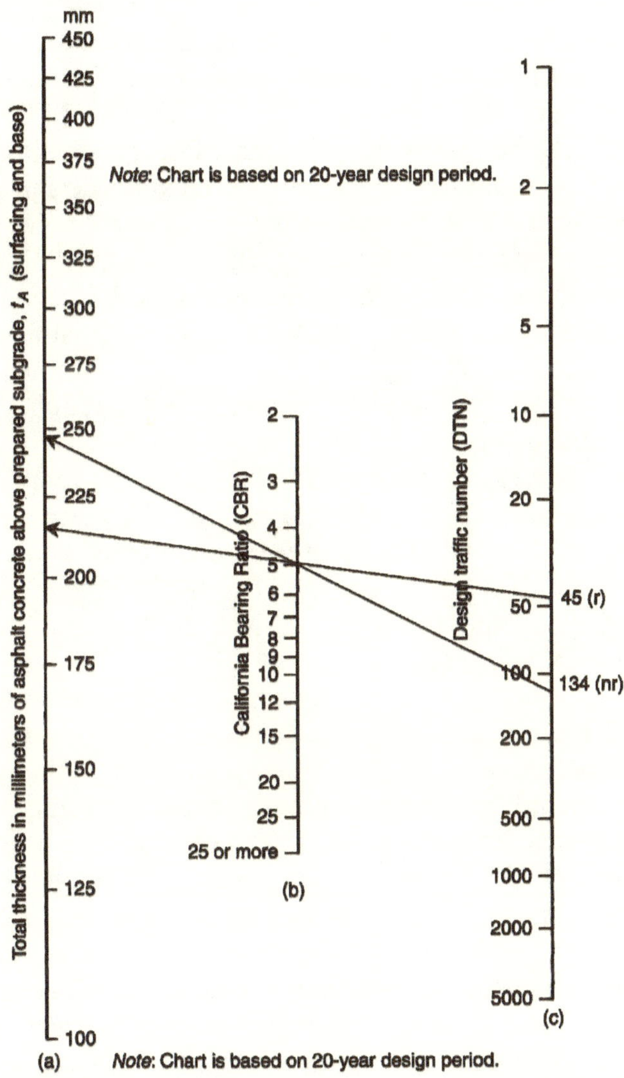

Figure 2.54 Thickness requirements for asphalt pavement structures using unsoaked subgrade soil CBR. (After Asphalt Institute [128])

1. Determine the soil subgrade strength value as represented by its *CBR* value.
2. Determine the initial traffic number (*ITN*) as discussed in reference 128. This is a combination of each vehicle's weight and respective number of load repetitions based on traffic counting.
3. Determine the adjustment factor for the desired design period and estimate traffic annual growth rate, as described in reference 128.
4. Multiply the *ITN* by the adjustment factor to obtain the design traffic number *DTN* for use in the thickness design chart.
5. Use figure 2.54 (or equivalent) to determine the full-depth asphalt-to-pavement thickness, t_{An}, needed for the design subgrade strength value, the *DTN*, and the selected design period.
6. Determine the effective thickness, t_e, of the existing pavement as discussed in [128].
7. The thickness of asphalt concrete overlay required, then, is equal to $t_{An} - t_e$.
8. This process is repeated for the geotextile-reinforced case using equation 2.74, which results in a thickness t_{Ar}.
9. The resulting two thicknesses (nonreinforced and geotextile-reinforced) are then compared ($t_{An} - t_{Ar}$) to note the savings in asphalt overlay thickness Δt since the base thickness is the same in both cases.

Example 2.29:

An interurban two-lane highway carries an average of 4000 vehicles per day, 400 (10%) of which are heavy trucks of 135 kN average gross mass. The single-axle load limit is 80 kN. Traffic growth rate is 4% annually. The existing pavement consists of 75 mm of asphalt concrete surface and 200 mm of crushed stone base on a soil whose *CBR* = 5.0. The pavement is in generally good condition, but visual evaluation indicates the need for an overlay. Find the overlay thickness for a twenty-year-design period (a) if you do not use a geotextile, (b) if you use a geotextile with *FEF* = 3.0, and (c) compare the two overlay thicknesses. This problem (without geotextile reinforcement) is from [128].

Solution: (a) Determine that the initial traffic number = 90, and the adjustment factor = 1.49, resulting in a DTN for the nonreinforced fabric case of

$$\begin{aligned} DTN_n &= 90 \times 1.49 \\ &= 134 \end{aligned}$$

Using this and a $CBR = 5$, figure 2.54 results in a full-depth nonreinforced asphalt pavement thickness (t_{An}) of

$$t_{An} = 245\text{mm}$$

The existing pavement effective thickness (t_e) calculation uses a weighing factor of 0.8 on the existing asphalt and 0.4 on the existing stone base.

$$\begin{aligned} t_e &= 75(0.0) + 200(0.4) \\ &= 140\text{mm} \end{aligned}$$

Therefore, the required overlay thickness (t_{on}) without using a geotextile (i.e., nonreinforced) is

$$\begin{aligned} t_{on} &= t_{An} + t_e \\ &= 245 - 140 \\ &= 105\text{mm} \end{aligned}$$

(b) The solution varies for the case of using a geotextile-reinforcement layer as follows:

$$\begin{aligned} DTN_r &= \frac{DTN_n}{FEF} \qquad (2.74) \\ &= \frac{134}{3.0} \\ &= 45 \end{aligned}$$

which, from figure 2.54, results in

$$t_{Ar} = 215 \text{ mm}$$

and

$$\begin{aligned} t_{or} &= 215 - 140 \\ &= 75\text{mm} \end{aligned}$$

(c) Thus, the savings in asphalt overlay thickness using a geotextile layer (and based on a reinforcement hypothesis) is

$$\begin{aligned} \Delta t_o &= t_{on} - t_{or} \\ &= 105 - 75 \\ \Delta t &= 30\text{mm} \end{aligned}$$

Note that this same result comes about from using the values of t_{An} and t_{Ar} directly, i.e.,

$$\begin{aligned} \Delta t &= t_{An} - t_{Ar} \\ &= 245 - 215 \\ \Delta t &= 30\text{mm} \end{aligned}$$

Waterproofing-Based Design. Again, using the Asphalt Institute's techniques [128], we can develop an alternate design procedure for geotextiles used in asphalt overlay situations; this time one that is based on a waterproofing hypothesis. This concept should not come as a surprise since adequate subgrade drainage of pavements has long been suspected as being the key factor for extending conventional pavement lifetimes. Cedegren [129], clearly illustrates this type of improved pavement lifetime. The particular procedure we will use, adopted from Bell [130], utilizes field-measured rebound deflections of the existing pavement system along with the design guide of figure 2.55. The individual steps in the design are as follows:

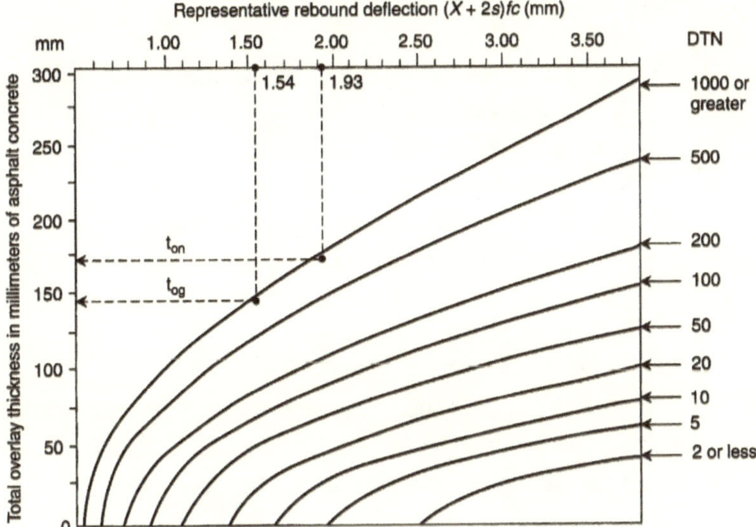

Figure 2.55 Asphalt overlay thickness required to reduce pavement deflection from a measured to a design deflection value (a rebound test). (After Asphalt Institute, [128])

1. Determine the representative rebound deflection as discussed in [128], which is based on Benkelman beam field deflection tests.
2. Determine the *ITN* as discussed in [128]. It is a combination of each vehicle's weight and respective number of load rejections based on traffic counting.
3. Determine the initial traffic number adjustment factor for the desired design period as described in [128].
4. Multiply the ITN by the ITN adjustment factor to obtain the DTN for use in the overlay thickness chart.
5. Enter the overlay thickness chart (figure 2.55) at the representative rebound deflection and move down vertically to the curve representing the DTN (interpolate if necessary). Move horizontally to the overlay thickness scale and read the thickness of overlay required.
6. For the case of a geotextile included in the pavement cross-section and of it being suitably waterproofed, one can appropriately modify the representative rebound deflection (RRD) equation as follows:

$$RRD = (\overline{X} + 2s)fc \tag{2.75}$$

where

$\overline{X}$ = arithmetic mean of measured Benkelman beam deflection values,
s = standard deviation,
f = temperature adjustment (see [128]), and
c = critical period adjustment factor, which is largely influenced by moisture in the subgrade system (this is the term that will be empirically adjusted in design example 2.30).

7. The design process is then repeated as with the nongeotextile case, and the resulting two thicknesses are compared to note the savings in asphalt overlay using the geotextile.

Example 2.30:

A four-lane interurban highway carries an average of 16,000 vehicles per day, 2400 (15%) of which are heavy trucks with an average gross mass of 145 kN; the design lane is estimated to carry 45% of the heavy trucks; the traffic growth rate is 5% annually, and the legal single-axle load limit is 80 kN. Some cracking of the pavement surface is evident. High deflections indicate the need for an overlay. Find the overlay thickness required for a twenty-year-design period (a) if you do not use a geotextile, (b) if you use a geotextile that changes the value of c in equation (2.75) from 1.25 to 1.00, and (c) compare the two overlay thicknesses. In the analysis use $ITN = 590$ and an adjustment factor of 1.67. This problem (without geotextile waterproofing) is also taken from [128].

Solution: (a) First calculate the representative rebound reflection from equation (2.75) using field-gathered data of

$$\overline{X} = 1.55\text{mm},$$
$$s = 0.10 \text{ mm},$$
$$f = 0.88,$$
$$c = 1.25, \text{ and}$$
$$RRD = (1.55 + 0.20)0.88 \times 1.25$$
$$= 1.93\text{mm}$$

The design traffic number is also needed:

$$DTN = ITN \times \text{adjustment factor}$$
$$= 590 \times 1.67$$
$$= 985$$

Using figure 2.55, the required overlay thickness without a geotextile is

$$t_{on} = 170 \text{ mm}$$

(b) For the case with a geotextile as a waterproofing layer, the constant c is changed from 1.25 to 1.00 and the process is repeated:

$$RRD = (1.55 + 0.20)0.88 \times 1.00$$
$$= 1.54\text{mm}$$

Using this value and a $DTN = 985$, the required overlay waterproofed geotextile thickness from figure 2.55 is

$$t_{og} = 140 \text{ mm}$$

(c) Based on these asphalt overlay thickness values, the saving using a geotextile is as follows:

$$\Delta t = t_{on} - t_{og}$$
$$= 170 - 140$$
$$\Delta t = 30\text{mm}$$

Note that the equivalent thickness of the existing pavement system is the same in both cases, and the resulting saving if a constant value were added to both t_{on} and t_{og} is still 30 mm.

Commentary. As mentioned in section 2.10.1, multiple-function geotextile applications are difficult to analyze since a clear-cut primary function cannot easily be identified. This topic of crack reflection prevention in bituminous pavement overlays illustrates the dilemma perfectly. Using two completely different hypotheses (one based on reinforcement and the other based on waterproofing), two completely different designs can be developed. It simply begs the question of where the truth actually lies. Considering these two extremes, it might be that a combination of the two phenomena are working together! Further, a separation interlayer may be another function that is yet to be quantifiably developed.

Clearly, well-instrumented, well-monitored, well-analyzed, and well-reported case histories are needed in this application area. Perhaps, then, a decision as to which is the primary function will evidence itself, pointing the way to the correct design methodology. The case histories that are currently available on the topic not only do not give this identification but cast doubt on where the technique can be used. In general, however, it is felt that reports are definitely on the positive side. Based on experiences to date, the following recommendations are offered:

- For those states that are using proprietary or competition-limiting specifications for geotextiles, the adaptation of Texas specifications for state use is encouraged.
- Prior to placement of a geotextile overlay system, the condition of the existing roadway should be documented. When an unstable roadway is suspected, deflection tests are recommended. While limiting deflection values have yet to be established for geotextile systems, it is important that data be obtained that could assist in their eventual evaluation.
- Since all geotextiles presently being marketed are not equivalent in physical properties, agencies should conduct the tests identified in the Texas specification, including the

asphalt retention test, so as to develop documentation that may be useful later in assessing relative geotextile performance.

- Rather than placing the geotextile on the cracked existing pavement, construct an asphalt leveling course first so as to provide a relatively unblemished surface for applying the tack coat and the geotextile. This will assure more complete and uniform impregnation of the geotextile by the tack coat and will also assist in determining the type of tack coat to be used and the proper application rate.
- For pavement rehabilitation projects that include pavement widening with new asphaltic concrete overlays, geotextiles placed longitudinally over the shoulder-pavement and/or widening joint should be considered. The state of Maine has had success in this regard [131]. Both longitudinal and transverse joints were greatly retarded by 380% and 320%, respectively, when using a high-strength geotextile directly spanning the cracks in question.
- Over jointed portland-cement concrete pavements, no evidence has been provided to support placing a geotextile system across the full roadway width in a continuous mat. Instead, the use of heavy-duty geotextile materials in strips over transverse and pavement edge joints and cracks is presently recommended.

A summary report by the US Army Corps of Engineers on this particular application of geotextiles has arrived at similar conclusions [132]. Of importance insofar as current research and development is concerned is a series of RILEM conferences focused specifically on reflective cracking in pavements [133].

2.10.3 Railroad Applications

Geotextiles are often used in railroads beneath the stone ballast on which the wooden or concrete tie system is placed. As will be discussed, a critical aspect of the design is the depth at which the geotextile is placed beneath the bottom of the tie (i.e., the thickness of overlying ballast). First, however, it is necessary to gain a perspective of the possible functions of the geotextile under various possible conditions.

Overview. It is virtually impossible to identify a unique, primary function for geotextile use in railroad applications. Site-specific conditions will vary the primary function among a number of possibilities. In listing these possibilities, it is important to keep in mind whether the railroad is being newly constructed or being rehabilitated. If new construction, the material beneath the geotextile will probably be the in-situ soil; if rehabilitation, the material beneath the geotextile will be previously placed ballast (now contaminated with soil), which has migrated into the soil over the working history of the railroad. Considering both of these situations, the possible geotextile functions are the following:

- Separation in new railroads, between in-situ soil and new ballast.
- Separation in rehabilitated railroads, between old, contaminated ballast and new clean ballast.
- Lateral confinement-type reinforcement in order to contain the overlying ballast stone from lateral movement.
- Filtration of soil pore water rising up from the soil beneath the geotextile due to rising water conditions, or the dynamic pumping action of the individual wheel loads across the plane of the geotextile.
- Lateral drainage from water entering from above or below the geotextile moving within the geotextile.

Irrespective of the difficulty of identifying a single function of the geotextile (it is obviously a multifunction application), the acceptance of geotextiles by railroad companies is reasonable and increasing. Newby [134] reports that as far back as 1982 the Southern Pacific Railroad had used geotextiles in over 1600 km of trackage.

Specific Design. A review of the geotextile literature on railroad applications shows some inconsistency. Railroad specifications seem to favor relatively heavy nonwoven needle-punched geotextiles because of their high flexibility and in-plane drainage (transmissivity) characteristics. The logic for high flexibility is apparent, since the geotextiles must deform around the relatively large ballast and not fail or form a potential slip plane. In-plane drainage, however, is not itself a dominant function, because any geotextile acting as a proper separator and filter will preserve the integrity of the drainage of the ballast stone itself, where ample void volume is always present. Nevertheless, geotextile drainage has been emphasized in railroad specifications [134].

Laboratory work, even large-scale model testing, seems to be directed at a reinforcing and stiffening function provided for by the geotextile. Work by Eisenmann and Leykauf [135] (which also included the filtration function), Saxena and Wang [136], and Bosserman [137] clearly illustrates the reinforcement benefits of using a geotextile beneath the ballast stone. In the author's opinion, however, such membrane-type reinforcement can be gained only after quite high deformation of the subgrade soils. For the majority of existing railroads, as in rehabilitation work, such deformations do not seem possible or desirable, since densification of the subsoil has usually occurred many years before rehabilitation is necessary. Similarly, new railroads would never be placed on deformable soil subgrades.

For these reasons, the design of geotextiles beneath ballast in railroad applications can be addressed in the following steps:

1. Design the geotextile as a *separator*, this function is always required. The procedures of section 2.5 have direct applicability here. High grab tensile, puncture, impact, and burst strengths all have significance in this particular application and lead to class 1 types of geotextiles, recall table 2.2a, or stronger.
2. Design the geotextile as a *filter* since this function is also usually required. The general procedures illustrated in section 2.8, in particular those illustrated for walls and underdrains, are relevant in this application. The general requirements of adequate permeability, soil retention, and long-term soil-to-geotextile flow equilibrium are needed as in all filtration designs. A note of caution, however: railroad loads are dynamic; thus pore pressures must be rapidly dissipated. For this reason high factors of safety are required in the permittivity design part of the process and the opening size would be designed using figure 2.4b for dynamic flow conditions.
3. Consider geotextile *flexibility* if the cross-section is raised above the adjacent subgrade. Here a very flexible geotextile is an advantage in laterally containing the ballast stone in its proper location. Quantification of this type of lateral

confinement reinforcement is, however, very subjective (recall section 2.3.2).

4. Consider the *depth* of the geotextile beneath the bottom of the railroad tie. The very high dynamic loads of a railroad acting on ballast imparts accelerations to the stone that are gradually diminished with depth. If the geotextile is not deep enough, it will suffer from abrasion at the points of contact with the ballast. Raymond [138] has evaluated a number of geotextiles beneath Canadian and US railroads and found many that are pockmarked with abrasion holes. In fact, there were so many cases that he has quantified the situation (see figure 2.56). Here it is seen that major damage occurs within 250 mm of the tie, and only deeper than 350 mm does the situation become acceptable. From this curve it can be concluded that the minimum depth for geotextile placement is 350 mm plus 50 mm for track settlement for a total 400 mm. If this depth is considered excessive in view of the large amount of ballast stone required, a highly abrasion-resistant geotextile must be used. Raymond [138] recommends a resin-dipped nonwoven needle-punched geotextile that has been forced-air dried to reestablish its porosity.
5. The last step in the design is to consider the geotextile's *survivability* during installation. To compact ballast under ties the railroad industry customarily uses a series of vibrating steel prongs forced into the ballast. Considering both the forces exerted and the vibratory action, high-geotextile puncture strength is required. Hence, and in keeping with the step 4, it is necessary to keep the geotextile deep or to use a special high-puncture resistant geotextile (recall section 2.3.5).

Commentary. Geotextile use in railroads beneath the ballast offers a number of possible benefits. These are separation, lateral confinement-type reinforcement, filtration, and drainage. It cannot be categorically stated that one function dominates over the others in all cases. This application is indeed a multifunctioned one that must be handled on a site-specific basis. Yet some functions (e.g., separation and filtration) are always present. A recommended design procedure has been outlined.

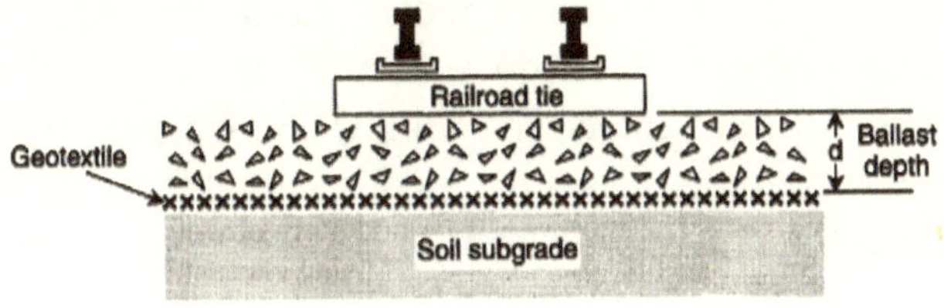

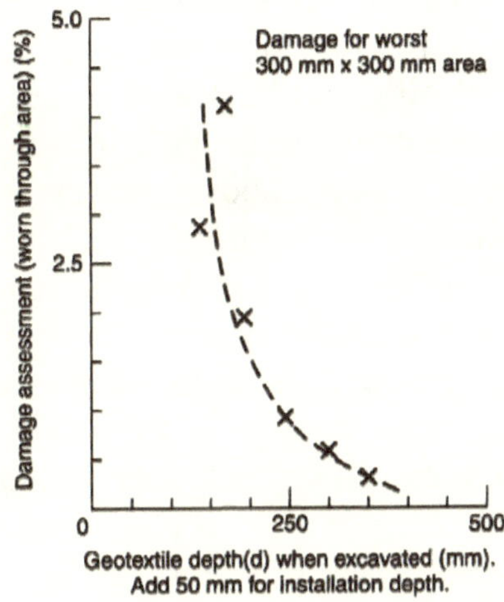

Figure 2.56 Observed geotextile abrasion damage as a function of depth beneath bottom of railroad tie. (After Raymond [138])

It must be emphasized that a geotextile beneath a railroad is a very demanding application. Avoiding both puncture during installation and abrasion during service lifetime requires deep placement beneath the bottom of the tie or specially treated geotextiles designed for high abrasion resistance and puncture resistance.

In closing, more well-planned, well-documented, and well-reported case histories of the type reported by Chrismer and Richardson [139] are recommended. Only with quantified results under specific conditions can a definitive function be associated with the geotextile. At that point the design process can advance in a position of strength.

2.10.4 Flexible Forming Systems

The traditional method of formwork for concrete or grout is to use wood or metal. These rigid forms are properly positioned and fixed in location until the material placed in them adequately cures and has sufficient strength of its own. While the constraint of a rigid form is an obvious advantage in building a concrete wall or footing to exact line and grade, it is a decided disadvantage in a number of other applications. These situations, which can capitalize on the use of flexible forms made from geotextiles, are explored in this section.

Overview. It is easy to visualize that a geotextile—in the form of a bag, container, or tube—could be used as a flexible form into which concrete, grout, or soil could be placed or pumped. Such a system would work as well under water (where water is displaced from within the fabric) as it would above ground (where air is displaced from within the fabric). Upon curing or stabilizing, the shape of the solidified mass takes the shape of the expanded geotextile; the number of possibilities is enormous. Additionally, the geotextiles, being flexible, can be inserted in difficult-to-reach locations and filled after proper positioning. This concept of flexible-forming systems with geotextiles can be used in many practical situations.

As a historical note, Terzaghi was the first to use geotextiles as flexible forms at Mission Dam (now Terzaghi Dam) in British Columbia, Canada, in 1955 [140]. A seepage cutoff wall was being placed within an existing dam. Since the surfaces on each side of cutoff wall were not parallel (one side was a multicurved concrete surface and the other was steel piling), a pipe or beam closure was not possible. Instead, Terzaghi used a nylon-reinforced canvas tube (today a geotextile) of 21 m length and 450 mm diameter and wrapped it around a 25 mm diameter grout pipe. The tube-surrounded grout pipe was placed in its proper position and then inflated with cement-bentonite grout. To be sure the system would work properly, a 6 m prototype was first constructed to see whether the gap would be completely filled and whether the fabric was sufficiently strong. The success was immediately obvious, and the system was used accordingly. Since that first use, the area of using geotextiles as flexible forms has grown rapidly and in many different applications.

Columns for Mine and Cavern Stability. Abandoned mines and caverns are obvious problems when any type of structure is to be founded on or near them. The obvious preconstruction remedy is to fill them before subsidence can occur; however, they often have no access or access is blocked and unavailable. Under these conditions, alternative methods are hydraulic backfilling, grouting, or the construction of conventional caissons to adequate depth, all of which are quite expensive. Geotextiles, however, can be used as flexible forms without the necessity of entering the mine or cavity. The technique consists of drilling 100 to 150 mm diameter holes to intercept the roof of the mine and then carrying these holes to the floor of the mine, penetrating the floor approximately 300 to 450 mm. A prefabricated and sewn tube of geotextile of approximately 500 mm diameter is wrapped around a grout pipe and is inserted down the drill hole into the keyhole at the floor of the opening. Then fine-aggregate concrete is injected under controlled pressure as the grout pipe is withdrawn. Expansion occurs to the full tube diameter where no resistance is met, that is, where voids exist (see figure 2.57). Thus, support shoulders are created beneath the competent rock strata. The tube of geotextile has to be supported at the surface or through rings at the top of the geotextile with a cable system to the ground surface. Each application requires a determination of how much pressure the geotextile can withstand. It may be necessary to pump the tube in multiple lifts. If necessary, fiber (steel or plastic) reinforcement can be utilized. The critical point in this application is to get maximum support of the column of concrete under the roof of the individual rock strata. Where the cavity or mine is dry, it is feasible to observe the expanded column with a television camera from an adjacent hole. In areas where the opening is fully or partially filled with soil or other compressible or objectionable materials, it is possible to jet out an opening in this material by having the grout pipe extend through the bottom of the tube and, while jetting through this pipe, maintain an adequate head of bentonite or grout in the geotextile form. This technique has the advantage over other methods of forming grout columns in that a positive form can be installed with a relative economy, as large quantities of concrete are not lost in a wasteful base that would be needed to build up the angle of repose of the concrete.

Using this concept, it is also possible to create a bulkhead in an underground mine. By drilling holes on a predetermined line, we pump alternate columns initially and allow for curing. The secondary or intermediate locations are then pumped and expanded with concrete to interlock between the originally placed columns. Two parallel walls can be created to form a bulkhead or cutoff. This technique has been used in relatively shallow mine and limestone applications. See

Koerner and Welsh [141] for several case histories. The method is very similar to the construction of secant-pile or tangent-pile cutoff walls in conventional foundation engineering.

As with other topics in this section, the geotextile to be used serves multiple functions (filtration, separation, and reinforcement through containment). The design centers on the following steps.

1. The opening size of the geotextile must be designed as a filter with emphasis on grout retention. A recommended criterion in this regard is the following:

$$0_{95} \leq 2d_{50} \tag{2.76}$$

where

O_{95} = 95% opening size of the geotextile (mm), and
d_{50} = average (50%) particle size of the cement used in the grout (mm).
Note that some "bleeding" of the grout will occur, but this should not significantly decrease its strength. It is necessary to have an open geotextile when construction is underwater so as to expel water within the fabric.

2. The required amount of shoulder area underneath the roof of the mine or cavern must be estimated on the basis of the surface loads to be imposed. This, together with the diameter of the geotextile column, will allow for a rough estimate of the support strength. The tensile strength must not be exceeded by the grout pressure, yet sufficient pressure must be generated to obtain the maximum possible shoulder area.

3. Since the geotextile must be longitudinally seamed, the sewing procedure is critical. In this application the seam strength must be equal to, or greater than, the required strength.

4. A numeric example on the use of geotextiles as flexible forms will be given at the end of the section.

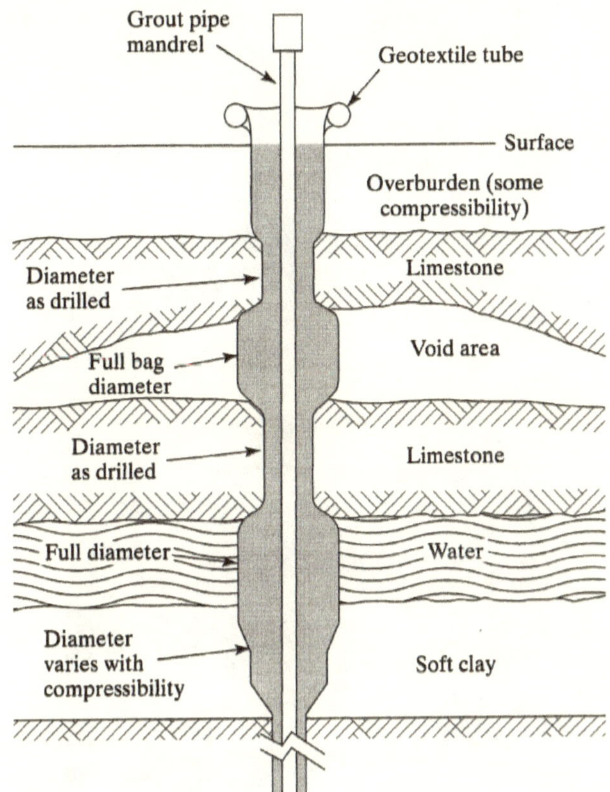

Figure 2.57 Idealized cross-sectional sketch of mortar piles showing grout pipe within a geotextile sleeve placed in a previously drilled 150 mm diameter hole. (After B. A. Lamberton, Intrusion Prepakt)

Geotextile Bags, Containers, and Tubes. The use of geotextiles as flexible forms in the context of bags, containers and tubes is growing at an incredible rate. The primary applications are erosion control and dewatering. The fill in erosion control applications is usually cohesionless sand or gravel, while dewatering applications focus on fine-grained dredge spoils or industrial waste sludges; not the least of which is agricultural waste.

Geotextile bags manufactured with UV-stabilized antioxidants are being successfully used for erosion control in numerous situations. Figure 2.58 illustrates such a situation where the beach erosion on both sides of the protected section is very obvious. A composite

(a) Fabricated bags being filled

(b) Seamed bags being barge transported

(c) Geotextile bag protecting the shoreline

Figure 2.58 Geotextile bag protection of Historic Farm Building along North Sea, Germany. (Compliments NAUE GmbH).

woven (340 g/m²) and nonwoven (610 g/m²) fabric was used for this particular application. With a veneer of soil cover placed over the completed bag system both UV degradation and accidental or intentional damage are essentially avoided.

Geotextile containers represent an extension of the previous geobags, however, the application is often removing river and harbor bottom sediments from shipping channels and navigable waterways. The technique generally utilizes high-strength woven geotextiles (greater than 50 kN/m tensile strength) and bottom dump barges, see figure 2.59. The geotextile is placed in the empty barge and then filled

(a) Geotextile within closed barge

(b) Geotextile being filled

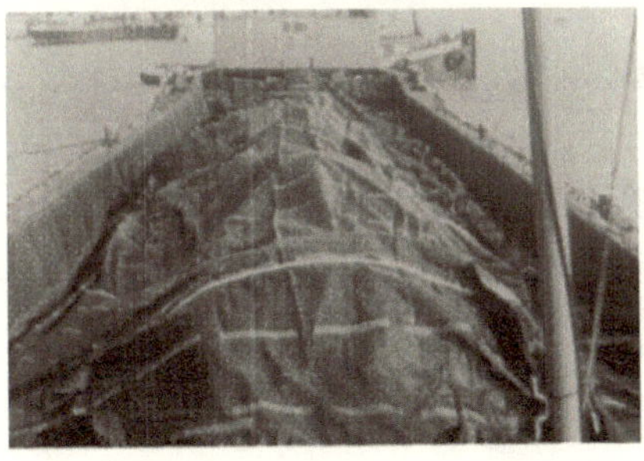

(c) Geotextile sewn over sediment fill forming

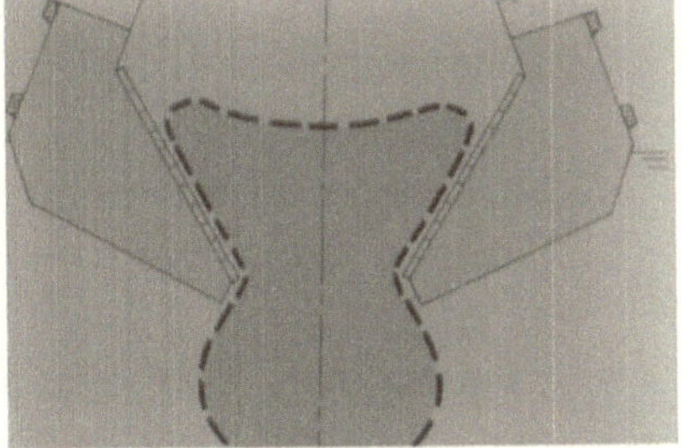

(d) Filled container being dropped out of bottom

Figure 2.59 Geotextile containers for removal of river and harbor sediment (sometimes contaminated) to construct underwater containment areas using bottom-dump barges. (Compliments TenCate Geosynthetics Co.)

with the bottom sediments. When full, the geotextile ends are folded over the top and sewn together, thereby completing the enclosure. The barge is towed to the disposal area, and when properly positioned, the split hull of the barge is opened and the sediment-filled container drops to the bottom. Subsurface embankments are being formed by this technique, which has the significant advantage that the sediments (particularly when contaminated) never leave the estuary or harbor. Furthermore, additional storage is possible behind the embankments, now reinforced by the filled containers themselves. Particularly intriguing is the design of the geotextile when considering its deformed shape, which includes the dynamic placement of the dredged sediment, its squeezing through the barge opening, and the impact stresses of its final positioning, see Leschinsky and Leshchinsky [142].

Geotextile tubes form the logical extension to bags and containers and have been increasing in size and length since their inception. Efforts to form flexible sand-filled tubes were made as early as 1957, but were not very successful. Eventually, in 1967, a patent was granted to a Danish firm, Aldek A. S., in conjunction with the Danish Institute of Applied Hydraulics. Since that time the technology has rapidly grown to the use of tubes as large as 20 m circumference and an unlimited length, [143]. Figure 2.60 shows a large geotextile tube filled with dredged sand acting as a beach erosion control system. There are a number of important design issues, among which are the following:

- Beach-side anchor tube dimensions, location, and connection to main tube.
- Water-side scour tube dimensions, location, and apron connection to main tube.
- Subsidence and/or rolling of main tube during severe storms.
- Opening size of fabric vis-à-vis particle size of soil being used as infill.
- Required tensile strength of the fabric and the seams, of which the seams are critical since they are invariably the weaker element.
- Geometric dimensions of the main tube after final pumping and filling.

(a) Method for sand slurry filling of prefabricated tube

(b) Filled tube showing longitudinal seam

(c) Tube after backfilling with recently planted dune grass

Figure 2.60 Sand filled geotextile tubes for beach erosion protection.

Progress on these last two issues has been made and is provided by means of a design nomograph after Pilarczyk [144], see figure 2.61, where example 2.31 illustrates its use.

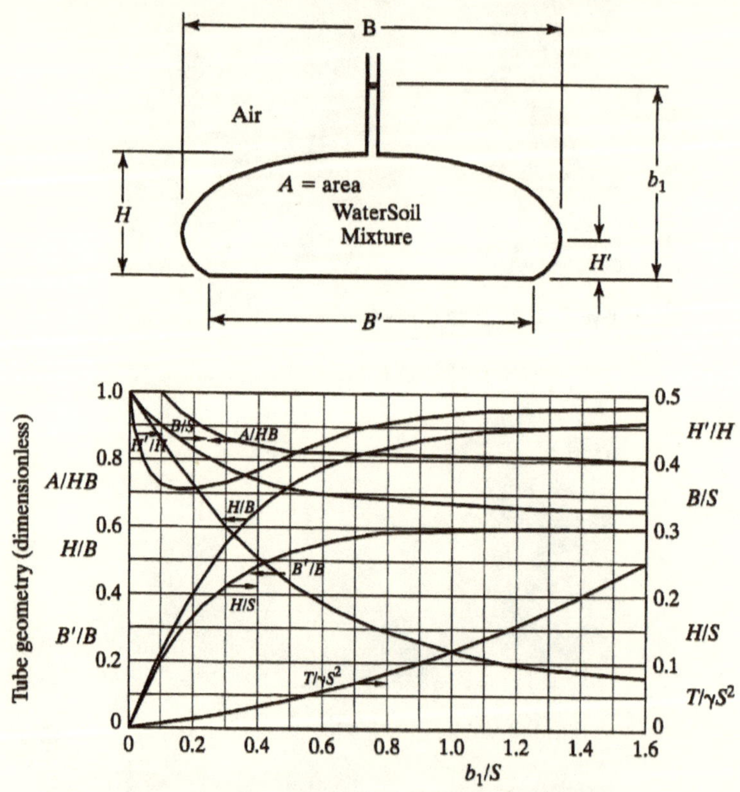

Figure 2.61 Geotextile tube definition sketch and design curves for fabric strength and final dimensions. (After Pilarczyk [144])

Example 2.31

A geotextile tube is filled with sand for the purpose of beach erosion control. (a) Determine the factor of safety of a 200 kN/m allowable strength geotextile used under the following conditions; pressure head, b_1 = 10.5 m (at 100 kPa) and circumference, S = 14 m and (b) what are the dimensional properties of the filled tube?

Solution: Using figure 2.61, the procedure is as follows:

(a) For b_1/S = 10.5/14 = 0.75
$T/\gamma S^2$ = 0.085
T_{reqd} = $(0.085)(9.81)(14)^2$
= 163 kN/m
FS = T_{allow}/T_{reqd}
= 200/163
FS = 1.2; which is marginally acceptable

(b) Figure 2.61 is again used for the final dimensional properties of the geotextile tube for a (b_1/S) value of 0.75:

B/S = 0.34; therefore, B = (0.34)(14) = 4.76 m
H/S = 0.28; therefore, H = (0.28)(14) = 3.92 m
H'/H = 0.45; therefore, H' = (0.45)(3.92) = 1.76 m
B'/B = 0.27; therefore, B' = (0.30)(4.76) = 1.43 m
A/HB = 0.81; therefore, A = (0.81)(3.92)(4.76) = 15.1 m^2
As a check on the curves and respective values:
H/B = 3.92/4.76 = 0.82 vs. 0.82, which is appropriate.

The extension of geotextile tubes into the dewatering of fine-grained soils, industrial sludges, sewage treatment sludges (biosolids), and agricultural farm waste is logical, but only to a certain degree (see Heibaum [145]). Clearly, the pumping of such sludges into a geotextile tube is possible, but the time for dewatering depends on the composition of the filter cake that is formed on the inside of the fabric. Figure 2.62 shows a field of such tubes with harbor bottom sediment being dewatered and an example of the low-permeability filter cake on the inside of the fabric as was just mentioned. The pumping must be carefully adjusted so as not to cause fabric or seam failure, yet optimize the tube's capacity and minimize stabilization time. In this regard a coagulant (also called a flocculant) is often added to the mixture as it enters the pump or pumping station. Coagulants that have been successful are iron hydroxide, aluminum hydroxide as well as polymer focculants such as polyacrylanide, polydmdaac, polyamine, and others. In all cases, the purpose is to decrease the thickness of the filter cake and/or increase its permeability.

Figure 2.62 Dewatering of contaminated harbor bottom sediment using geotextile tubes (upper photograph) and the subsequent filter cake on the inside of the fabric (lower photograph)

When the infilled sediment or sludge has hazardous chemical contained within it, however, the situation calls for additional additives. The goal in this regard is to either attach the contaminants to the infilled soil particles or embed them within the filter cake. Whatever the case, the pollutants must not become fugitive in the effluent. Figure 2.63 shows the concept graphically wherein high soluability and low sorption pollutants are the most troublesome in this regard. Huang and Koerner [146] have selected activated carbon and charcoal for immobilizing organic pollutants, and phosphoric rock for precipitating heavy metals. Their preliminary design indicates that adding a small fraction (0.5 wt%) of charcoal can cause an order of magnitude reduction of aqueous phase concentrations for many EPA-prioritized organic pollutants, and that a small quantity of phosphoric rock added can reduce aqueous phase concentrations of lead, zinc, copper, etc., to their respective EPA drinking water standards. Work is ongoing in this regard.

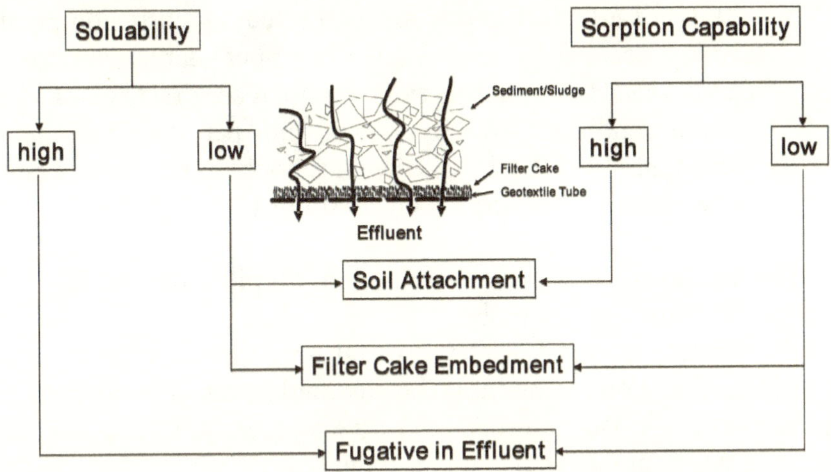

Figure 2.63 Alternative pathways regarding decontamination of polluted sediments and sludges using geotextile tubes; Huang and Koerner [146]

Restoration of Piles (Pile Jacketing). All piles in a marine environment suffer deterioration at varying rates. The deterioration is caused by normal marine exposure, wet-dry cycles, freeze-thaw cycles, and chemical, industrial, and sanitary wastes. Moreover, each type of pile has its own particular problems.

1. Wood piles in a marine environment can be subject to attack by bores. There are three basic types: teredo and bankia (both mollusks), and the arthropod limnoria. Although the limnoria is a surface eroder and the mollusks are internal borers, it is extremely difficult to detect damage to wood piles by any of them by visual inspection until serious deterioration has occurred.
2. Concrete piles are subject to deterioration caused by the wet-dry, freeze-thaw cycles in the water and splash fluctuation zone. Some concrete piles (both precast and cast-in-place) deteriorate below the mudline because of poor original placement techniques. This comes about from permeable or cracked concrete, which allows corrosion of the reinforcing steel and subsequent expansion and spalling of the concrete. It is also sometimes caused by a sulfite reaction of the concrete.

3. All types of steel pipes are subjected to corrosion, with average corrosion rates being 0.13 mm per year under normal conditions. However, this value can increase drastically under certain conditions. Values of 0.36 mm per year have been measured [147]. Still higher values could conceivably result under certain stress corrosion conditions.

The methods used for rehabilitation of piles are varied and constantly increasing in number.

The oldest technique is to use metal forms, such as corrugated steel in half sections joined together by angles, attached to each of the half sections. Dockworkers and divers place, join, and seal the sections, which are used to contain cement grout, which bonds to the deteriorated pile section. This highly labor-intensive operation has led to the development of other, more economical techniques. Rigid plastic forms have been proposed as not only an economical form but as high-strength envelope to prevent further deterioration to the piling system. The annular space between the form and pile can be filled with either concrete grout or specially formulated epoxy. Another pile jacketing technique utilizes bituminized fiber forms. However, all these systems have a problem with bottom closure and sealing when they are acting as a form, and the pile is not to be jacketed down to the firm subsoils. Also, complex configurations are all but impossible to form when using rigid enclosures.

In the 1960s, a technique was developed that utilizes geotextiles as a concrete-forming system [141]. Basically, this concept uses a geotextile jacket as a concrete form, the lengthwise seam of which is joined by a heavy industrial zipper prefabricated onto the geotextile. The ends of the geotextile above and below the deteriorated pile zone are banded to the pile. These flexible geotextile forms possess economic advantages over other concrete-forming systems because of their light weight, ease of installation, adaptability to any configuration, relatively low cost, and ease of connection onto the piles at any location below or above the mudline. The geotextile is so designed that when concrete is injected into it, the excess water bleeds through the voids of the geotextile without allowing the cementatious portion to escape. This lowering of the water/cement ratio produces a dense surface of concrete to resist further deterioration of the pile. Typical installation procedures are illustrated in figure 2.64.

(a) Cleaning pile and placing reinforcement

(b) Banding geotextile to lower, sound pile

(c) Preparing geotextile for filling

(d) Completed pile restoration

Figure 2.64 Procedure for using geotextiles as flexible forms in pile jacketing rehabilitation work.

The design procedure for this application is similar to other cases of using geotextiles as flexible-forming systems and will be described later. One departure worth noting, however, is that elongation of the geotextile under load should be kept to a minimum. Thus, high-strength woven geotextiles are often used. As with other situations where grout is being pumped, high-strength seams comparable to the required design strength are necessary.

Bridge Pier Underpinning. An estimate of foundation soil scour depth beneath shallow foundation bridge piers in rivers and streams is extremely difficult [148]. Even for bridge piers founded on rock, the rock often deteriorates and is scoured away during floods that are accompanied by high-velocity water situations. The problem is so severe that divers sometimes find that they can swim beneath the bridge pier itself [141]. (One has to wonder about the structural factor of safety under such circumstances!)

In a related problem, the Ambursen Hydraulic Co. constructed a number of hollow-core, reinforced concrete slab-and-buttress dams throughout the eastern and midwest USA from 1910 until 1940. The dams consisted of flat concrete slabs at 45°, which were supported by vertical buttresses at 3.0 to 4.5 m spacing. Both the slabs and buttresses were relatively thin (e.g., 300 to 600 mm of reinforced concrete). Today, many of these dams are in need of repair, particularly at their buttress footing regions where compressive stresses are the highest.

Using fabrics as flexible forms, some very clever solutions to these difficult problems have been developed. Shown in figure 2.65 is a solution used by Welsh [149] for a number of scoured bridge piers. A geotextile tube is prefabricated to fit around the perimeter of the pier between the top of the stable foundation material and the bottom of the pier foundation. As grout inflation of the geotextile proceeds, pipes are placed to communicate from the outside of the pier to within the enclosure. After curing of the perimeter tube, injection of

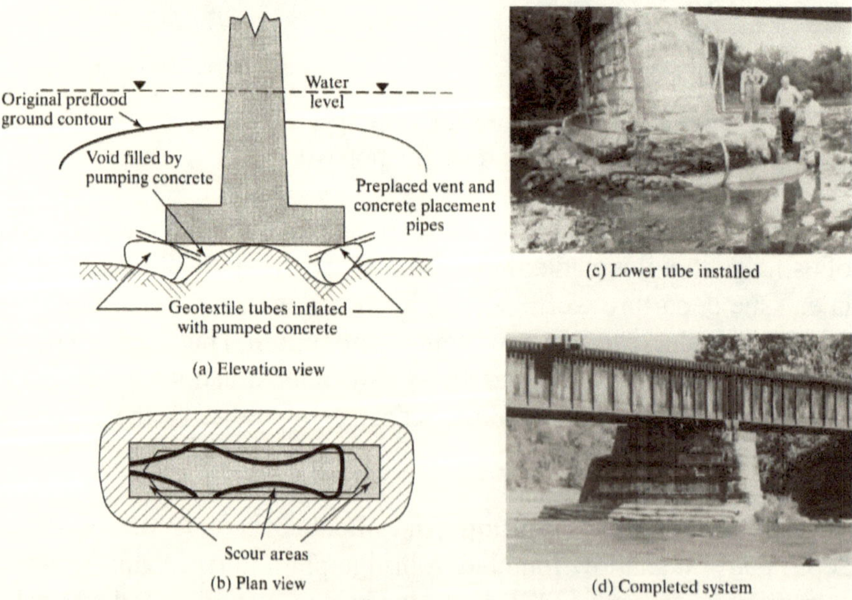

Figure 2.65 Underpinning of scoured bridge pier using grout-filled geotextile forms. (After Welsh [149])

high-strength grout into the inside of the perimeter tube reestablishes the bearing capacity of the pier foundation. The previously installed pipes serve the dual function of allowing grout to enter the enclosure and allowing entrapped water to be displaced. The concrete cures as does typical tremie concrete placed under water.

Erosion Control Mattresses. By taking two sheets of geotextile and joining them at discrete points, a form will result that can be pumped with grout to field-fabricate a mattress that will conform to essentially any subsoil shape or grade. The thickness and geometry are controlled by internal spacer threads woven between the upper and lower sheets of geotextile. Thicknesses of up to 500 mm have been made with various configurations. Such mattresses must have firm soil subgrades (after curing they have no flexibility) and essentially no water to dissipate (most of the surface is covered). The articulating block mattresses shown in figure 2.49b have significant advantages in this regard.

Design Procedures for Flexible Forming Systems. As shown in the preceding subsections, there are many applications for soil, grout or concrete-filled geotextile forms. There are subtle differences between the designs, yet the basic design philosophy is quite similar. Regarding the geotextile design, the following points must be considered:

- Sufficient permeability or permittivity must be available to allow removal of water either within the forms or from within concrete or grout during its curing.
- Proper opening size of the geotextile must be present to prevent excessive loss of the soil, cement or grout.
- Adequate strength must be available to prevent rupture of the geotextile and its seams particularly under high-pressure pumping.
- Adequate elongation must be available to fulfill the specific need (e.g., columns for mine or cavern stability) without failure of the geotextile.

Example 2.32 illustrates several of the above points.

Example 2.32:

Design a geotextile for pile jacketing in 3 m of water as shown in the diagram below:

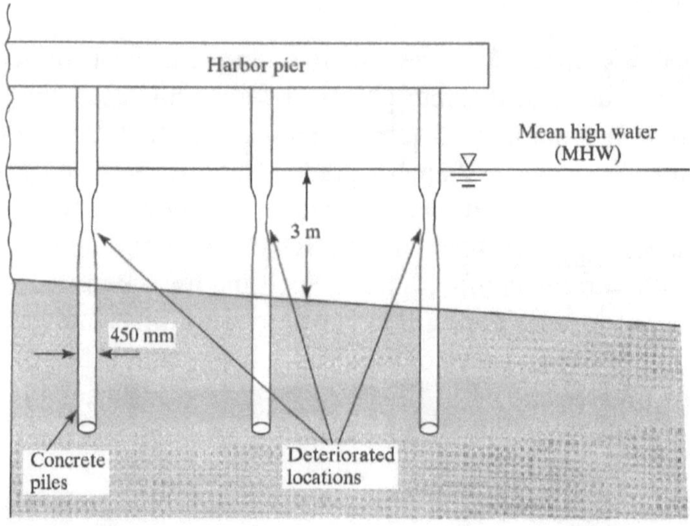

Solution: The three most important design issues follows: (a) For adequate permeability/permittivity, an estimate of the flow of water out of the geotextile during grouting is necessary. Using an estimated value of 0.2 m^3/min, you can obtain the following:

$$q = kiA = \frac{k\Delta hA}{t}$$

$$\psi = \frac{k}{t} = \frac{q}{\Delta hA}$$

$$= \frac{0.2}{(3.0)\pi(0.45)(3.0)(60)}$$

$$\psi = 2.6 \times 10^{-4} \text{ sec}^{-1}$$

Since most geotextiles are in the range of 10^{-1} to 10^{-3} sec^{-1} for their actual permittivity, there are many commercial geotextiles available.

(b) For retention of grout mix, assume that d_{50} of sand is 0.4 mm and that it has a $CU = 4.0$; use

$$0_{95} < 9(d_{50})/CU$$
$$< \frac{(9)(0.40)}{4.0}$$
$$0_{95} < 0.9\,\text{mm}$$

Most geotextiles are in the AOS range of 0.1 to 0.5 mm, thus a wide range is available. If the cement paste is to be retained, the design is straightforward but will result in a geotextile of much smaller openings.

(c) For strength and elongation, the stress-versus-strain curve of the candidate geotextile must be known so as to resist the grouting pressure and required elongation. This is very much a site-specific situation. Note that woven geotextiles with grab strength of 1000 N and elongation at failure of less than 15% are often used for geotextile flexible forms of this type.

Commentary. To be sure, the use of geotextiles as flexible forming systems is an exciting concept simply waiting for new and innovative applications. Contrasted to other geotextile uses, the geotextile is sacrificial in most cases (e.g., when grout or concrete is placed within the geotextile form). Thus, UV degradation is no problem in these cases. For sand-filled bags, containers, and tubes, however, it is a definite problem that must be considered during polymer selection.

The design of the geotextile follows nicely along lines of other geotextile systems. Strength and elongation considerations are invariably necessary along with proper filtration; it is a multiple-function situation. Thus, the topic is being considered in this section. A good deal of future activity will undoubtedly be seen in this application of geotextiles, particularly in the areas of geotextile bags, containers, and tubes.

2.11 CONSTRUCTION METHODS AND TECHNIQUES USING GEOTEXTILES

2.11.1 Introduction

Although this book is devoted primarily to design issues, it is critically important to consider the constructability of the final design. All too often adequate designs have been negated by the inability to construct them or by use of improper construction methods. Of course, either situation can be disastrous as far as the final system is concerned.

Construction with geotextiles is not particularly difficult as long as it is remembered that the textile product being dealt with typically has a mass per unit area from 150 to 600 g/m^2—that is, it is not a steel-wire blasting mat! Most building contractors, heavy-construction contractors, land developers, and federal, state, and local construction forces that deal with other types of construction materials are well-equipped for handling geotextiles. In fact, it is most interesting to note the adaptability of these groups in devising new and clever geotextile deployment and installation procedures. Freely available literature by manufacturers is also extremely helpful in this regard. There are, however, certain areas where new and unique techniques are required, and these have been discussed in their specific sections. Most important are the reinforcement area (walls, embankments, and foundations), the reflective crack prevention area, and flexible forming systems.

One area that does require constant vigilance is that of ultraviolet UV-light susceptibility. Contractors often fail to recognize that geotextiles can be literally destroyed by exposing them to prolonged sunlight, especially in southern climates. The work of Koerner et al. [57], originally commented on in section 2.3.6, is reemphasized here to bring attention to this susceptibility. In some cases more than 50% of the strength and elongation was lost within a few months. Clearly, there is the utmost need for the contractor to keep the geotextile in its protective plastic cover as long as possible and even perhaps to keep it in an enclosure. Once the roll is opened and the geotextile is placed in its final position, it must be backfilled *in a timely manner*. Unused portions of rolls or sampled rolls must be rerolled and suitably protected. The specification must be clear and the inspection rigid in this regard. There are numerous references available, e.g., Hsuan et al. [150], as well as agency guidelines. The latter guidelines are often two weeks for polypropylene and four weeks for polyester geotextiles as being maximum length of exposure unless the geotextile is of a unique type and/or especially stabilized. When the geotextile is UV-stabilized (generally using carbon black and/or a chemical inhibitor in the polymer mixture) UV durability is increased, but certainly not indefinitely.

2.11.2 Geotextile Installation Survivability

Geotextile survivability refers to the ability of the geotextile to withstand handling, installation, and backfilling stresses. It is related to construction equipment, construction technique, substrate material, substrate condition, backfill material, backfill size and shape, and so on. Recall the levels of installation damage shown in figure 2.18 and discussed in section 2.3.5. Table 2.18 considers these features and rates general geotextile requirements of survivability in categories of low, moderate, high, very high, or not recommended. Depending on, and keyed into, these categories are a set of survivability requirements that are considered to be the minimum geotextile properties for necessary placement in the intended and final position. The numeric values of these survivability properties are given in table 2.2a. The classes corresponding to the requirements in table 2.18 are as follows:

- Class 1 (highest values) $\simeq$ "very high" to "high" survivability
- Class 2 (intermediate values) $\simeq$ "high" to "moderate" survivability
- Class 3 (lowest) values $\simeq$ "moderate" to "low" survivability

While a class 0 is not provided in table 2.2a, its values would be proportionately higher than class 1. It should be emphasized that if the values in table 2.18 exceed those calculated on the basis of functional design (as they sometimes will), the values in the table must be used. Thus, calculated design properties do not always prevail.

2.11.3 Cost and Sustainability Considerations

Of prime importance to all involved with the candidate geotextile is its installed cost. Although this has been noticeably absent in the book because of changing price indices, site and climate variations, type and quantity of geotextile, and so on, a few comments are in order.

TABLE 2.18 REQUIRED DEGREE OF SURVIVABILITY AS A FUNCTION OF SUBGRADE CONDITIONS AND CONSTRUCTION EQUIPMENT*

Subgrade Conditions	Low ground-pressure equipment (≤ 25 kPa)	Medium ground-pressure equipment (> 25 kPa, ≤ 50 kPa	High ground-pressure equipment (> 50 kPa)
Subgrade has been cleared of all obstacles except grass, weeds, leaves and fine wood debris. Surface is smooth and level so that any shallow depressions and humps do not exceed 450 mm in depth or height. All larger depressions are filled. Alternatively, a smooth working table may be placed.	Low (Class-3)	Moderate (Class-2)	High (Class-1)
Subgrade has been cleared of obstacles larger than small to moderate-sized tree limbs and rocks. Tree trucks and stumps would be removed or covered with a partial working table. Depressions and humps should not exceed 450 mm in depth or height. Larger depressions should be filled.	Moderate (Class-2)	High (Class-1)	Very high (Class-0)
Minimal site preparation is required. Trees may be felled, delimbed, and left in place. Stumps should be cut to project not more than ± 150 mm above subgrade. Fabric may be draped directly over the tree trunks, stumps, large depressions and humps, holes, stream channels, and large boulders. Items should be removed only if placing the fabric and cover material over them will distort the finished road surface.	High (Class-1)	Very High (Class-0)	Not recommended

*Recommendations are for 150 to 300 mm initial lift thickness. For other initial lift thicknesses:
300 to 450 mm: reduce survivability requirement one level;
450 to 600 mm: reduce survivability requirement two levels;
>600 mm: reduce survivability requirement three levels
For special construction techniques such as prerutting, increase the fabric survivability requirement one level.
Placement of excessive initial cover material thickness may cause bearing failure of the soft subgrade.
Source: After Christopher, Holtz, and DiMaggio [152]

The cost of the geotextile itself is reasonably related to its mass per unit area. Heavier geotextiles cost proportionately more than lighter ones. Note, however, that the installation cost may not be significantly higher for the heavier geotextiles. The type of manufacture is also a factor, with woven slit-film types generally being the least expensive, then nonwoven heat-bonded and nonwoven needle-punched, and then woven monofilament types, which are the most expensive on the basis of an equivalent mass per unit area. These comments, however, should in no way sway a design toward preference of one geotextile over another. They are offered only to give a feeling for the costs involved. As of this writing, these costs ranged from $0.60 to $2.00 per square meter for geotextiles in the range 150 to 500 g/m^2, with installation costs being an additional $0.10 to $0.25 per square meter depending on the site conditions, quantity involved, and particular application.

The issue of sustainability, as measured by the carbon footprint of various alternative solutions, is important and becoming more so as time progresses. For example, the replacement of 300 mm thick sand filter by a geotextile filter will reduce the amount of CO_2 generated by 60 to 80% depending on site-specific conditions. An entire conference has been devoted toward comparing natural soil versus geosynthetic alternatives in this regard, Koerner et al. [151]. It is quite possible that designers will have to go beyond cost and benefit/cost proposals for projects and provide carbon footprint information as well.

2.11.4 Summary

As with all construction materials, failures sometimes occur. In my personal investigations of geotextile-related failures (also recall table 2.16), they primarily fall into the following groups:

1. *Construction related.* This is the largest group of failures with excessive ultraviolet degradation, installation damage, poorly constructed seams, and the lack of intimate contact being the major problems. This last issue is particularly important for filtration and erosion-control applications.
2. *Design/specification related.* Design failures per se have been relatively few. Some embankments on soft soils have had excessive deformations, but they have generally been repaired on-site without insurance or litigation problems. Retaining walls backfilled with low-permeability soils have been problematic (excessive deformations and collapse) particularly when adequate drainage has not been provided. Specifications have often been lax and, in many instances, contributed to some of the construction related failures mentioned before.
3. *Testing related.* There have been a number of testing-related problems, but no serious failures to my knowledge. The major concern in this regard is the use of literature values for interface shear strengths, instead of properly simulated direct shear tests. This practice must cease before major problems arise.

In spite of the concern over the lack of true performance tests, conservative reduction factors probably compensate for the shortcoming. In this regard, a designer should never apologize for using high-reduction factors or high factors of safety.

4. *Product related.* Other than supplying the wrong product, and then accepting and installing same, product failures are sparse. The use of the MARV concept (but only for geotextiles) is certainly warranted and provides the needed safeguard against product variability. It also challenges the manufacturer to decrease product variability to the maximum extent possible.

While geotextile failures have indeed occurred, their number is small in light of the number of installations. We certainly have an engineering material capable of being considered in a comparable manner with other materials conventionally used by civil engineers and related professions. Field performance to date, after more than thirty-five years of service life in a multitude of applications, has been excellent.

REFERENCES

1. Broms, B. B., "Triaxial Tests with Fabric-Reinforced Soil," *C. R. Coll. Inst. Soils Text.* Paris, Vol. 3, 1977, pp. 129-133.
2. Taylor, D. W., *Fundamentals of Soil Mechanics*, John Wiley & Sons Inc., 1948.
3. Christopher, B. R. and Fischer, G. R., "Geotextile Filtration Principles, Practices and Problems," *Jour. of Geotextiles and Geomembranes*, Elsevier, Vol. 11, Nos. 4-6, 1992, pp. 337-354.
4. *Report on Task Force 25*, Joint Committee Report of AASHTO-AGC-ARTBA, American Association of State, Highway and Transportation Officials, Washington, DC, January 1991.
5. Carroll, R. G., Jr., *Geotextile Filter Criteria*, TRR 916, Engineering Fabrics in Transportation Construction, Washington, DC, 1983, pp. 46-53.
6. Luettich, S. M., Giroud, J. P. and Bachus, R. C., "Geotextile Filter Design Guide," Jour. of Geotextiles and Geomembranes, Vol. 11, No. 4-6, 1992, pp. 19-34.
7. Haliburton, T. A. and Wood, P. D., "Evaluation of US Army Corps of Engineers Gradient Ratio Test for Geotextile Performance,"

Proc. 2nd Int. Conf. on Geotextiles, Aug. 1-6, 1982, IFAI, pp. 97-101.

8. Halse, Y., Koerner, R. M. and Lord, A. E. Jr., "Filtration Properties of Geotextiles Under Long Term Testing," Proc. ASCE/PennDOT Conf. on Advances in Geotechnical Engineering, Hershey, PA, Apr. 1987, pp. 1-13.
9. Williams, N. D. and Abouzakhm, M. A., "Evaluation of Geotextile/Soil Filtration Characteristics Using the Hydraulic Conductivity Ratio Analysis," *Jour. Geotex. and Geomemb.*, Elsevier, Vol. 8, No. 1, 1989, pp. 1-26.
10. Koerner, R. M. and Koerner, G. R., "Geotextile Filter Failures: Examples and Lessons Learned," *Proc. 25th Central Pennsylvania Geotechnical Conference*, Hershey, PA, 2011, (on CD).
11. McGown, A., "The Properties of Nonwoven Fabrics Presently Identified as Being Important in Public Works Applications," *Index 78 Programme*, University of Strathclyde, Glasgow, Scotland, 1978.
12. Heerten, G., "A Contribution to the Improvement of Dimensioning Analogies for Grain Filters and Geotextile Filters," *Proc. Intl. Conf. on Filters, Filtration and Related Phenomena*, Karlsruhe, Germany, 1992, pp. 110-122.
13. Giroud, J. P., "Granular Filters and Geotextile Filters," *Proc. Geofilters '96*, J. LaFleur and A. Rollin, Eds., Montreal, Canada, 1996, pp. 565-680.
14. Maiser, M. and Myles, B., "Possible Culpability of Filter Geotextile in the Failure of a Sea Wall," Proc. 1st Pan American Geosynthetics Conference, Cancun, Mexico, 2008, pp 833-841.
15. Gerry, G. S. and Raymond, G. P., "The In-Plane Permeability of Geotextiles," *Geotech. Test. J.*, ASTM, Vol. 6, No. 4, Dec. 1963, pp. 181-189.
16. Kaswell, E. R., *Handbook of Industrial Textiles*, West Point Peperell, New York, 1963.
17. Booth, J. E., *Principles of Textile Testing*, Newnes-Butterworths Publ., London, 1968.
18. Morton, W. E. and Heart, J. W., *Physical Properties of Textile Fibers*, John Wiley & Sons, Publ., 1975.
19. Haliburton, T. A., Fowler, J. and Langan, J. P., "Design and Construction of a Fabric Reinforced Test Section at Pinto Pass, Mobile, Alabama," *Trans. Res. Record #79*, Washington, DC 1980.

20. Myles, B. and Carswell, I., "Tensile Testing of Geotextiles," *Proc. 3rd Intl. Conf. Geotextiles*, IFAI Publ., 1986, pp. 713-718.
21. McGown, A., Andrawes, K. Z. and Kabir, M. H., "Load-Extension Testing of Geotextiles Confined in Soil," *Proc. 2nd Intl. Conf. Geotextiles*, Aug. 1-6, 1982, IFAI Publ., pp. 793-796.
22. McGown, A., Andrawes, K. Z. and Murray, R. T., "The Load-Strain-Time-Temperature Behavior of Geotextiles and Geogrids," *Proc. 3rd Intl. Conf. Geotextiles*, IFAI Publ., 1986, pp. 707-712.
23. Wilson-Fahmy, R. F., Koerner, R. M. and Fleck, J. A., "Unconfined and Confined Wide Width Testing of Geosynthetics Used in Reinforcement Applications, S. C. J. Cheng, Ed., ASTM STP 1190, 1993, pp. 49-63.
24. Wayne, M. H., Carey, J. E. and Koerner, R. M., "Epoxy Bonding of Geotextiles," *Jour. Geotex. and Geomemb.*, Vol. 9, No. 4-6, 1990, pp. 559-564.
25. Ashmawy, A. K. and Bourdeau, P. L., "Geosynthetic Reinforced Soils Under Repeated Loadings: A Review and Comparative Study," *Geosynthetics International*, Vol. 2, No. 4, 1995, pp. 643-678.
26. Raumann, G., "A Hydraulic Tensile Test with Zero Transverse Strain for Geotechnical Fabrics," *Geotech. Test. J., ASTM*, Vol. 2, No. 2, June 1979, pp. 69-76.
27. Andrejack, T. L. and Wartman, J., "Development of a Multi-Axial Test for Geotextiles," *Proc. GeoAmerican 2008 Conf.*, Cancun, Mexico, 2008, (on CD).
28. Koerner, R. M., Monteleone, M. J., Schmidt, R. K. and Roethe, A. T., "Puncture and Impact Resistances of Geosynthetics," *Proc. 3rd Intl. Conf. on Geotextiles*, IFAI Publ., 1986, pp. 677-682.
29. Alfheim, S. L. and Sorlie, A., "Testing and Classification of Fabrics for Application in Road Construction," *C. R. Coll. Int. Soils Tex.*, Paris, Vol. 2, 1977, pp. 333-338.
30. Murphy, V. P. and Koerner, R. M., "CBR Strength (Puncture) of Geosynthetics," *J. Geotech. Engr. Testing, ASTM*, Vol. 11, No. 3, Sept. 1988, pp. 167-172.
31. Cazzuffi, D. and Venesia, S., "The Mechanical Properties of Geotextiles: Italian Standard and Interlaboratory Test Comparison," *Proc. 3rd Int. Conf. Geotextiles*, IFAI Publ., 1986.

32. Ingold, T. S., "Some Observations on the Laboratory Measurement of Soil-Geotextile Bond," Geotech. Test. Jour., GTJODT, ASTM, Vol. 5, No. 3/4, 1982, pp. 57-67.
33. Koerner, R. M., "A Recommendation to Use Peak Shear Strengths for Geosynthetic Interface Design," GFR, Vol. 21, No. 3, April 2003, pp. 28-30 [including Letters-to-the-Editor, GFR, Vol. 21, No. 6, August 2003, pp. 14-15].
34. Marr, W. A. and Christopher, B., "Recommended Design Strength for Needlepunched GCL Products," GFR, Vol. 21, No. 8, Oct./Nov. 2003, pp. 18-23.
35. Martin, J. P., Koerner, R. M., and Whitty, J. E., "Experimental Friction Evaluation of Slippage between Geomembranes, Geotextiles and Soils," *Proc. Int. Conf. Geomembranes*, June 20-24, 1984, IFAI Publ., pp. 191-196.
36. Collios, A., Delmas, P., Gourc, J. P. and Giroud, J. P., "Experiments of Soil Reinforcement with Geotextiles," *Proc. Symp. Use of Geotextiles for Soil Improvement*, ASCE, Apr. 14-18, 1980, pp. 53-73.
37. Gourc, J. P., Faure, Y., Rollin, A. and LeFleur, J., "Standard Tests of Permittivity and Application of Darcy's Formula," *Proc. 2nd Int. Conf. Geotextiles*, Aug. 1, 1982, IFAI Publ., Vol. 1, pp. 149-154.
38. Holtz, R. D., *Mercury Intrusion Characterization of Geotextile Pore Size Distribution*, Report to Hoechst-Celanese Corp., Spartanburg, SC, 1988.
39. Bhatia, S. K., Smith, J. L. and Christopher, B. R., "Geotextile Characterization and Pore Size Distribution: Part III. Comparison of Methods and Applications to Design," *Geosynthetics International*, 1996, Vol. 3, No. 3, pp. 301-328.
40. "Designing with Terram," Design Brochure from ICI Fibres Ltd., Gwent, Great Britain.
41. McGown, A. W., "The Properties and Uses of Permeable Fabric Membranes," Residential Workshop on Materials and Methods for Low Cost Roads and Reclamation Works," Leura, Australia, 1976, pp. 663-710.
42. Dierickx, W. and Myles, B., "Wet Sieving as a European EN-Standard for Determining the Characteristic Opening Size of Geotextiles," *Recent Developments in Geotextile Filters and Prefabricated Drainage Geocomposites*, ASTM STP 1281,

Shobha K. Bhatia and L. David Suits, Eds., American Society for Testing and Materials, 1996, pp. 54-64.

43. Hausmann, M., *Engineering Principles of Ground Engineering*, McGraw-Hill Publishing Co., 1991, 632 pgs.
44. Koerner, R. M. and Bove, J. A., "In-Plane Hydraulic Properties of Geotextiles," *Geotech. Test. J.*, ASTM, Vol. 6, No. 4, 1983, pp. 190-195.
45. Koerner, R. M., Bove, J. A. and Martin, J. P., "Water and Air Transmissivity of Geotextiles," *J. Geotextiles and Geomembranes*, Vol. 1, 1984, pp. 57-73.
46. Koerner, G. R. and Koerner, R. M., "The Installation Survivability of Geotextiles and Geogrids," *Proc. 4th IGS Conference on Geotextiles, Geomembranes and Related Products*, The Hague, Balkema, Rotterdam, 1990, pp. 597-602.
47. Shrestha, S. C. and Bell, J. R., "Creep Behavior of Geotextiles Under Sustained Loads," *Proc. 2nd Int. Conf. Geotextiles*, Aug. 1-6, 1982, IFAI Publ., pp. 769-774.
48. den Hoedt, G., "Creep and Relaxation of Geotextile Fabrics," *Jour. Geotextiles and Geomembranes*, Vol. 4, No. 2, 1986, pp. 83-92.
49. Thornton, J. S. and Baker, T. L., "Comparison of SIM and Conventional Methods for Determining Creep-Rupture Behavior of a Polypropylene Geotextile," *Proc. 7th ICG*, Nice, France, A. A. Balkema Publ., 2003 pp. 1545-1552.
50. Koerner, R.M., Hsuan, Y. and Lord, A. E., Jr., "Remaining Technical Barriers to Obtain General Acceptance of Geosynthetics," Inaugural Mercer Lecture, *Jour. of Geotextiles and Geomembranes*, Vol. 12, No. 1, 1993, pp. 1-52.
51. Ingold, T. S., Montanelli, F. and Rimoldi, P., "Extrapolation Techniques for Long Term Strengths of Polymeric Geogrids," Proc. IGS 5, Singapore, 1994, pp. 1117-1120.
52. Koerner, R. M. and Ko, F. K., "Laboratory Studies on Long-Term Drainage Capability of Geotextiles," *Proc. 2nd Int. Conf. Geotextiles*, Aug. 1-6, 1982, IFAI Publ., pp. 91-95.
53. Wayne, M. H. and Koerner, R. M., "Correlation Between Long Term Flow Testing and Current Geotextile Filtration Design Practice," *Proc. Geosynthetics '93*, IFAI Publ., 1993, pp. 501-517.
54. Williams, N. D. and Abouzakhm, M. A., "Evaluation of Geotextile/Soil Filtration Characteristics Using the Hydraulic

Conductivity Ratio Analysis," *Jour. Geotex. and Geomemb.*, Vol. 8, No. 1, 1989, pp. 1-26.

55. Luettich, S. M. and Williams, N. D., "Design of Vertical Drains Using the Hydraulic Conductivity Ratio Analysis," *Proc. Conf. on Geosynthetics*, IFAI Publ., 1989, pp. 95-103.
56. Van Zanten, R. V. ed., *Geotextiles and Geomembranes in Civil Engineering*, Balkema, Rotterdam, 1986, 658 pgs.
57. Koerner, G. R., Hsuan, Y. and Koerner, R. M., "Photo-Initiated Degradation of Geotextiles," *Jour. Geotechnical and Geoenvironmental Engineering*, ASCE, Vol. 124, No. 12, December 1998, pp. 1159-1166.
58. Hsuan, Y. G., Koerner, R. M. and Lord, A. E., Jr., "A Review of the Degradation of Geosynthetic Reinforcement of Materials and Various Polymer Stabilization Methods," ASTM STP 1190, S.C. J. Cheng, Ed., ASTM, 1993, pp. 228-244.
59. Halse, Y., Koerner, R. M. and Lord, A. E., Jr., "Effect of High Alkalinity Levels on Geotextiles—Part I $Ca(OH)_2$ Solutions," *Jour. Geotextiles and Geomembranes*, Elsevier Publ. Co., Vol. 5 No. 4, 1987, pp. 261-282, and Part II NaOH Solutions," *Jour. Geotextiles and Geomembranes*, Vol. 6, No. 4, 1987, pp. 295-305.
60. Hsieh, C. W., Lin, C. K. and Chiu, Y. F., "The Strength Properties of Geotextiles in Ocean Environments," *Proc. EuroGeo3*, Munich, Germany, 2004, pp. 377-382.
61. Koerner, G. R. and Koerner, R. M., "Leachate Flow Rate Behavior Through Geotextile and Soil Filters and Possible Remediation Methods," *Jour. Geotextiles and Geomembranes*, Vol. 11, No. 4-6, 1992, pp. 401-430.
62. Koerner, G. R., Koerner, R. M. and Martin, J. P., "Geotextile Filters Used for Leachate Collection Systems: Testing, Design of Field Behavior," *Jour. Geotechnical Engr. Div.*, ASCE, Vol. 120, No. 10, Oct. 1994, pp. 1792-1803.
63. US Nuclear Regulatory Commission (NRC) Workshop Proceedings, "Engineered Barrier Performance Related to Low Level Radioactive Waste, Decommissioning, and Uranium Mill Tailings Facilities," Aug. 2-5, 2010, 148 pgs.
64. Gourc, J.-P. and Faure, Y.-H., "Soil Particles, Water and Fibers—A Fruitful Interaction Now Controlled," *Proc. 4th IGS Conf.*, The Hague, The Netherlands, 1990, pp. 949-972.
65. Geosynthetics Magazine, IFAI, Roseville, MN (a monthly publication).

66. Voskamp, W. and Risseeuw, P., "Method to Establish the Maximum Allowable Load Under Working Conditions of Polyester Reinforcing Fabrics," *Jour. Geotex. and Geomemb.*, Vol. 6, 1988, pp. 173-184.
67. GSI White Paper #4, "Reduction Factors (RF's) Used in Geosynthetic Design," GSI, Folsom, PA, 2005, 14 pgs.
68. Giroud, J. P., "Designing with Geotextiles," *Mater. Const.* (Paris), Vol. 14, No. 82, 1981, pp. 257-272, and *Geotextiles and Geomembranes, Definitions, Properties and Designs*, IFAI Publ., 1984.
69. Koerner, G. R., "Long-Term Benefit Cost Performance and Analysis of Geotextile Separators in Pavement Systems," *Proc. Geosynthetics '97*, IFAI Publ., 1997, pp. 701-713.
70. Suits, L. D. and Koerner, G. R., "Site Evaluation/Performance of Separation Geotextiles," *Proc. Geosynthetics '01 Conference*, IFAI Publ., 2001, pp. 451-468.
71. Hausmann, M. R., "Fabric Reinforced Unpaved Road Design Methods - Parametric Studies," *Proc. 3rd Intl. Conf. Geotext.*, Vienna, Austria, 1986, IFAI, pp. 19-24.
72. Giroud, J. P. and Noiray, L., "Design of Geotextile Reinforced Unpaved Roads," *Jour. Geotech. Eng. Div.*, ASCE, Vol. 107, No. GT9, Sept. 1981, pp. 1233-1254.
73. Barenberg, E. J. and Bender, D. A., "Design and Behavior of Soil-Fabric-Aggregate Systems," 57th Transp. Research Board Meeting, Washington, DC, June 1978.
74. Holtz, R. D. and Sivakugan, K. "Design Charts for Roads with Geotextiles," *Jour. Geotextiles and Geomembranes*,Vol. 5, No. 3, 1987, pp. 191-200.
75. Diaz, V., "Thread Selector for Geotextiles," *Geotechnical Fabrics Rep.*, Vol. 3, No. 1, Jan.-Feb., 1985, IFAI Publ., pp. 15-19.
76. —, *Field Seaming of Geotextiles*, IFAI Publ., 1989, 29 pgs.
77. Perkins, S. W. et al., "Geosynthetics in Pavement Reinforcement Applications," *Proc. 9th ICG*, Guaruja, Brazil, 2010, pp. 115-164.
78. Hoffman, G. L. and Shamon, M. E., *Premature Failure of Permeable Subbase Pavement Sections Incorporating Geotextiles*, Paper 6, PennDOT, Harrisburg, PA, Apr. 5-6, 1984.

78. Broms, B. B., "Design of Fabric Reinforced Retaining Structures," *Proc. Symp. Earth Reinforcement,* ASCE, 1978, Pittsburgh, PA, pp. 282-303.
80. Steward, J. E., Williamson, R., and Mohney, J., "Earth Reinforcement," chapter 5 in *Guidelines for Use of Fabrics in Construction and Maintenance of Low Volume Roads*, US Forest Service, Portland, OR, June 1977.
81. Whitcomb, W. and Bell, J. R., "Analysis Techniques for Low Reinforced Soil Retaining Walls," *Proc. 17th Eng. Geol. Soils Eng. Symp.*, Moscow, ID, Apr. 1979, pp. 35-62.
82. Lee, K. L., Adams, B. D. and Vagneron, J. M. J., "Reinforced Earth Retaining Walls," *J. Soil Mech. Fdtn. Eng. Div.*, ASCE, No. SM10, Oct. 1973, pp. 745-764.
83. Bell, J. R., Stilley, A. N., and Vandre, B., "Fabric Retained Earth Walls," *Proc. 13th Eng. Geol. Soils Eng. Symp.*, Moscow, ID, Apr. 1975.
84. Jarrett, P. W. and McGown, A. (Eds.), *Proc. on Appl. of Polymeric Reinforcement in Soil Retaining Structures*, Royal Military College, Kingston, Ontario, Canada, 1988.
85. Leshchinsky, D., "Geosynthetic Reinforcement: Is it Magic?" Geosynthetics Magazine, Vol. 28, No. 3, June/July, 2010, pp. 16-24.
86. NAVFAC, DM-7.2, Bureau of Yards and Docks, US Navy, Apr. 1982.
87. Broms, B. B., "Polyester Fabric as Reinforcement in Soil," *C. R. Coll. Int. Soil Text.*, Paris, 1977, Vol. 1, pp. 129-135.
88. Koerner, R. M., *Construction and Geotechnical Methods in Foundation Engineering*, McGraw-Hill Book Co., 1984.
89. Barrett, R. K., "Geotextiles in Earth Reinforcement," *Geotechnical Fabrics Report*, Vol. 3, No. 2, March/April 1985, pp. 15-19.
90. Allen, T. M., Christopher, B. R. and Holtz, R. D., "Performance of a 12.6 m High Geotextile Wall in Seattle, Washington," *Proc. Geosynthetic Reinforced Retaining Walls*, J. T.H. Wu, Ed., A. A. Balkema, 1992, pp. 81-100.
91. Stevens, J. B. and Souiedan, B., "Geotextile Wall Aids Bridge Construction," *Geotech. Fabrics Rpt.*, Vol. 8, No. 3, 1990, pp. 10-15.

92. Richardson, G. N. and Behr, L. H. Jr., "Geotextile-Reinforced Wall: Failure and Remedy," *Geotechnical Fabric Report*, Vol. 6, No. 4, 1988, IFAI Publ., pp. 14-18.
93. Koerner, G. R., Koerner, R. M. and Elias, V., "Geosynthetic Installation Damage Under Two Different Backfill Conditions," ASTM STP 1190, S. C. J. Cheng, Ed., ASTM, 1993, pp. 163-184.
94. Kaniraj, S. R., "Direction and Magnitude of Reinforcement Force in Embankments on Soft Soil," Earth Reinforcement, Ochiai,Yasufuku and Omine, Eds., A. A. Balkema Publ., 1996, pp. 221-225.
95. Guler, E. and Dosay, S., "Analysis and Behavior of Geotextile Reinforced Slopes With Different Foundations, Backfills and Inclinations Using Finite Element Method," *4th European Geosynthetics Conf.*, Scotland, 2008, Paper 89.
96. Haliburton, T. A., Fowler, J. and Langan, J. P., *Design and Construction of a Fabric Reinforced Test Section at Pinto Pass, Mobile, Alabama*, Trans. Res. Rec. 79, Washington, DC, 1980.
97. Fowler, J., "Theoretical Design Considerations for Fabric Reinforced Embankments," *Proc. 2nd Intl. Conf. Geotex.*, 1982, IFAI Publ., pp. 665-676.
98. Daniel, D. E. et.al, *The New Orleans Hurricane Protection System: What Went Wrong and Why*, ASCE, 2007, 84 pgs.
99. Sprague, C. J. and Koutsourais, M., "The Evolution of Geotextile Reinforced Embankments," Geotechnical Spec. Publ. 30, Borden, R. H., Holtz, R. D. and Juran, I., Eds., ASCE, 1992, pp. 1129-1141.
100. Koerner, R. M., Hwu, B-L., Wayne, M. H., "Soft Soil Stabilization Designs Using Geosynthetics," *Jour. Geotex. and Geomemb.*, Vol. 6, No. 1-3, 1987, pp. 33-52.
101. Koerner, R. M. and Uibel, B. L., "Hydraulic Fill Embankments Utilizing Geosynthetics," *Proc. ASCE Conf. on Hydraulic Fill Structures '88*, Colorado State Univ., ASCE, 1988.
102. Koerner, R. M., Fowler, J. and Lawrence, C. A., *Soft Soil Stabilization Study for Wilmington Harbor South Dredge Material Disposal Area*, US Army Engineer Waterways Experiment Station, Misc. Paper GL-86-38, Dec. 1986.
103. Humphrey, D. N., Discussion "Current Design Methods," *Jour. Geotex. and Geomemb.*, Vol. 6, No. 1-3, 1987, pp. 89-92.

104. Guglielmetti, J. L., Koerner, G. R. and Battino, F. S., "Geotextile Reinforcement of Soft Landfill Process Sludge to Facilitate Final Closure," *Journal of Geotextiles and Geomembranes*, Vol. 14, Nos. 7/8, July/August 1996, pp. 377-392.
105. British Standard Code of Practice, "Strengthened/Reinforced Soils and Other Fills," BS 8006, 1995, London, England, Section 8, pp. 97-121.
106. Filz, G. M. and Smith, M. E., "Design of Bridging Layers in Geosynthetic-Reinforced, Column-Supported Embankments," Report No. VTRC 06-CR-12, Virginia Transportation Research Council, Charlottesville, Virginia, 2006, 48 p.
107. Han, J. and Akins, K., "Use of Geogrid Reinforced and Pile Supported Earth Structures," *Proc. Intl. Deep Foundations Congress*, ASCE, Feb. 14-16, 2002, pp. 668-679.
108. Alexiew, D., Brokemper, D. and Lothspeich, S., "Geotextiles Encased Columns (GEC): Load Capacity, Geotextile Selection and Pre-Design Graphs," *Proc. GeoFrontiers*, Austin, Texas, 2005, (on CD).
109. Koerner, R. M. and Wong, W.-K., "Geosynthetic Supported Basal Reinforcement Over Deep Foundations," *Proc. 23rd Central PA Geotechnical Conf.*, Hershey, PA, 2008, 26 pgs.
110. Guido, V. A., Biesiadecki, G. L. and Sullivan, M. J., "Bearing Capacity of a Geotextile Reinforced Foundation," *Proc. 11th ISSMFE*, San Francisco, 1985, Vol. 3, pp. 1777-1780.
111. Shen, C. K., Bang, S. and Herrman, L. R., "Ground Movement Analysis of Earth Support System," *J. Geotech. Eng. Div., ASCE*, Vol. 107, No. GT12, Dec. 1981, pp. 1609-1624.
112. Koerner, R. M., "Slope Stabilization Using Anchored Geotextiles: Anchored Spider Netting," *Proc. Spec. Geotech. Eng. for Roads and Bridges Conf.*, PennDOT, Harrisburg, PA, 1984, pp. 1-11.
113. Koerner, R. M., "In-Situ Soil Slope Stabilization Using Anchored Nets," *Proc. Conf. Low Cost and Energy Saving Construction Materials*, Rio de Janeiro, Brazil, July 1984, Envo Publ. Co., Bethlehem, PA, pp. 465-478.
114. Lambe, T. W. and Whitman, R. V., *Soil Mechanics*, J. Wiley and Sons Inc., 1969.
115. Koerner, R. M. and Robins, J. C., "In-Situ Stabilization of Soil Slopes Using Nailed Geosynthetics," *Proc. 3rd Conf. on Geotextiles*, IFAI Publ., 1986, pp. 395-399.

116. Ghiassian, H., Hryciw, R. D. and Gray, D.H., "Laboratory Testing Apparatus for Slopes Stabilized by Anchored Geosynthetics," *Geotechnical Testing Journal*, ASTM, Vol. 19, No. 1, March 1996, pp. 65-73.
117. Ghiassian, H., Gray, D. H. and Hryciw, R. D., "Seepage Considerations and Stability of Sandy Slopes Reinforced by Anchored Geosynthetics," *Proc. Geosynthetics '97*, IFAI Publ., 1997, pp. 581-593.
118. Cedegren, H. R. *Seepage, Drainage and Flow Nets*, John Wiley and Sons Inc., 1967.
119. Bell, J. R. and Hicks, R. G., *Evaluation of Test Methods and Use Criteria for Geotechnical Fabrics in Highway Applications*, Final Report, FHwA, Contract No. DOT-FH-119353, Oregon State Univ., Corvallis, OR, 1984.
120. Richardson, G. R. and Middlebrooks, P., "A Simplified Design Method for Silt Fences," Geosynthetics '91 Conference, IFAI Publ., 1991, pp. 879-885.
121. Koerner, R. M. and Koerner, G. R., "Geotextile Filter Failures: Examples and Lessons Learned," *Proc. 25th Central PA Geotechnical Conference*, Hershey, PA, 2011, 27 pgs.
122. Anderson, O., "The Use of Plastic Fabric for Pavement Protection During Frost Break," *Proc. Int. Conf. Use of Fabrics in Geotechnics*, Paris, Apr. 20-22, 1977, Vol. 1, pp. 143-149.
123. Gamski, K. and Rigo, J.M., "Geotextile Soil Drainage in Siphon or Siphon-Capillarity Conditions," *Proc. 2nd Int. Conf. Geotextiles*, Aug. 1-6, 1982, IFAI Publ., Vol. 1, pp. 145-452.
124. Clough, I. R. and French, W. J., "Laboratory and Field Work Relating to the Use of Geotextiles in Arid Regions," *Proc. 2nd Int. Conf. Geotextiles*, Aug. 1-6, 1982, IFAI Publ., pp. 447-452.
125. Button, J. W., Epps, J. A., Lytton, R. L. and Harmon, W. S., "Fabric Interlayer for Pavement Overlays," *Proc. 2nd Int. Conf. Geotextiles*, Aug. 1-6, 1982, IFAI Publ., pp. 523-528.
126. Murray, C. D., "Simulation Testing of Geotextile Membranes for Reflection Cracking," *Proc. 2nd Int. Conf. Geotextiles*, Aug. 1-6, 1982, IFAI Publ., Vol. 2, pp. 511-516.
127. Majidzadeh, K., Luther, M. S. and Skylut, H., "A Mechanistic Design Procedure for Fabric-Reinforced Pavement Systems," *Proc. 2nd Int. Con. Geotextiles*, Aug. 1-6, 1982, IFAI Publ., pp. 529-534.

128. *Asphalt Overlays and Pavement Rehabilitation*, The Asphalt Institute Manual Series No. 17 (MS-17), College Park, MD, Nov. 1977.
129. Cedergren, H. R., "Seepage, Drainage and Flow Nets," John Wiley & Sons Inc., 1989, New York, NY.
130. Bell, J. R., Jr., "Designing with Geosynthetics," unpublished course notes of R. M. Koerner and J. R. Bell, 1983 to 1985.
131. Personal communication from Maine Department of Transportation, 1994.
132. Ahlrich, R. C., "Evaluation of Asphalt Rubber and Engineering Fabrics as Pavement Interlayers," US Army Corps of Engineers, Vicksburg, MS, GL-86-34, 1986.
133. —, "Reflective Cracking in Pavements" Conference Proceedings, RILEM Intl. Union of Testing and Research Laboratories for Materials and Structures, Paris, France, 1989, 1993 and 1996.
134. Newby, J. E., "Southern Pacific Transportation Co. Utilization of Geotextiles in Railroad Subgrade," *Proc. 2nd Int. Conf. Geotextiles*, Aug. 106, 1982, IFAI Publ., pp. 467-472.
135. Einsenmann, J. and Leykauf, G., "Investigation of a Nonwoven Fabric Membrane in Railway Track Construction," *Proc. Int. Conf. Use of Fabrics in Geotechnics*, Paris, Apr. 20-22, 1977, pp. 41-45.
136. Saxena, S. K. and Wang, S., "Model Test of a Rail-Ballast-Fabric-Soil System," *Proc. 2nd Int. Conf. Geotextiles*, Aug. 1-6, 1982, IFAI Publ., pp. 495-500.
137. Bosserman, B., "Reviewing Geotextiles at FAST," *Railroad Track and Structures*, June 1981, pp. 42-58.
138. Raymond, G., "Geotextiles for Railroad Bed Rehabilitation," *Proc. 2nd Int. Conf. Geotextiles*, Aug. 1-6, 1982, IFAI Publ., pp. 479-484.
139. Chrismer, S. M. and Richardson, G. R., "In-Track Performance of Geotextiles at Caldwell, Texas," Transportation Research Board 1071, Transportation Research Board, Washington, DC, 1986, pp. 72-80.
140. Terzaghi, K. and Lacroix, Y., "Mission Dam: An Earth and Rockfill Dam on a Highly Compressible Foundation," *Geotechnique*, Vol. 14, Mar. 1964, pp. 13-50.

141. Koerner,R. M. and Welsh, J. P., *Construction and Geotechnical Engineering Using Synthetic Fabrics*, John Wiley & Sons Inc., 1980.
142. Leshchinsky, D. and Leshchinsky, O., "Geosynthetic Confined Pressurized Slurry (GeoCoPS)," Technical Report, CPAR-GL-96-1, US Army Corps of Engineers, Washington, DC, September 1996, 45 pgs.
143. Lawson, C., "Geotextile Containment—International Perspectives," *Proceedings GRI-17. Hot Topics in Geosynthetics IV*, GII, Folsom, PA, 2003, pp. 178-201.
144. Pilarczyk, K. W., "Application and Design Aspects of Geocontainers," *Proc. Geosynthetics '97*, IFAI Publ., pp. 147-160.
145. Heibaum, M., Editor, "Special Issue on Geotextile Containers," *Journal of Geotextiles and Geomembranes*, Vol. 20, No. 5, October, 2002, pp. 279-342.
146. Huang, W. and Koerner, R. M., "An Amendment Strategy for Enhancing the Performance of Geotextile Tubes Used in Decontamination of Polluted Sediments and Sludges," *Proc. of GRI-18 Conference at GeoFrontiers*, Austin, Texas, January 26, 2005, 6 pgs.
147. Escalante, E. M. and Iverson, R. J., "Corrosion of Steel Piles," *Mater. Perform.*, Vol. 17, No. 10, ASTM, Oct. 1978, pp. 9-15.
148. *Scour at Bridge Waterways*, NCHRP Publ. 5, Highway Research Board, National Academy of Engineering, Washington, DC, 1970.
149. Welsh, J. P., "PennDOT Uses a New Method for Solving Scour Problems beneath Bridge Piers," *Highway Focus*, Vol. 9, No. 1, May, 1977, pp. 72-81.
150. Hsuan, Y. G., Koerner, R. M. and Soong, Y.-T., "Behavior of Partially Ultraviolet Degraded Geotextiles," Proc. 5th IGS Conference, Singapore, Sept. 5-9, 1994, pp. 1209-1212.
151. Koerner, R. M., Koerner, G. R., Hsuan, Y. G., Ashley, M. V. and Koerner, J. R. (2011), Eds., *Enhancing Sustainability Using Geosynthetics*, Proc. GRI-24 Conference, GII Publ., Folsom, PA, 150 pgs.
152. Christopher, B. R., Holtz, R. D. and DiMaggio, J. A., *Geotextile Engineering Manual*, US DOT, FHWA Contract No. DTFH 61-80-C-00094, Feb. 1984.

PROBLEMS

2.1 A shopping center site developer has \$15,000 available to purchase a geotextile to be used as a separator between subgrade soil and stone base. The total area to be covered is 2.5 ha. Which of the following would the developer probably use?

a. 165 g/m^2 fabric at \$0.75/m^2
b. 200 g/m^2 fabric at \$0.85/m^2
c. 250 g/m^2 fabric at \$0.95/m^2

2.2 Given available funds of \$7,500, how many linear meters of pipe underdrain trench could be covered with geotextile if it is 2.5 m wide and costs \$0.95/m^2 in place?

2.3 In the Pennsylvania Department of Transportation specification given in table 2.1

a. For erosion-control geotextiles, when would you use type A versus type B?
b. For sediment-control geotextiles, when would you use type A versus type B?

2.4 Regard the statistics involved in the discussion of figure 2.1:

a. Define mean and standard deviation.
b. For a normally distributed statistical behavior what percentage of total occurrences falls between $\bar{x} \pm s$; $\bar{x} \pm 2s$; and $\bar{x} \pm 3s$?

2.5 Assuming that the values listed in the specifications of table 2.1 are MARV and that one standard deviation of a particular geotextile is 8%, convert them (with exception of AOS and UV strength retention) to as follows:

a. Average (or mean) values
b. Absolute minimum values at $\bar{x} - 3\sigma$.

2.6 Each class of geotextile in table 2.2a was subdivided according to the elongation at break of 50%. Why are all the required woven geotextile strengths greater than the nonwovens?

2.7 Compare the Pennsylvania Department of Transportation Specification values given in table 2.1 against the AASHTO M288 class 2 values listed in table 2.2a for the following:

a. Subsurface filtration vs. table 2.2b
b. Separation vs. table 2.2c
c. Erosion control—type A vs. table 2.2e
d. Sediment control—type A vs. table 2.2f

2.8 Using the M288 Specifications of table 2.2, determine what geotextile properties are needed for the following applications under severe installation conditions.

a. Woven monofilament erosion control geotextile under stone rip-rap for soil subgrade of 65% passing the 0.075 mm sieve size

b. Unsupported woven geotextile silt fence

c. Nonwoven needle-punched geotextile for prevention of reflective cracking as a paving fabric

2.9 In the description of designing-by-function (section 2.1.3), what would you use for a range of required factors of safety for the following situations:

a. A highway underdrain in the shoulder area of a secondary road

b. A filter fabric behind a retaining wall, above which is a paved parking lot

c. A reinforcing fabric for an access road for construction equipment

d. A filter to protect a chimney drain within an earth dam where a small town is located downstream

2.10 In designing-by-function (recall section 2.1.3), the focus is on the primary function. How does one know what is primary? (Hint: Would the resulting factors of safety from various aspects of the problem be indicative in any way?)

2.11 In section 2.2, separation was distinguished from reinforcement. In the case of geotextiles placed on soil subgrades beneath stone base courses for highways, when does the geotextile act as a separator vis-a-vis reinforcement?

2.12 Regarding liquid flow situations using geotextiles:

a. In section 2.2, filtration was distinguished from drainage. Describe these two functions and how they are different.

b. In handling flow for these two different functions, ψ and θ were defined. Why is this necessary and why is the thickness of the geotextile included in each term?

2.13 Using figure 2.4a for steady-state flow conditions, what is the appropriate opening size formula for the following conditions?

- **a.** Gravel with $C'_U = 5$ in a dense condition.
- **b.** Stable sand favoring retention with $C'_U = 2.5$ in a loose condition.
- **c.** Sandy silt that is nonplastic favoring permeability, widely graded $C'_U > 15$ and in a medium-density state.

2.14 Using figure 2.4b for dynamic flow conditions, what is the appropriate opening size formula for the following conditions?

- **a.** Clayey (plastic) soil that is nondispersive.
- **b.** Silty sand that is widely graded ($C_U = 13$) under mild water current.
- **c.** Gravel under severe wave attack

2.15 In geotextile testing:

- **a.** What is an index test?
- **b.** What is a performance test?
- **c.** How can typical laboratory-test values be made into allowable values for design-by-function procedure?

2.16 Calculate the compressibility modulus and the compressibility coefficient of the four geotextiles shown in figure 2.6. (Note that the thickness change must be converted to strain using original fabric thickness measured at 2.0 kPa.)

2.17 Determine the strength, elongation at break (strain), toughness, and tangent modulus of fabrics A, B, C, and E shown in figure 2.7 in both geotextile units and standard engineering units.

2.18 Concerning the different geotextile tensile test specimen patterns shown in figure 2.8:

- **a.** Estimate the tensile strength in units of kN/m for a woven geotextile for the various shapes shown if the grab strength is 30 kN.
- **b.** Estimate the tensile strength in units of kN/m for a nonwoven needle-punched geotextile for the various shapes shown if the grab strength is 20 kN.

2.19 Your answer to problem 2.18 should have comparable strength for the different patterns for woven geotextiles, but greater strengths for nonwoven geotextiles. Why is this generally the case? (Stated differently, what happens in the testing of nonwovens that does not happen with wovens?

2.20 Concerning the type of puncture probe used in evaluating puncture resistance of geotextiles as described in section 2.3.3, why should (a) a large-diameter probe be used for nonwovens, (b) a beveled probe used for wovens, and (c) a tapered probe not used for neither?

2.21 You have been given the following set of data from a soil-geotextile friction test:

Normal Stress (kPa)	Shear Strength (kPa)
17	8.6
35	20
70	36
140	75

a. Plot the Mohr failure envelope.
b. Obtain the friction angle.
c. Calculate the fabric efficiency based on a soil friction angle of 38 deg.

2.22 Calculate the porosity of the following geotextiles:

Geotextile	Mass/unit area (g/m^2)	Unit Weight (kN/m^3)	Density (kg/m^3)	Thickness (mm)
A	135	4.7	480	0.33
B	200	6.0	610	0.38
C	350	5.5	560	0.63
D	600	6.0	610	1.32

2.23 What is an "image analyzer," and how could it be used to determine fabric porosity?

2.24 Given the projection of a woven monofilament geotextile shown below, determine its percent open area (POA). (Hint: you will have to enlarge the figure considerably. A photocopier used a number of times will get you to a convenient size to use cross-section paper or a planimeter.)

2.25 The apparent opening size test described in section 2.3.4 is very controversial. There are three versions described in the literature; dry sieving (ASTM D4751), wet sieving (ISO 12956) and hydrodynamic sieving (used in Germany). Make a table describing the advantages and disadvantages.

2.26 If a series of glass bead sieve tests (either dry or wet) gave the following 0_{95} values in mm, what would be the "next highest" AOS sieve size value?

Geotextile	0_{95} (mm)
A	0.87
B	0.415
C	0.079

2.27 You have been given the following data for constant-head cross-plane flow of water through a 50 mm-diameter, 0.30 mm thick geotextile. Calculate the permittivity ψ (in units of s^{-1}) and coefficient of permeability k (in units of cm/s).

Δh (mm)	q (cm^3/min.)
31	300
62	680
125	1010
250	1400

2.28 You have been given the following constant-head data for planar flow of water in a 1.50 mm thick geotextile that is 300 mm wide by 600 mm long. Calculate the transmissivity θ in cm^3/min-cm of fabric and then the planar coefficient of permeability k in cm/sec.

Δh (mm)	q (cm³/min.)
75	21
150	41
225	60
300	79

2.29 You have been given the following constant head data set for radial flow of water in a 1.02 mm thick geotextile that has a 57 mm outer radius and a 28.7 mm inner radius. Calculate the transmissivity in cm³/min-cm of fabric and the planar coefficient of permeability in cm/sec.

Δh (mm)	q (cm³/min.)
150	1400
300	2900
450	4500
600	6000

2.30 Assume the data from problem 2.29 were taken at normal pressure on the fabric of 7 kPa and the test was repeated at 14, 28, and 56 kPa, giving the additional data given below. Plot the transmissivity (cm³/min-cm) versus applied normal pressure (kN/m²) response curve.

Pressure (kPa)	Δh (mm)	q (cm³/min.)
14	150	820
	300	1730
	450	2500
	600	3500
28	150	570
	300	1220
	450	1730
	600	2250

56	150	540
	300	1080
	450	1650
	600	2200

2.31 What is the range of factors of safety for installation damage for the data shown in figure 2.18b in regions A, B, and C?

2.32 Calculate the strain of the polyester fabric shown in figure 2.19 at the end of 30 years assuming a linear extension of the data for both 20% and 60% of failure loads.

2.33 Review the long-term flow tests in section 2.3.5:

- **a.** Describe the physical phenomena that accompany the various portions of the long-term flow behavior shown in figure 2.20b.
- **b.** Which of these phenomena is most important as to the possibility of the response shutting off flow completely?

2.34 Regarding the gradient ratio tests in section 2.3.5:

- **a.** In the gradient ratio test, head is measured via plastic tubing attached to a manometer board. Since it is this value that is most important in calculating the gradient ratio, what experimental problems could you envision?
- **b.** The value 3.0 is used to separate acceptable and nonacceptable fabrics. Why is this the case?

2.35 Concerning clogging of geotextiles in the field:

- **a.** What are the main soils and conditions where excessive clogging of geotextiles is likely to occur?
- **b.** In such instances, what is the logical recommendation?

2.36 The type of clogging described in sections 2.3.5 had to do with various soils upstream of the fabric and water as the permeant. If the permeant was a liquid with a large amount of suspended solids and microorganisms in it (e.g., agriculture runoff or landfill leachate), what possible mechanisms of clogging could occur?

2.37 How would you evaluate the chemical resistance of a geotextile to a hazardous waste leachate of which there is no standard test data (recall section 2.3.6)?

2.38 Ultraviolet light exposure of geotextiles is known to cause degradation over time (recall section 2.3.6). What specific wavelength is most harmful to the common resins used to manufacture geotextiles, i.e., PE, PET, and PP?

2.39 How do manufacturers avoid, or minimize the harmful effects of ultraviolet degradation to exposed geotextiles?

2.40 How would you predict the *exposed* lifetime of a geotextile on the basis of accelerated weathering (ultraviolet, temperature, and moisture) laboratory-test data?

2.41 Regarding lifetime prediction of geosynthetics that are covered in a timely manner:

a. Devise an experiment for evaluating the long-term lifetime of a PP geotextile to be used for a reinforced wall application

b. Same question but now for a PET geotextile

2.42 If the ultimate strength of a geotextile from an index-type test is 45 kN/m, what would be the allowable strength for the design purposes according to table 2.8a for the following:

a. Separation?

b. Reinforced embankment?

c. Increased bearing capacity?

d. A flexible forming system?

2.43 If the ultimate flow rate of a geotextile from an index-type test is 18 l/min-m, what would be the allowable flow rate for design purposes according to table 2.8b for the following:

a. Gravity drainage problems?

b. Pressure drainage problems?

2.44 If the ultimate permittivity of a geotextile from an index-type test is 1.2 sec^{-1}, what would be the allowable value for design purposes according to table 2.8b for the following:

a. Retaining-wall filter?

b. Highway-underdrain filter?

c. Filter beneath rock riprap?

d. Filter above leachate collection system?

2.45 What are the mechanical properties of a geotextile that are of most importance when using it as a separator in an unpaved road situation with only stone base above it and relatively firm soil below it?

2.46 Regarding geotextiles used in roadway separation:

a. What is the required burst pressure of a geotextile supporting 75 mm maximum-size stone and heavy trucks with a tire inflation pressure of 1000 kPa? Use $p \simeq 0.75\ p_a$, a cumulative reduction factor of 2.0, and a factor of safety of 2.0?

b. What is the required burst pressure of a geotextile under the same conditions in part (a) except that now the road will haul only light vehicles of tire inflation pressure of 500 kPa.

2.47 What would be the influence on a burst analysis if well-graded stone base were used instead of poorly graded aggregate materials?

2.48 Redo example 2.9 (in section 2.5.3) on the basis that slippage does occur between the stone and geotextile. In your analysis assume that the strains mobilized result in values of f(ε) = 0.75, 0.52, 0.30, and 0.10.

2.49 What is the factor of safety for a geotextile with allowable puncture strength of 250 N according to ASTM D 4833, considering a 600 kPa tire inflation pressure and relatively large stone of 30 mm size? The stone comes from different quarries, as described in table 2.9.

a. Angular

b. Subrounded

c. Rounded

2.50 What is the factor of safety for a geotextile with ultimate puncture strength of 350 N according to ASTM D 4833 (use ΠRF = 2.5) considering a 700 kPa tire inflation pressure and relatively small stone of 15 mm size considering the stone comes from different sources, as described in table 2.9?

a. Angular

b. Subrounded

c. Rounded

2.51 The data of table 2.9 for protrusion factors, scale factors, and shape factors is limited in scope and quite subjective.

a. What other variables besides shape and size of the puncturing object might be important?

b. What type of laboratory experiment could be developed to authenticate the values?

2.52 Regarding impact resistance, what energy is mobilized by rock of 150 mm size falling out of a dump truck 1.5 m to the geotextile, if the geotextile rests on:

a. A "soft" soil of unsoaked $CBR = 3$?

b. A "firm" soil of unsoaked $CBR = 9$?

c. A "hard" soil of unsoaked $CBR = 16$?

2.53 What energy is mobilized by a 250 N jackhammer falling 2.0 m out of a dump truck onto a geotextile on a "hard" soil of unsoaked $CBR = 20$? Would this situation cause a problem?

2.54 Regarding geotextiles for use in unpaved roads:

a. When using geotextiles as reinforcement for unpaved roads on soft subsoils, do they also act as separators?

b. If the answer to part (a) is yes, of what benefit is it?

c. From product literature, or simply gut feeling, what does this separation function amount to as far as thickness of stone base saved (if any) is concerned?

2.55 Regarding soil subgrade characteristics for roadway construction:

a. Describe the details of the CBR test (both unsoaked and soaked versions).

b. Using table 2.10 for a soil whose $CBR = 1.0$, what is the equivalent in:

- **i.** Shear strength?
- **ii.** *R* value (California)?
- **iii.** *S*, soil support value?
- **iv.** Group index?
- **v.** *R* value (Washington)?
- **vi.** Cone index (320 mm^2 probe)?
- **vii.** Bearing value (300 mm plate, 5 mm deflection)?
- **viii.** Bearing value (760 mm plate, 2.5 mm deflection)?
- **ix.** Modulus of subgrade reaction?

c. In table 2.10 why do the ASTM, AASHTO, and FAA classifications not go down as far as CBR = 1.0?

2.56 You have been given a 80 kN axle-load vehicle with tire inflation pressure of 480 kPa undergoing a settlement depth of 0.3 m for an unpaved road. The base course thickness is to be designed without a geotextile, then with a geotextile using figure 2.28, and ultimately resulting in a thickness of stone aggregate to be saved:

a. What is the response for 340 passages without and with a 90 kN/m modulus geotextile? Draw the response curve to varying soil *CBR* values from 0.5 to 4.0.

b. What is the response for 340 passages and a *CBR* = 1.0? Draw the response curve for varying geotextile moduli from 450 kN/m to 10 kN/m.

c. What is the resonse for a soil of *CBR* = 1.0 and a geotextile modulus of 90 kN/m? Draw the response curve to a varying number of vehicle passages from 10,000 to 10.

2.57 Evaluate the sensitivity of varying settlement depths (the value S in equation 2.38) on the required thickness of stone base using typical values for unpaved road problems, and plot your results. (Hint: It would help to go to the original reference by Giroud and Noiray [72] to gain insight into this aspect of the development.)

2.58 Using equation (2.39) in example 2.14 of section 2.6.1, plot the required thickness of stone base of an unpaved road as a function of *CBR* varying from 0.50 to 10. Use 25 kN equivalent single wheel loads for 10,000 coverages on a 300 × 450 mm tire contact area in your solution.

2.59 Regarding the sewing of geotextile seams:

a. What type of test would you use to evaluate the strength of a sewn seam?

b. In writing a nonnumeric procedural guide for sewing of seams, what items would you include?

c. If you specified a minimum tensile strength for a geotextile seam, what percentage of the minimum tensile strength of the geotextile would you recommend?

d. How would you conduct field tests for strength of sewn seams?

2.60 For geotextiles to be used to reinforce paved roads on firm soil subgrades, the geotextile must somehow be prestressed (recall section 2.6.2).

a. Why is this the case?

b. Sketch some methods for prestressing geotextiles for such an application.

c. In light of your answer above, consider creep and possible stress relaxation, and postulate the road system's long-term performance.

2.61 The concept of flexible wall systems, as opposed to rigid concrete and masonry walls, is very much in style. List advantages and disadvantages of each type.

2.62 Mechanically stabilized earth (MSE) walls and slopes are stabilized by so-called inextensible and extensible reinforcement. What do these terms imply and what are the subsets of each category?

2.63 What are the major long-term concerns over an exposed geotextile wraparound wall?

2.64 What actions can be taken to alleviate the long-term concerns mentioned in the previous problem?

2.65 From a geotechnical perspective, this book's wall design has focused on lateral pressures as calculated by a Rankine design procedures. What are some limitations in this procedure?

2.66 Regarding geotextile-reinforced walls (this is an extremely long problem):

a. Design an 5.5 m high wraparound geotextile wall carrying a road consisting of 300 mm stone base (γ = 22 kN/m^3) and 150 mm asphalt (γ = 24 kN/m^3) for 180 kN dual-tandem-axle loads whose wheel pattern is shown in example 2.16 (section 2.7.1). The wall is to be backfilled with SW-ML soil of γ = 18 kN/m^3, φ = 35 deg., and c = 0. The geotextile to be used is a nonwoven heat-bonded fabric of 200 g/m^2 with an ultimate tensile strength of 50 kN/m. Use reduction factors 1.2, 2.5, and 1.26 for installation damage, creep, and chemical/biological degradation respectively.

b. Check and/or modify your answer for external stability considering that the foundation soil is

ML-CL with $\gamma = 19$ kN/m^3, $\varphi = 15$ deg., $\delta = 0.95\varphi$, $c = 24$ kPa, and $c_a = 0.90c$.

2.67 Develop a design chart similar to figure 2.36 for the wheel load and backfill of problem 2.66, for vertical faced walls (β = 90 deg.) and for wall heights varying from 1.5 to 8.0 m and geotextile allowable tensile strengths of 20, 40, 60 and 80 kN/m. (Note: this is another extremely long problem and even requires a computer program to be generated.)

2.68 For examples 2.18 and 2.19 (section 2.7.2) using geotextiles to stabilize embankments, the fabric's allowable tensile strength is 40 kN/m. Repeat this problem using a single fabric whose allowable strength varies from 20, 40 (the example), 80, 150, and 300 kN/m and plot the resulting *FS* value against allowable strength.

2.69 Concerning the allowable geotextile tensile strength to be used in reinforcement problems (both for walls and embankments):

a. What test method should be used?

b. What considerations enter into your choice of reduction factors?

c. How does creep enter into the situation?

2.70 A 15 m high embankment has a slope angle of $\beta = 40$ deg. The soil strength parameters are $\varphi = 22$ deg. and $c = 15$ kPa in both the embankment and foundation sections. The unit weight is 16 kN/m^3. For a failure circle located at coordinates of (+3, +18) with respect to the toe at (0, 0) (see figure 2.38) and a radius of 21 m, what is the factor of safety? How many layers of geotextiles spaced 300 mm apart and having an allowable tensile strength of 55 kN/m placed at the interface of the foundation and the embankment are required to raise this factor of safety to 1.40?

2.71 For problem 2.70, find the minimum factor of safety for both the nonreinforced and reinforced conditions. (Note that this is an extremely long problem requiring a search of both variation in radius and center of circle. A computer program is necessary.)

2.72 In using high-strength geotextiles for areal stabilization projects as described in section 2.7.3, how do you sew the transverse seams after the larger panels are already deployed?

Furthermore, how does one sew the intersection where the four panels come together?

2.73 In placing fill on a high-strength geotextile used to stabilize very soft soils by the linear fill method, section 2.7.3 mentioned the edges being placed first in the Wilmington Harbor South Disposal Area project. Why is this necessary? What would happen if the entire fill were advanced together? What would happen if the center were advanced first?

2.74 Describe by means of a sketch how you would instrument a high-strength geotextile of the type mentioned in section 2.7.3 for the linear fill to verify the various design models shown in figure 2.40.

2.75 What is the required wide-width strength of a geotextile used for basal reinforcement (recall equations 2.60 a and b) spanning between stone column deep foundations under the following conditions.

- Column center-to-center distance = 4.0 m
- Column diameter = 1.5 m
- Embankment height over columns = 2.0 m
- Embankment soil unit weight = 18 kN/m^3
- Surcharge load above embankment = 20 kN/m^2
- Allowable strain in geotextile = 10%
- Use ultimate conditions for partial load factors

2.76 There is considerable appeal to the use of geotextiles to improve the bearing capacity of shallow foundations (section 2.7.5). Most efforts are aimed at showing improvement in bearing capacity as indicated in figure 2.42.

a. Why are the improvements low at low deformations and considerably better at high deformations?

b. How could the low-deformation behavior be improved?

c. By observing these data, would you consider using only one high-strength geotextile instead of a number of lower-strength layers?

d. If the answer to part (c) is yes, how would you join the geotextile ends and sides?

2.77 Regarding improved bearing capacity:

a. Would a geotextile placed under a large mat foundation measuring 30 × 30 m on a compressible soil prevent (total) settlement?

b. If the answer to part (a) is no, would it be of any help insofar as reinforcement is concerned?

2.78 Regarding the "spider netting" of section 2.7.6:

a. Illustrate the various mechanisms that soil nails can provide in soil slope stabilization?

b. What are the high-stress regions of the net as currently configured?

c. Why must the nails be continually driven into the ground even after initial installation?

2.79 The geotextile is exposed to the surface of the slope in spider netting. What advantages and disadvantages occur?

2.80 What are the three essential features that must be addressed in filtration design?

2.81 What is the difference in total active earth pressure between the two retaining wall cases shown below? One has good drainage; the other has no drainage and full hydrostatic head.

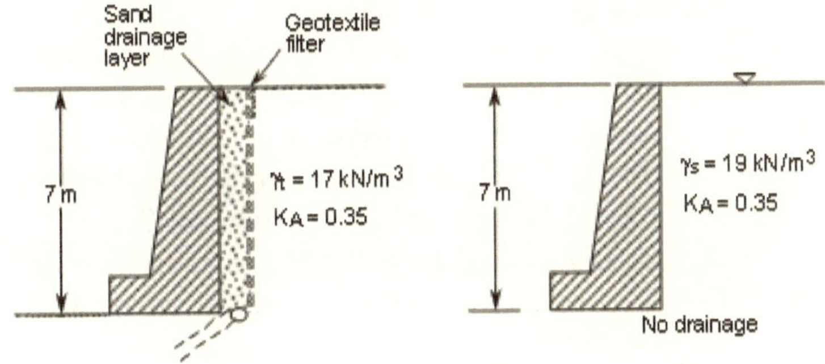

2.82 a. In considering hydraulic designs involving geotextiles, why is permittivity used in filtration and transmissivity used in drainage rather than just the respective permeability (hydraulic conductivity) coefficients?

b. In the laboratory determination of a specific geotextile's permittivity or transmissivity, at what pressure should the geotextile's thickness be measured?

2.83 A stone aggregate has been placed around a highway underdrain with a geotextile:

a. What is the major function of the geotextile?
b. If properly designed, what should be the long-term condition of the stone base?
c. If the stone base has sufficient open space to transmit the entering water, of what necessity is the perforated pipe?
d. What is a French drain?

2.84 A geotextile filter is being considered to protect the stone aggregate drain behind a cantilever retaining wall as shown in figure 2.47a, b. The wall stem is 7.5 m high, retaining an ML soil with k = 2.5×10^{-4} m/s, d_{50} = 0.05 mm, CU = 4.8, and D_R = 85%. The candidate geotextile is a heat bonded nonwoven with a permittivity (ψ) = 0.01 sec^{-1} and an AOS = no. 70 sieve. What is the factor of safety against flow and the adequacy as far as soil retention is concerned?

2.85 If the required permittivity of a sandy soil beneath a rock riprap-protected slope is 0.052 sec^{-1} slope and riprap covers 75% of the geotextile's surface, would a geotextile of k = 3.5 $\times$ 10^{-4} m/s and 0.65 mm thick be adequate? What is the factor of safety in this case?

2.86 Concerning geotextile silt fences:
a. While geotextile silt fences function as filters, section 2.8.6 focused mainly on strength considerations. Why is strength so important?
b. How do geotextile silt fences filter the turbid water and retain the suspended soil particulates?
c. Discuss UV degradation of silt fences in light of their use as erosion and sedimentation control systems.

2.87 Recalculate example 2.24 of section 2.8.6 using the same values, except vary the storm intensity from 10, 50, 100 (the example problem), and 200 mm/hour to determine the following:
a. Height of the silt fence
b. Strength of the geotextile
c. Type of posts to be used for support

2.88 Regarding the drainage capability of geotextiles:
a. Which manufactured style is best suited to convey water in its plane?

b. What conditions are required to satisfy Darcy's formula?

c. What is the driving mechanism for water flow in gravity drainage situations?

d. What are some driving mechanisms for water flow in pressure drainage situations?

2.89 Geotechnical engineers are generally reluctant to use drainage geotextiles as chimney drains and drainage galleries as shown in the earth dam example 2.25 of section 2.9.3. What are some reasons for this reluctance?

2.90 For the 8 m high concrete cantilever retaining wall of example 2.26 (section 2.9.3), recalculate the soil's permeability to determine what value is required to have the $\theta_{allow} = 0.00015$ m^2/min. be adequate with $FS = 4.0$ (i.e., work the problem backward).

2.91 Repeat example 2.27 illustrating pressure drainage of the consolidating soil beneath a surcharge fill (section 2.9.4) with a $B = 30$ m. Vary the time for the surcharge fill to be placed from 1 day to 1 year. Plot the results and show on the graph the acceptable zone based on $FS = 5.0$.

2.92 A geotextile is being used as a capillary migration break, as in section 2.9.5:

a. What is to prevent the water from continuing across the barrier as though it were not there?

b. Does the fact that geotextiles are hydrophobic play a role in this situation?

c. If soil particles become clogged in the fabric structure and accumulate, does the situation change?

d. How is the situation in part (c) prevented from occurring?

2.93 There are some areas where full-pavement-width geotextiles should not be used in reflective crack prevention as per section 2.10.2. Describe why this is so for each of the following cases.

a. Where subsoil foundation problems exist beneath the stone base.

b. For reinforced or nonreinforced concrete pavements.

c. In areas of rapid and harsh cyclic freeze-thaw temperature conditions.

2.94 Regarding the proper quantity of sealant for geotextile in reflective cracking applications:

a. Why is sealant in excess of saturation a problem?

b. Why is sealant less than saturation a problem?

2.95 In reflective crack prevention of section 2.10.2, assuming reinforcement to be the primary function, the *FEF* is of paramount interest.

a. What is *FEF* ?

b. How is it determined?

c. What geotextile property is it mainly dependent on?

d. What are the dangers in taking laboratory-generated data and using it to project field performance?

2.96 Regarding geotextiles in reflective cracking assuming reinforcement as the primary function as described in section 2.10.2.

a. Using $DTN_N = 200$ and *CBR* value varying from 2 to 20, what is the nonreinforced value of T_A using figure 2.54?

b. Using $DTN_N = 200$ and a geotextile resulting in $FEF = 4.0$, what is T_A if the *CBR* varies from 2 to 20?

c. Plot the two resulting curves on a graph of T_A versus *CBR* and comment on the results and the differences between the curves.

2.97 Postulate on the mechanism(s) that might occur in assuming waterproofing to be the major function in using geotextiles to prevent reflective cracks in asphalt pavements.

2.98 Regarding geotextiles in reflective cracking assuming moisture barriers as the primary function as described in section 2.10.2:

a. Using $DTN_N = 500$ and $x = 1.57$ mm, $s = 0.15$ mm, $f = 0.80$, and $c = 1.25$, determine the asphalt overlay thickness according to figure 2.55.

b. Using a fabric assumed to function as a waterproofing barrier, redo part (a) for "c" varying from 2.5 to 0.5 in equation 2.75.

c. Plot the results on a graph of T_o versus c and comment on the curves and the differences.

2.99 Design and sketch a field experiment for determining whether reinforcement or waterproofing is the major function in using fabrics as preventing reflective cracks in asphalt overlays.

2.100 For reflective-crack prevention using narrow strips of high-strength fabric reinforcement:

- **a.** Fiberglass geotextiles have some distinct advantages and potential disadvantages over polymeric materials. What are they?
- **b.** How does one anchor the sides of the geotextile strip on each side of the crack?
- **c.** Which are the more troublesome cracks in old pavements: transverse or longitudinal, and why?

2.101 List the functions (in order of priority) that you feel are acting when geotextiles are placed beneath railroad ballast in the following situations:

- **a.** New railroad track construction.
- **b.** Remediation of existing railroad trackage.

2.102 Comment on why most railroad specifications call for thick needle-punched nonwovens and most laboratory generated research papers use relatively thin wovens or heat-bonded nonwoven fabrics.

2.103 Regarding geotextiles used as railroad ballast separators:

- **a.** In exhuming geotextiles within railroad ballast one often sees small holes (a few millimeters) and large holes (centimeters in size). What was the most likely cause of each type?
- **b.** What is the minimum depth at which a geotextile should be placed beneath the bottom of a railroad tie to prevent abrasion problems?

2.104 In the use of geotextiles as flexible forms for columns in mine stabilization or pile jacketing (section 2.10.4), will the resulting shape be circular? If not, why?

2.105 For the fabric used as a flexible form in mine stabilization as shown in figure 2.57, what type, dimensions and properties of geotextile should be considered?

2.106 As in section 2.10.4, a geotextile tube has dimensionless parameters b_1 = 10.5 m and S = 26.2 m (thus, b_1/S = 0.40) in figure 2.61.

a. What would be the necessary allowable strength of the geotextile for a $FS = 1.25$?

b. Using ΠRF = 1.50, what would be the necessary ultimate strength?

c. Determine the geotube's dimensions B, H, H', B' and A

2.107 In the geotextile tube shown in figure 2.60b, is the fabric tensile strength critical? If not, estimate the longitudinal seam's efficiency as a percent of nonseamed fabric tensile strength. [Hint: Recall figure 2.10].

2.108 In the dewatering of fine materials using geotextile tubes as shown in figure 2.62, what could be done to reduce the negative impact of the low-permeability filter cake?

2.109 If the fine sediment pumped into a geotextile tube for dewatering (as in the photograph of figure 2.62) was contaminated, what two characteristics of the pollutant(s) are critical in assessing whether or not the pollutants escape with the effluent water?

2.110 Regarding the use of geotextiles as flexible forms:

a. Is there any limit to the depth at which tremie concrete or grout can be placed under water?

b. When placed in geotextile forms, how is the water displaced and the concrete or grout kept in?

c. After the concrete or grout is set (hardened), what is the function of the geotextile?

2.111 Concerning the geotextile survivability concepts discussed in section 2.11.2 and table 2.18, what minimum mechanical properties would be required in the following cases using the specification of table 2.2a?

a. 35 kPa construction equipment using average site preparation

b. Same as in (a), except no site preparation has been provided

c. 20 kPa construction equipment on a dredged site with no vegetation or growth

2.112 What general and specific methods does the design engineer have with respect to the contractor regarding the proper care and handling of geotextiles during the installation process?

Chapter 3

Designing with Geogrids

3.0 INTRODUCTION

The geotextiles discussed in chapter 2 and the geogrids discussed in this chapter compete for use in most reinforcement applications. They are also designed by similar methods but they differ in their manufacture, appearance, properties and placement. It should be understood that geogrids are not used for separation, filtration, drainage, or barrier functions unless they are composites formed with other geosynthetics. Such geocomposites will be treated in chapter 8. Therefore, a geogrid can be defined as follows:

> Geogrid: A geosynthetic reinforcement material consisting of connected parallel sets of polymeric tensile ribs with apertures of sufficient size to allow strike-through of surrounding soil, aggregate, or other particulate material.

Thus, geogrids are matrixlike materials with large open spaces called *apertures*, which are typically 10 to 100 mm between the intersecting *ribs.* These ribs are called *longitudinal* and *transverse* respectively. The ribs themselves can be manufactured from a number of different materials, and the rib cross-over joining or junction-bonding methods can also vary. Since the primary function of geogrids is reinforcement, the sections within this chapter are organized not by function, but by type of reinforcement application. In those applications where the direction of the major stresses are known, as in walls and slopes, *unidirectional*, or *uniaxial*, *geogrids* are used. In those applications where the applied stresses come from random directions, as in pavements and foundations, *bidirectional*, or *biaxial*, *geogrids* are used.

Figure 3.1 illustrates the three categories of geogrids that are currently available: (a) unitized PE or PP polyolefins, (b) coated PET or PVA yarns, and (c) PET or PP bonded straps (or rods). These categories will be explained in some detail, as will the testing related to their method of manufacture and performance.

Figure 3.1a shows the original type of geogrids, which are characterized as being unitized insofar as the continuity of the intersecting longitudinal and transverse ribs are concerned. Both unidirectional and bidirectional products are available. Each style begins as a polyolefin polymer sheet (that is, a thick geomembrane) that subsequently has a uniform and controlled pattern of holes punched in it. The punched sheet is then sent over and under a number of rollers, each going faster than the one before it, thus inducing longitudinal stretching of the sheet. The elongated material between holes becomes the geogrid's ribs. In the unidirectional deformed products, circular holes punched in high-density polyethylene (HDPE) sheet become elongated ellipses with stretched longitudinal ribs in the machine direction and unstretched transverse ribs in the cross machine direction. The eventual draw ratio is as high as 15 to 1. The molecular structure in the longitudinal ribs is highly elongated and the strength, modulus, and resistance to creep are increased significantly over the original nondeformed material. A number of different styles with different strength properties are available. In the bidirectional products, squares are punched in polypropylene (PP) sheet, which is then drawn longitudinally (using rollers) as before, then transversely (using a stretcher), forming near-square or rectangular apertures. This process increases strength in both longitudinal and transverse directions in the bidirectional product. Unidirectional products are for applications in which the major principal stress direction is known (such as walls and slopes) and the bidirectional products are for applications in which mobilized stresses are essentially random (such as pavements and foundations).

A recently developed variation is called a triaxial geogrid. Here the stretching of the punched PP sheet is such that three sets of ribs are formed (about 120° apart) and the strength is balanced along three axes.

Figure 3.1b shows a variety of coated yarn geogrids. There are more products in this category of geogrids than any other. Most often, the yarns are bundles of high tenacity polyester (PET) filaments. (The side walls of automobile and truck tires are reinforced with similar filaments—that is, the so-called *tire chords*). Polyvinyl alcohol (PVA) and fiberglass (FG) filaments have also been used, but PET filaments predominate. The yarn bundles are then woven or knit on conventional textile machinery into the desired grid pattern. Strength can easily be varied using more or fewer filaments per yarn in both directions, giving rise to unidirectional and bidirectional products. Yarn spacing can also be varied. The entanglement of the yarns at their intersections is an important issue and varies from product to product. There is obviously a selvedge on both edges of the manufactured material. As

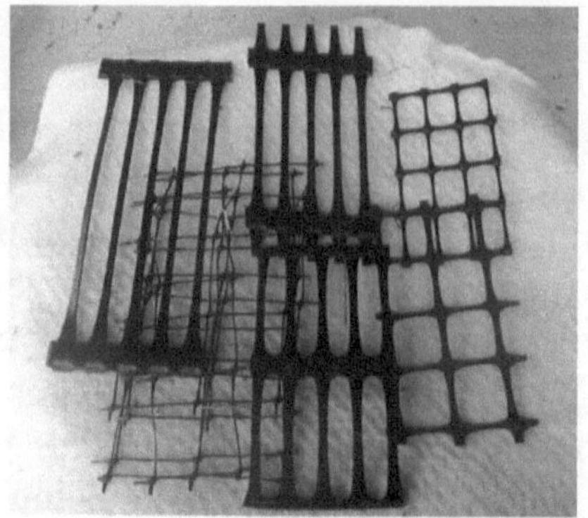

(a) Unitized PE and PP polyolefin geogrids

(b) Coated woven or knit PET and PVA yarn geogrids

(c) PET and PP bonded strap (or rod) geogrids

Figure 3.1 Various categories of geogrids.

a secondary step, the geogrids are coated, usually by spraying and then dipping in bitumen, latex, or polyvinyl chloride (PVC). The purpose of the coating is to maintain geometric stability of the product and to protect the filaments from damage during installation and service.

Figure 3.1c shows geogrids made from high tenacity PET or polypropylene (PP) straps or rods. These are similar to packaging and bonding materials used for shipping purposes. The individual straps are approximately 10 mm wide and 1.0 mm thick and are manufactured by overlapping the longitudinal ribs over and/or under the transverse ribs. The crossover locations, called *junctions* or *nodes*, are then either ultrasonically or laser bonded to provide junction strength. Different rib layout patterns give rise to different styles of unidirectional and bidirectional products.

3.1 GEOGRID PROPERTIES AND TEST METHODS

In contrast to the entire range of test methods described in chapter 2, only those involved in reinforcement applications will be addressed here. Geogrid tests are unique in a number of aspects when compared with geotextiles. Properties relating to separation, filtration, drainage, and barrier applications are not included since geogrids always serve the primary function of reinforcement.

3.1.1 Physical Properties

Many of the physical properties of geogrids—including the type of structure, rib dimensions, junction type, aperture size, and thickness—can be measured directly and are relatively straightforward. Other properties that are of interest are mass per unit area, which varies over a tremendous range from 200 to 1000 g/m^2, and percent open area (POA), which varies from 40 to 95%. Such large POA values give rise to large apertures such that almost all soils will communicate, or *strike-through*, the plane of the geogrids.

Density. The density or specific gravity of a geogrid depends on the polymer from which it is made. Homogeneous geogrids are made from HDPE or PP and density can be measured using ASTM D792 or D1505. Values will be less than unity. Rod or strap geogrids made from PET or PVA can use the same test methods, and the resulting values will be greater than unity. Coated yarn geogrids are difficult to evaluate since the coating cannot be readily removed. In addition, the very fine filaments are troublesome to measure in their own right.

Out-of-Plane Bending Stiffness. Bending stiffness is a physical property of geogrids that is of direct interest insofar as constructability is concerned. This can be measured using ASTM D1388, a test for flexural rigidity. This test method slides a geogrid test specimen hanging over an inclined plane measuring an angle of 41.5° with the horizontal. When the geogrid bends and eventually touches the surface of the inclined plane, its distance is measured and then related to the mass per unit area. The test is described in section 2.3.2. The unitized and strap geogrids are quite *stiff* and are characterized by having flexural rigidity values significantly greater than 1000 g-cm in this test. The woven or knit yarn geogrids are quite *flexible* and are characterized by having flexural rigidity values less than 1000 g-cm in this test.

In-Plane Torsional Stiffness. Kinney [1] has proposed clamping a square bidirectional geogrid test specimen in a rigid frame and firmly gripping the central node. A torque is applied and the angular rotation versus the geogrids's resistance is measured. The test is formalized as GRI-GG9. For stiff geogrids, the resulting plot shows a near linear performance. For flexible geogrids, the response is initially low, but after a 5° to 10° rotation, the resisting force increases markedly. The test has its greatest applicability for bidirectional geogrid reinforcement in pavement base courses and perhaps soft soil foundation stabilization as well.

3.1.2 Mechanical Properties

The mechanical properties of geogrids covered in this section all relate directly to their use in tensile reinforcement applications. Some are index tests, while others are clearly performance oriented.

Single Rib and Junction (Node) Strength. The initial tendency toward assessing a geogrid's tensile strength is to pull a single rib in tension until failure and note its behavior. A secondary tendency is to evaluate the in-isolation junction strength by pulling a longitudinal rib away from its transverse rib's junction. It is important to state *in-isolation* since there is no normal stress on the junction; thus the test will not represent performance conditions. A performance junction strength test must be done with the entire geogrid structure contained within soil embedment. This is a much more complicated test and will be covered in this section under anchorage strength from soil pullout.

A *single rib tension strength* test merely uses a constant rate-of-extension testing machine to pull a single rib to failure, as described in ASTM D6637. For unidirectional geogrids, this would most likely be a longitudinal rib. For bidirectional geogrids, both longitudinal and transverse ribs require evaluation. By knowing the repeat pattern of the ribs, an equivalent wide-width strength can be calculated. Alternatively, a number of ribs can be tested simultaneously to obtain a more statistically accurate value for the wide-width strength (see below).

An *in-isolation junction or node strength* test can also be performed. The test method uses a clamping fixture that grips the transverse ribs of the geogrid immediately adjacent to and on each side of the longitudinal rib (see figure 3.2). The lower portion of the longitudinal rib is gripped in a separate clamp, and each clamp is mounted in a tensile testing machine, where the test specimen is pulled apart. The strength of the junction, in force units, is obtained. The test is standardized as GRI-GG2. Also, note that the individual rib strength can also be evaluated as described previously. Having both sets of data, a junction strength efficiency can be calculated.

In general, the unitized geogrids give junction efficiencies from 90 to 100; the bonded strap geogrids from 40 to 70; and the woven or knit geogrids from 10 to 25%. It is important to note that these results have the junction in an unconfined status. The GRI-GG2 test method has an alternative that constrains junction rotation and results in improved strength efficiencies for the woven or knit geogrids. That said, both of the alternative clamping methods still result in an index, rather than performance, test result.

Wide-Width Tensile Strength. Clearly, the wide-width tensile strength of a geogrid, in its machine direction for unidirectional geogrids and in both machine and cross-machine directions for bidirectional and tridirectional geogrids, is of prime importance.

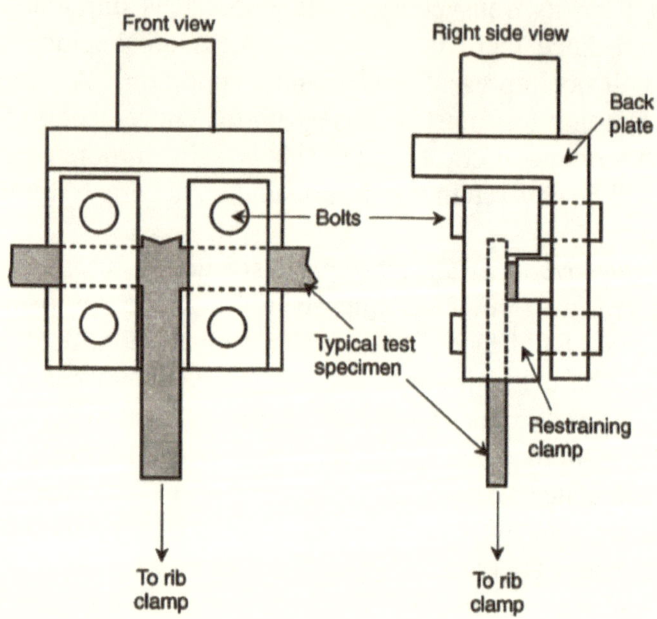

Figure 3.2 Test fixture for measuring geogrid junction strength.

When testing the products, the test clamps grip a larger test specimen than described above, including a number of repeating rib units in the width direction and perhaps repeating length sections as well. The resulting data gives strength values in units of force per unit width, that is calculated by using the repeat distance of the actual geogrid structure. Obviously there is an extremely wide range in product behavior depending on type of polymer, its structure, spacing of ribs, and so on. The strength of geogrids with respect to geotextiles is also of interest. Geogrid strengths fall at an intermediate point between conventional geotextiles and those geotextiles specifically made for high-strength applications.

There are two procedural test methods used to evaluate the wide-width tensile strength of geogrids. One choice is to use ASTM D6637, which has provision for measuring single rib strength (described previously), or multiple rib strength. The width of the test specimen, its length, its clamping mechanism, its strain rate, and the method for measuring deformation are all important considerations. If this test is selected, these items must be agreed on by the parties involved. The second choice is to use ISO 10319 for wide-width strength testing of geogrids. In its procedure both the width and length of the test specimen are prescribed. The type of clamps used should be roller or capstan grips as shown in figure 2.9e. Deformation monitoring in almost all situations must be based on an external measurement system—such as optical, laser, or transducer.

The information gained from a wide-width tension test on a geogrid comprises several factors: the tensile strength at which the test specimen fails (kN/m); the tensile elongation at which the test specimen fails—i.e., its failure strain (%); the tensile stress at different elongations prior to specimen failure—for example, stresses (kN/m) at 1%, 2%, 5%, strain; and the tensile modulus (kN/m) taken from the initial portion of the strength-versus-elongation curve, or possibly other defined modulus values. Manufacturers' literature is available on the these data for their particular products and styles that are in current production. Some compilations of comparative data are available in the literature, but the situation changes regularly and direct inquiries to the manufacturer are recommended.

In the tension testing just discussed, the geogrid is evaluated in isolation—that is, with no soil adjacent to or surrounding it. With soil pressure adjacent to the geogrid, the material may show an improvement in its strength characteristics, although the effect is felt to be nominal at best. This comment is based on the work of Wilson-Fahmy et al. [2] in which a variety of geosynthetic materials were tested without, then with, lateral confinement; only the nonwoven needle-punched geotextiles showed significant improvement under confining pressure. Unfortunately, geogrids were not evaluated in that study. If such tests are conducted on geogrids under confined conditions, it is important to eliminate friction-induced contributions by the pressurizing medium (usually soil) by careful lubrication of both surfaces of the geogrid test specimen (see Wilson-Fahmy et al. [2]).

Shear Strength. One type of performance test that is used regularly on geogrids is an adapted form of a conventional geotechnical engineering direct shear test. In such a test (its analog for geotextiles is discussed in section 2.3.3), the geogrid is fixed to a block and is forced to slide over stationary soil in a shear box while being subjected to normal stress, as shown in figure 3.3a. The maximum shear stress, i.e., its shear strength, is obtained accordingly (see figure 3.3b). Then a new test with a replicate geogrid specimen and soil (but now at a different normal stress) is conducted. This process is repeated sufficiently often to develop a set of strength-versus-normal stress points, which are plotted in figure 3.3c. The resulting best-fit line defines what is known as the failure envelope, properly called the Mohr-Coulomb failure envelope. From this graph the shear strength parameters of the geogrid to the particular soil are obtained—that is, the values of friction angle (δ) and apparent cohesion or adhesion (c_a). Note that if a strain-softening response of the stress-versus-displacement curves occurs, two sets of values result, peak and residual. If the shear strength parameters of the soil by itself, friction angle (φ) and cohesion (c), are also determined in their own separate tests with soil in both halves of the shear box, a comparison or *efficiency* can be calculated as follows:

$$E_\varphi = (\tan\delta / \tan\varphi)\times 100 \quad (3.1)$$
$$E_c = (c_a / c)\times 100 \quad (3.2)$$

where

E_φ = efficiency on friction,
E_c = efficiency on cohesion,
δ = friction angle of soil-to-geogrid,
φ = friction angle of soil-to-soil,
c_a = adhesion of soil-to-geogrid, and
c = cohesion of soil-to-soil.

A large shear box must be used for geogrid testing in order to minimize scale effects. A rule of thumb used in soil testing is that the shear testing device must be more than 10 times the size of the largest soil particle. If an analogy is made to the geogrid's apertures, this would generally require a 300 × 300 mm box or larger for geogrid shear testing. Both ASTM D5321 and ISO 12957 for direct shear testing of geosynthetics require at least this size of test device.

Test results using a 450 × 450 mm shear box are shown in table 3.1. The soil in all cases consisted of a well-graded angular sand (SW) in the dry condition and in a dense compaction state (≅ 90% relative density). The cohesion of the soil was zero; hence the adhesion is also zero. The resulting peak friction angle of the soil by itself was

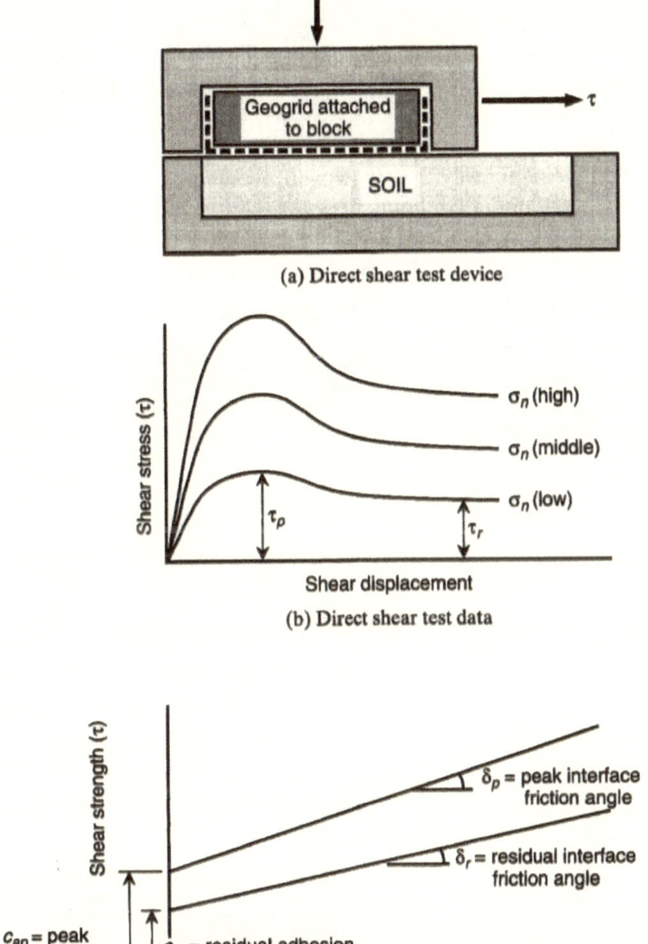

Figure 3.3 Test setup and procedure to assess interface shear strengths involving geogrids.

44°. The efficiencies were calculated on the basis of equation 3.1. Note that the efficiencies are all quite high, with many as high as the soil itself. This is understandable since the general configuration of geogrids (with their rather large apertures and relatively thick ribs) forces the failure plane into the soil itself. If any reduction in the soil's strength occurs, it is only along the surface of the geogrid's ribs.

Conversely, improvement in the soil's shear strength might be gained by having bearing stresses occur against the edges of the transverse ribs of the geogrid. While it is generally prudent to do product-specific and soil-specific shear testing, the shear strength of most soils, small enough to fit into the geogrid's aperatures, will be fully mobilized by most geogrids.

An investigation of the influence of aperture size versus soil particle size on the frictional efficiency of a number of geogrids is available from Sarsby [3]. He finds that the optimum transfer of shear stress—that is, the highest efficiency—occurs when

$$B_{GG} > 3.5d_{50} \tag{3.3}$$

where

B_{GG} = the minimum width of geogrid aperture, and
d_{50} = the average particle size of the backfilling soil.

This is an important consideration when selecting the type of backfill to be used around geogrids. Fortunately, the criterion can readily be accommodated by a wide selection of soil types for backfilling purposes.

TABLE 3.1 RESULTS OF DIRECT SHEAR TESTS (PEAK STRENGTHS) USING VARIOUS GEOGRIDS*

Test Condition	Test #1		Test #2	
	Friction Angle (deg.)	Efficiency (%)	Friction Angle (deg.)	Efficiency (%)
Soil-to-Soil	44	100	44	100
Soil-to-Bidirectional Geogrid #1	43	96	44	100
Soil-to-Bidirectional Geogrid #2	45	103	45	103
Soil-to-Bidirectional Geogrid #3	46	107	46	107
Soil-to-Unidirectional Geogrid #1	35	72	37	78
Soil-to-Unidirectional Geogrid #2	37	78	39	84
Soil-to-Unidirectional Geogrid #3	42	93	43	96

*The geogrids were firmly attached to a wooden platen in the movable portion of the shear box and slid over the stationary soil in the bottom of the shear box.

Anchorage Strength from Soil Pullout. The intrinsic merit of geogrids comes about by their anchorage strength or pullout resistance, which can far exceed the direct shear strength that was just discussed. Interesting comparison tests between steel grids, steel plate, polymer geogrids, and polymer geonets are reported by Ingold [4]. This behavior comes about by virtue of the large apertures in the geogrid allowing for soil strike-through from one side of the geogrid to the other. Obviously, the soil particles must be sufficiently small to allow for full penetration; thus the d_{50} value in equation 3.3 represents the recommended maximum particle size for a particular geogrid's minimum aperture width.

The anchorage strength or pullout resistance is a result of three separate mechanisms, as illustrated in figure 3.4. The first is the shear strength along the top and bottom of the longitudinal ribs of the geogrid. The second is the shear strength contribution along the top and bottom of the transverse ribs. The third mechanism is the passive resistance against the front of the transverse ribs. In the last mechanism, the soil goes into a passive state and resists pullout by means of bearing capacity. It has been analytically shown that this bearing capacity can be a major contributor to the overall anchorage strength of geogrids [5]. It indeed is a geogrid's forte and can be used admirably in this behavioral mode. Experimental evidence follows the same trends. Example 3.1 describes this effect.

Example 3.1:

For an idealized relatively stiff geogrid, such as the one shown in figure 3.4, calculate the anchorage capacity of each of its three strength components and the percent contribution due to bearing capacity (BC). The arbitrary dimensions are L = 900 mm by W = 300 mm. The longitudinal ribs are at 50 mm spacings and the transverse ribs are at 100 mm spacings. All ribs are 15 mm wide by 3.5 mm thick. In the analysis use a shear strength of 14.4 kPa (= 25 tan 30°) and a bearing capacity 800 kPa based on a soil friction angle of 35°.

Solution: The formulation based on figure 3.4 is as follows. Note that no numeric deduction was taken at rib cross-over points. The anchorage force at failure is

$$
\begin{aligned}
A &= 2\,(\Sigma LR_s + \Sigma TR_s)\,\tau + (\Sigma TR_b)\,q_o \\
&= 2[(0.015 \times 0.900)6 + (0.015 \times 0.300)9] \\
&\quad 14.4 + [(0.0035 \times 0.300)9]800 \\
&= 2.33 + 1.17 + 7.56 \\
A &= 11.06 \text{ kN}
\end{aligned}
$$

and the percent contributed by bearing capacity is

$$BC = (7.56/11.06)100$$

$$BC = 68\% \text{ of the total anchorage force}$$

Note that the degrees of mobilization of the three components of anchorage resistance during pullout are functions of the load-extension properties of the longitudinal ribs and the flexibility and load-extension properties of the transverse ribs [5].

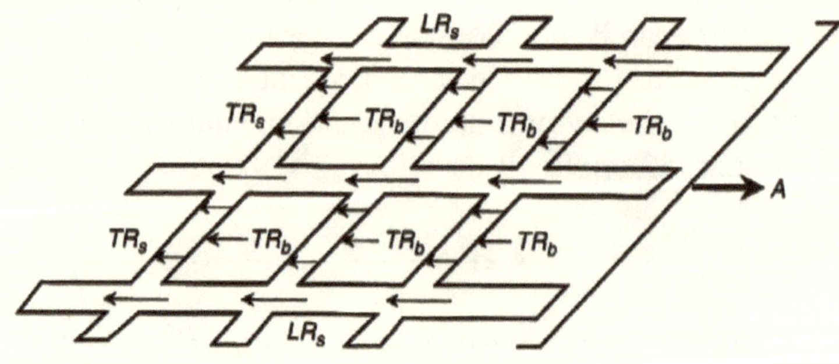

Legend

A = Total anchorage (pullout) strength
LR_s = Longitudinal rib shear strength
TR_s = Transverse rib shear strength
TR_b = Transverse rib bearing strength

Figure 3.4 Mechanisms involved in geogrid anchorage strength. (After Koerner et al. [5] and Wilson-Fahmy and Koerner [6])

The following considerations are important for a soil pullout test setup to determine anchorage strength:

- The test box must be deep enough to permit soil deformation above and below the geogrid as it pulls out of the soil mass. For gravel-size soils, this probably requires 300 mm of soil above and below the geogrid.
- The test box must be long enough to allow for the applied stress on the geogrid to dissipate fully. The number of transverse ribs that are required is dependent on both the geogrid structure and soil type. A box *at least* 1 m long is necessary.
- With such a large size test box, functioning at a high normal stress, the total forces involved can be enormous. This requires a very strongly braced and supported containment system.
- Using a sleeve insert, the geogrid must be gripped from within the encapsulated soil mass. If gripped from outside of the test box, passive pressure will be set up against the face of the box, which will impose additional (and quite unknown) resisting stresses on the front portion of the test specimen.
- Geogrids, being quite strong, will require a high-strength withdrawal system for the actual geogrid pullout (or tensile failure) to occur.
- To monitor the geogrid's deformation behavior, a number of deformation telltales on different parts of the embedded geogrid are necessary—that is, the incremental movement should be monitored. These telltales are often steel wires attached to the geogrid's nodes and extend out through the back of the box to which dial indicators are attached. Alternatively, strain gages attached to the longitudinal ribs can be used.

Figure 3.5 shows plan and elevation views of a soil pullout box for testing the anchorage behavior of geogrids. While the above described test is clearly not simple (the test is probably the most complex and expensive of all geosynthetic performance tests), many such tests have been conducted. Results from the above type of soil pullout testing are shown in figure 3.6.

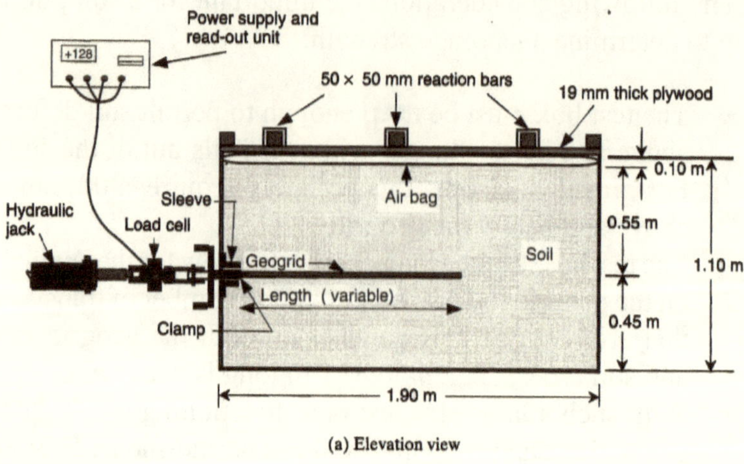

(a) Elevation view

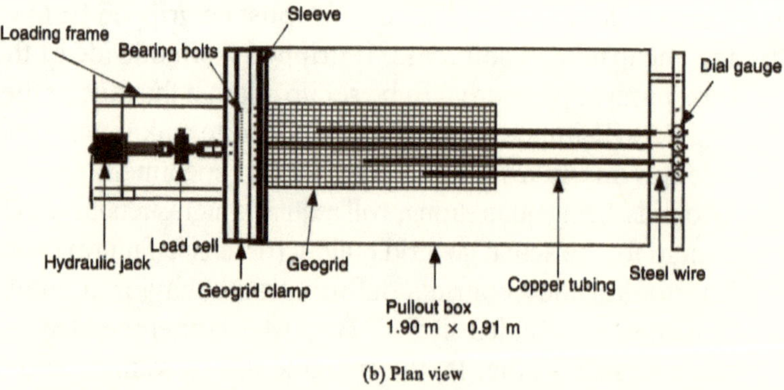

(b) Plan view

Figure 3.5 Diagrams of soil pullout box for evaluating the anchorage behavior of geogrids.

Analysis of the type of data shown in figure 3.6 leads to the determination of an interaction coefficient C_i, which can be used for the design of a specific type of geogrid embedded in the anchorage zone behind a potential failure plane. The value of C_i is soil type and test parameter specific. For example, if a geogrid anchorage test is conducted to failure by sheet pullout, the following equation can be formulated:

$$A = 2C_iL_e\sigma'_n \tan\varphi' \qquad (3.4)$$

where

A = anchorage capacity per unit width (kN/m),
C_i = interaction coefficient (dimensionless),
L_e = length of geogrid embedment (m),
σ_n' = effective normal stress in the geogrid (kPa), and
φ' = effective soil friction angle (deg.).

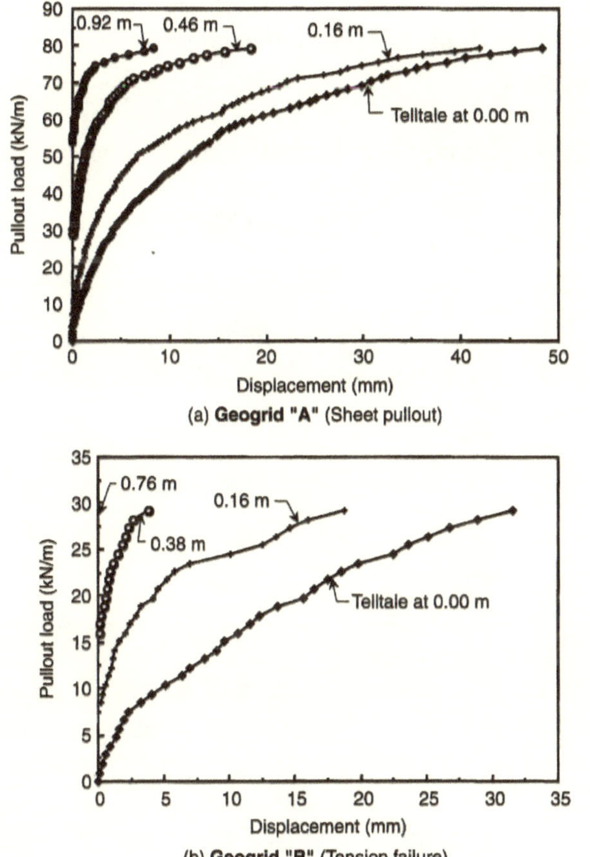

Figure 3.6 Results from selected geogrid pullout tests in a well-graded concrete sand at 69 kPa normal stress at a pullout rate of 1.5 mm/min. and geogrid length of 0.92 m. (After Wilson-Fahmy and Koerner [6])

Note that the value of φ' is for the soil alone and not the soil-to-geogrid value. Also equation 3.4 could be modified to handle cohesive soils, but usually granular soils are selected for backfill materials and if not, the omission of a cohesion term leads to a conservative design.

Example 3.2:

Given the pullout data of figure 3.6a, where the length was 0.92 m, the normal stress was 300 mm of 19 kN/m³ soil plus a 70 kPa surcharge, and the effective friction angle was 35°, what is the interaction coefficient for this geogrid in this particular type of well-graded sandy soil?

Solution: First effective normal stress is calculated:

$$\begin{aligned} \sigma_n' &= (0.3)\,(19) + (70) \\ &= 75.7 \text{ kPa} \end{aligned}$$

Then the interaction coefficient is calculated using equation 3.4 along with A_{ult} = 80 kN/m taken from figure 3.6:

$$\begin{aligned} 80 &= 2\,C_i\,(0.92)\,(75.7) \tan 35° \\ 80 &= 97.5\,C_i \\ C_i &= 0.82 \end{aligned}$$

Note that the consideration of a soil pullout test, such as described here, completely avoids the issue of in-isolation junction strength (recall the preceding discussion of single rib and junction strength). The geogrid junctions in this test are challenged the same way they are in-situ, by having a specific normal stress applied to them through soil embedment. If the junctions are inadequate, the system will fail at a low tensile stress, and this will be reflected by a relatively low value of interaction coefficient.

Wall Connection Anchorage Strength. When geogrids are used to construct reinforced retaining walls, the front edge generally terminates with a facing panel (mechanical connection) or modular block/welded wire facing (friction and/or mechanical connection).

The capability of the geogrid's connection to the wall facing should generally be evaluated. Based in part by research in the literature (e.g., Bathurst and Simac [7]) ASTM has developed the D6638 Test Method. It is primarily focused on modular concrete blocks with a geogrid in its proper location and orientation, at a predetermined normal stress on the blocks. The geogrid is tensioned until failure. Failure can come about in numerous ways—from geogrid tension and connection failure to geogrid slippage or block wall failure. The test nicely exposes the mode of failure and also the ultimate strength of the entire anchorage system. However, recognize that there are a large number of modular-block wall systems, many with a matching geogrid type and facing type. Thus, this test must use the specific materials that will be used in the actual wall construction. Furthermore, substitutions cannot be allowed at the time of bidding or during construction via value engineering or the like unless acceptable data is available.

Torsional Rigidity. The in-plane torsional rigidity (or rotational stability) of bidirectional or tridirectional geogrids can be evaluated using GRI-GG9. In this test method, an unsupported geogrid specimen is fixed on its four sides in a horizontally oriented containment box. Its central node is then clamped by a torquing device that has the capability of applying moment to the geogrid structure and of simultaneously measuring the resulting rotation. The modulus of the angular rotation versus moment curve is the desired value of geogrid torsional rigidity in units of mm-kg/deg. The focused application of this test method is the use of geogrids in paved or unpaved roadway base courses in such a manner as to challenge the entire geogrid structure, i.e., all sets of ribs and the junctions as well.

3.1.3 Endurance Properties

As geogrids are used in critical reinforcement applications, some of which require long service lifetimes, it is generally necessary to evaluate selected endurance properties. Installation damage, creep and accelerated test methods will be addressed.

Installation Damage. As with all geosynthetics, the placement of geogrids in the field requires a considerable degree of planning and care. As happens all too often with careless field construction crews and heavy machinery, installation damage of the geogrid can occur. Other uncertainties in this same area are coarse soil impingement, falling objects, and other accidents that may occur before the geogrid is covered. A few studies have been conducted whereby geogrids have been exhumed after installation and subsequently tested with comparisons being made to the as-received material. Loss in strength has often occurred. In-house investigations (recall figure 2.18, which had nine geogrids in the total study) show that strength reductions of 0 to 30% are possible (see Koerner et al. [8]). There is a formalized procedure available to assess installation damage, namely, ISO 10722, Hsieh and Wu [9] and Sprague and Allen [10] give details and results for several geogrid samples. Generally, the higher strength-loss values come about where large, poorly graded, quarried aggregate is used and heavy construction equipment performs the placement and compaction. If it is necessary to use such materials and methods, it is prudent to first place a cushioning layer of sand above and sometimes below the geogrid.

Tension Creep Behavior. A major endurance property involving geogrids is their sustained-load deformation or tension "creep." Since all polymers used in the manufacturing of geogrids consist of long-chain molecules arranged in crystalline regions with intersperced amorphous regions, the creep response reflects on the percent crystallinity and the glass transition temperature (T_g) (recall section 1.2). In general, this structure is reflected in the following manner:

- For nonoriented polyolefins (polypropylene and polyethylene), which function below T_g, the molecular chains slip along one another within the crystalline regions. Note that polyolefins are, in general, highly crystalline.
- For oriented polyolefins (also below T_g), the orientation creates a molecularly fibrous (or *affine*) structure, and when creep occurs it does so between fibers in the oriented, and stressed, direction.
- For polymers like polyester, polyamide and polyvinyl alcohol, which function above T_g, creep slippage hardly occurs in the crystalline region. Here the chains break at the interface of the crystalline and amorphous regions. In this regard creep rupture (described in the next subsection) is more likely to occur before a limiting deformation is reached.

Apart from the molecular structure and T_g values, creep is predominantly a function of stress level, time, temperature, and a number of environmental factors to be discussed later. Creep has been extensively evaluated on many geogrids, with results for a particular HDPE product given in figure 3.7a. Note that up to approximately 29.8 kN/m, which is 41% of the breaking load of this particular geogrid, the creep deformation is within a creep-limited strain value of 10%, which has been extrapolated as being equivalent to a 120-year design life. The allowable, or working, stress will reflect this type of information in that the reduction factor against creep will be the inverse of 41%, which is 2.4. Note in table 2.8a that PE and PP geotextile yarns require creep reduction factors that are generally higher (i.e., 3.0 to 4.0), due to the lack of geogrid orientation effects as occurs in the manufacture of homogeneous PE and PP geogrid materials. The tension creep test has been adopted as ASTM D5262 and ISO 13431. Creep test data on polyolefin geogrids can be portrayed as isochronous creep data, also shown in figure 3.7b. The creep behavior is readily observed in such a graph as is the 10% strain limit, which is generally considered as the maximum allowable amount.

Creep Rupture Behavior. A variation of the tension creep test just described is the creep rupture procedure presented by Ingold et al. [12]. In this procedure, higher stresses are imposed on the test specimens, causing failure to occur in a relatively short time. Upon performing a number of such tests, a graph of load-versus-log time can be generated. When extrapolated out to the desired service-lifetime, an acceptable load can be obtained. When normalized to the short-term value, the inverse of this ratio becomes the reduction factor to be applied on ultimate strength (see Miyata [13]).

Accelerated Testing Methods. Both tension creep and creep rupture can be evaluated more rapidly than conventionally done by recognizing that elevated testing temperature accelerates the relevant mechanisms. The concept utilizes time-temperature-superposition (TTS) to produce a series of curves at various elevated temperature steps. The resulting curves are then shifted along the horizontal axis (the log-time axis) to represent service lifetimes at the site-specific temperature. Based on field measurements, this is usually taken as 10°C (see Hsuan et al. [14]). Using TTS, the testing time is drastically reduced and projections far into the future can be obtained.

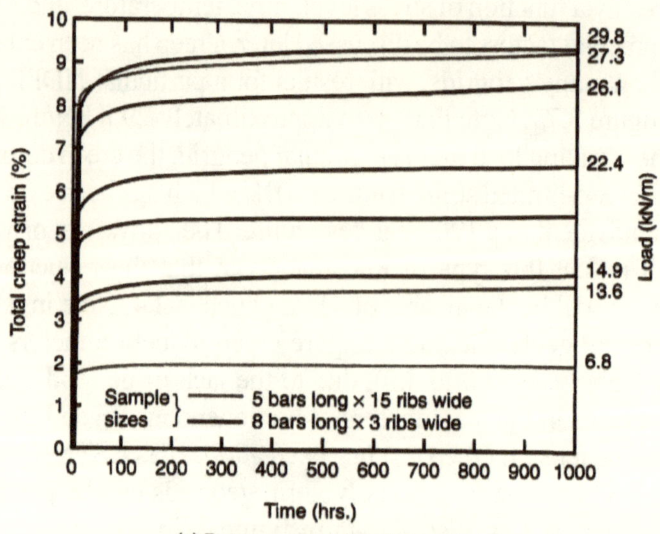

(a) Data generated directly from test

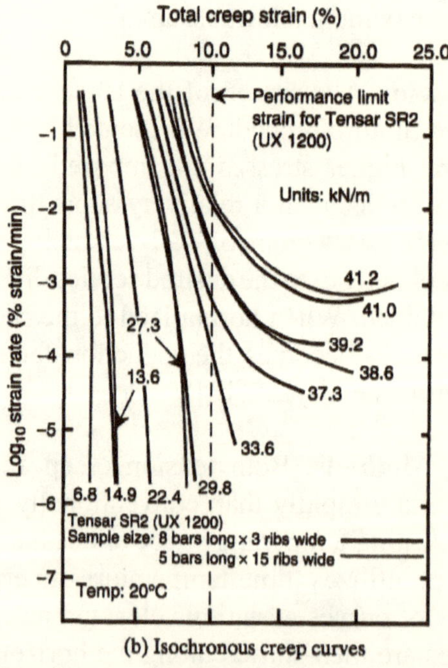

(b) Isochronous creep curves

Figure 3.7 Constant stress deformation (creep) test results on an HDPE geogrid of 72 kN/m ultimate strength. (After McGown, et al. [11]

There are two variations of TTS testing. The first (called standard TTS per ASTM D5292 appendix) uses individual test specimens for each temperature step and load increment (see Farraq and Shirazi [15] and Farraq [16]). The second (called the Stepped Isothermal Method, or SIM, per ASTM D6992) uses the same test specimen and cascades strain measurements at each temperature increment (see Thornton et al. [17] and Greenwood and Voskamp [18]). This has the desirable effect of eliminating test specimen variation in these quite sensitive tests. For different load increments, however, individual test specimens are still required.

The SIM method is particularly revealing when doing investigative research or when comparing products against one another. Yeo and Hsuan (19) show results of an HDPE unitized geogrid at 20% and 40% of ultimate tensile strength (see figure 3.8a) wherein the 20% stress level was less than 10% strain but the 40% stress level greatly exceeded it. Using an average of 30% this infers a RF_{creep} value of about 3.3. They also show results of a PET woven yarn geogrid at 30% and 50% of ultimate tensile strength (see figure 3.8b) wherein neither response exceeds a 10% strain. However, the 50% strain level resulted in creep failure. Again, using an average of 40% this infers a RF_{creep} value of about 2.5. The data nicely shows the influence of geogrids made from different polymers, and in this case the influence of the T_g values and whether creep strain or creep rupture is the predominate strength mechanism.

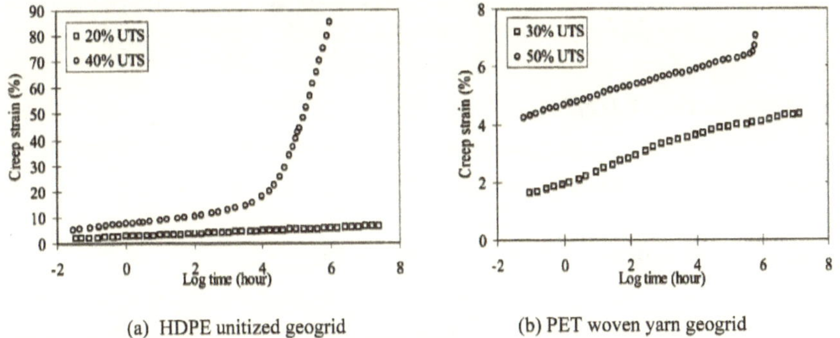

(a) HDPE unitized geogrid

(b) PET woven yarn geogrid

Figure 3.8 Creep master curves from SIM testing of two different geogrid types; Yeo and Hsuan [19])

3.1.4 Degradation Considerations

For all types of geogrids being used in permanent reinforcement applications, it is generally necessary to evaluate selected degradation considerations. This section briefly discusses some of these issues.

Temperature Effects. Given the temperature ranges of typical environments, temperature extremes (hot or cold) should have no serious adverse effects on geogrids. The one caution is that high temperatures can exacerbate strains arising from tension creep, creep rupture, and/or stress relaxation. This requires actual testing at the anticipated field temperatures or the use of TTS techniques as was just described.

Oxidation Effects. This long-term mechanism, which is applicable to polyolefin (HDPE and PP) degradation, was discussed in section 2.3.6.

Hydrolysis Effects. This long-term mechanism, which is applicable to polyester (PET) degradation, was discussed in section 2.3.6.

Chemical Effects. Polyolefins and polyesters have shown excellent resistance to a wide range of chemicals. If unusual conditions exist, however, the situation may dictate specific testing in the actual chemical environment—for example, regarding such conditions as landfill leachates. The laboratory incubation procedure can utilize ASTM D5322 followed by ASTM D6213 for the actual geogrid tests to be performed. Interpretation of the test results, however, is left to the designer.

Radioactive Effects. Unless high-level radioactive materials are in the immediate vicinity, low-level and mixed radioactive materials should pose no problem to geogrids.

Biological Effects. The discussion on the general lack of biological degradation of geotextiles in section 2.3.6 is applicable for geogrids, with the possible exception of the coatings on flexible geogrids. Latex, bitumen, or plasticizers in PVC may be sensitive to microorganisms, but no studies in this regard are available to my knowledge. Even if such attack on the coatings were possible, the high crystallinity PET or PVA fibers (to which these geogrids owe their strength) would remain unaffected.

Sunlight (UV) Effects. As with all polymeric materials, ultraviolet degradation can occur over time, and degradation of the polymer will follow. The discussion in section 2.3.6 applies to geogrids as well as with geotextiles. Regarding the issue of timely cover, the exposure time of geogrids can be considerably longer than geotextiles. This is due to the thickness of the ribs of the polyolefin geogrids and the good UV existence of PET and PVA geogrids. A specification should not be left open-ended, however, and a suggested thirty-day maximum exposure before covering is recommended.

The situation for exposed wrap-around geogrids at the face of MSE walls and steep slopes is quite another matter. These systems are very popular since they avoid the costs of special facings like modular masonry blocks. In such cases, the geogrid is required to have lifetimes of 75-100 years, and even longer for landfill berms. The method of time-temperature-superposition using ultraviolet weathering devices, followed by Arrhenius modeling of the resulting data, is ongoing (20). Lifetime details using this technique are in an advanced stage for geomembranes (buried and exposed) and will be presented in chapter 5.

Stress Crack Resistance. Highly crystalline polymers are sometimes sensitive to brittle cracking while under stress. The test used to evaluate this tendency is ASTM D5397, which describes the notched constant tensile load (NCTL) test and the single-point version (SP-NCTL) of the appendix to D5397. Both are explained in chapter 5, which addresses HDPE geomembranes. Only highly crystalline polyethylene geogrids are of concern and a review with respect to them is given by Wrigley [21]. It is not known to be a problem insofar as field failures are concerned.

3.1.5 Allowable Strength Considerations

The basis of the design-by-function concept is the establishment of a factor of safety. For geogrids, where reinforcement is the primary function, this factor of safety takes the following form:

$$FS = \frac{T_{allow}}{T_{reqd}} \tag{3.5}$$

where

FS = factor of safety; (to accommodate unanticipated loading conditions and uncertainties in design or testing),

T_{allow} = allowable tensile strength from laboratory testing, and

T_{reqd} = required tensile strength from design of the particular field situation.

The allowable value comes from a tensile test of the type described in section 3.1.2, where we must compare the laboratory test setup versus to the intended field situation. If the test method is not completely field-simulated, the laboratory value must be suitably adjusted. This will generally be the case. Thus, the laboratory-generated tensile strength is usually an ultimate value, which must be reduced before being used in design, thus $T_{allow} < T_{ult}$. As with geotextiles (recall section 2.4), we place reduction factors on each of the items not modeled in the laboratory test. For example, the following equation should be considered [22]:

$$T_{allow} = T_{ult}\left[\frac{1}{RF_{ID} \times RF_{CR} \times RF_{CBD}}\right] \tag{3.6}$$

where

T_{ult} = ultimate tensile strength from a standard in-isolation tensile test,

T_{allow} = allowable tensile strength to be used in equation 3.5 for final design purposes,

RF_{ID} = reduction factor for installation damage,

RF_{CR} = reduction factor for avoiding excessive creep or creep rupture over the duration of the structure's lifetime, and,

RF_{CBD} = reduction factor against long-term chemical and biological degradation.

Note that some of these values may be 1.0 or slightly above 1.0, and may therefore be inconsequential. Still others, not specifically mentioned in equation 3.6, may be included as the situation warrants. For example, reduction factors against ultraviolet degradation (RF_{UV}), field seams (RF_{seam}), or penetrations (RF_{pen}) may be included on a site-specific basis. Guidelines for the usual reduction factor values are given in table 3.2.

TABLE 3.2 RECOMMENDED REDUCTION FACTOR VALUES FOR USE IN EQ. (3.6) FOR DETERMINING ALLOWABLE TENSILE STRENGTH OF GEOGRIDS

Application Area	Reduction Factor Values		
	RF_{ID}	RF_{CR}	RF_{CBD}
Paved roads	1.2 to 1.5	1.5 to 2.5	1.1 to 1.7
Unpaved roads	1.1 to 1.6	1.5 to 2.5	1.0 to 1.6
Embankments	1.1 to 1.4	2.0 to 3.0	1.1 to 1.5
Slopes	1.1 to 1.4	2.0 to 3.0	1.1 to 1.5
Walls	1.1 to 1.4	2.0 to 3.0	1.1 to 1.5
Foundations	1.2 to 1.5	2.0 to 3.0	1.1 to 1.6
Veneer Covers	1.1 to 1.4	1.5 to 2.5	1.1 to 1.6

Also note that ranges are given rather than specific values. It is necessary to consider each item individually and make a conscious decision as to how important it is for the site-specific situation. For example, the largest is the creep reduction factor—hence its importance for proper evaluation. In examples 3.3 and 3.4 the values used are assumed on the basis of a hypothetical project and construction method.

Example 3.3:

What is the allowable geogrid tensile strength to be used in the construction of an unpaved road separating stone base from subgrade soil if the ultimate strength of the geogrid is 80 kN/m?

Solution: Using estimated values from table 3.2 in equation 3.6, the following results:

$$T_{allow} = T_{ult}\left[\frac{1}{RF_{ID} \times RF_{CR} \times RF_{CBD}}\right]$$

$$= 80\left[\frac{1}{1.3 \times 2.0 \times 1.5}\right]$$

$$= 80\left[\frac{1}{3.9}\right]$$

$$T_{allow} = 20.5\, kN/m$$

Example 3.4: ______

What is the allowable geogrid tensile strength to be used in the construction of a permanent wall adjacent to a major highway if the ultimate strength of the geogrid is 70 kN/m?

Solution: Using estimated values from table 3.2 in equation 3.6 gives:

$$T_{allow} = T_{ult}\left[\frac{1}{RF_{ID} \times RF_{CR} \times RF_{CBD}}\right]$$

$$= 70\left[\frac{1}{1.3 \times 2.5 \times 1.3}\right]$$

$$= 70\left[\frac{1}{4.22}\right]$$

$$T_{allow} = 16.6\, kN/m$$

Note that these examples could just as well have been framed so as generate an ultimate strength from a given allowable value. This would be the case if we were working from an analytical method that generated a design value. This design value (as with the allowable) would have to be *increased* by reduction factors to arrive at a required (or ultimate) tensile strength.

3.2 DESIGNING FOR GEOGRID REINFORCEMENT

The primary function of geogrids being reinforcement; this section will proceed from one reinforcement application area to another. The order will parallel that of sections 2.6 and 2.7 on geotextile reinforcement, with the addition of several areas that are unique to geogrids.

3.2.1 Paved Roads—Base Courses

The use of geogrids in paved road aggregate base courses is an area where the large aperture size of geogrids provide an excellent advantage. Here the geogrids are placed within the granular base course, typically crushed stone, with the intention of providing an increased modulus, hence a lateral confinement to the system. This lateral confinement is intended to resist the tendency for the base course aggregate to *walk out* from beneath the repetitive traffic loads imposed on the concrete—or bitumen-pavement surface. The situation is applicable for the ballast beneath railroad tracks as well, and perhaps even more so due to nature and intensity of the dynamic loads.

A number of laboratory tests have been conducted to assess the potential benefits and mechanisms involved, significant of which is the work of Haas [23] and Abd El Halim [24, 25]. In a large test setup measuring 4.0 m long by 2.4 m wide by 2 m deep and using 10 kN loads applied sinusoidally at a frequency of 10 Hz on a 300 mm diameter circular plate, five test series (called *loops*) were performed. Loop 1 compared the response of nonreinforced and reinforced sections using both dry (strong) and saturated (weak) subgrade conditions. Failure appeared in the nonreinforced sections earlier than the reinforced sections under both conditions. Loop 2 provided data that shows little difference in elastic deflection between the four trials. More significant was the angle of curvature and the elastic strain at the bottom of the asphalt pavement. Both indicate a 50% reduction for the reinforced sections, thereby indicating a significant load-spreading phenomenon. The permanent surface deformation of the reinforced section is substantially improved over the nonreinforced section. At a 20 mm failure assumption, the nonreinforced section carried 110,000 load repetitions, compared to 320,000 for the reinforced case. In the context of the discussion on geotextiles used in the control of reflective cracking of paved roadways, this would be called a geogrid effectiveness factor (*GEF*) equal to 2.9.

Loop 3 investigated the equivalent thickness that can be attributed to the reinforcement. The results indicate that the 150 mm reinforced section carried about 80,000 load cycles compared to only 34,000 load cycles for the 200 mm nonreinforced and 92,000 loads cycles for the 250 mm nonreinforced. In other words, 150 mm of reinforced asphalt nearly compared to 250 mm of nonreinforced asphalt. Loop 4 confirmed these results, in that reinforced sections result in a savings of 50 to 100 mm of nonreinforced asphalt. Loop 5 involved pressure cells in the soil subgrade and confirmed the load-spreading capability of the reinforcement.

Studies such as this typically indicate that a reinforcement function is provided to the pavement system by the geogrid, albeit by a rather complex set of mechanisms. Some possible contributors include the following: increasing initial stiffness, decreasing in long-term vertical deformation, decreasing long-term horizontal deformation, increasing tensile strength, reducing cracking, improving cyclic fatigue behavior, and simply holding the system together. This leads to difficulties as far as a specific design methodology is concerned. We could use a geogrid effectiveness factor and divide it into the design traffic number to determine a modified value and design accordingly, that is,

$$DTN_R = \frac{DTN_N}{GEF} \tag{3.7}$$

where

DTN_R = design traffic number for the geogrid-reinforced case,
DTN_N = design traffic number under standard (nonreinforced) conditions (e.g., using the Asphalt Institute's procedures), and
GEF = geogrid effectiveness factor ($\cong$ 3.0 for the unitized, homogeneous geogrid evaluated).

Carroll et al. [26] have further refined the technique using the same experimental data to calculate a structural number as per AASHTO [27]. Using the concept of a structural number, the nonreinforced (control) section is

$$SN = 25a_1d_1 + 25a_2d_2 \tag{3.8}$$

where

SN = structural number,
a_i = layer coefficients (0.40 for asphalt and 0.14 for granular stone base), and
d_i = thickness (mm) of each layer.

Using a soil subgrade support value S, obtained from a CBR test, the number of 80 kN single-axle equivalents for any cross section can be calculated. A load-correction factor is then calculated for geogrid-reinforced sections. An estimate of the reinforced-pavement SN is derived, and a ratio for reinforced-to-nonreinforced section is generated. When plotted against the actual reinforced base course thickness, this ratio is seen to be linear. Different values, but the same trend, are seen for geogrids placed in the middle and at the bottom of the base course. A design chart that enables a conventional nonreinforced base course thickness to be converted to a geogrid-reinforced section is given in figure 3.9. Note that a transition occurs at 250 mm, where

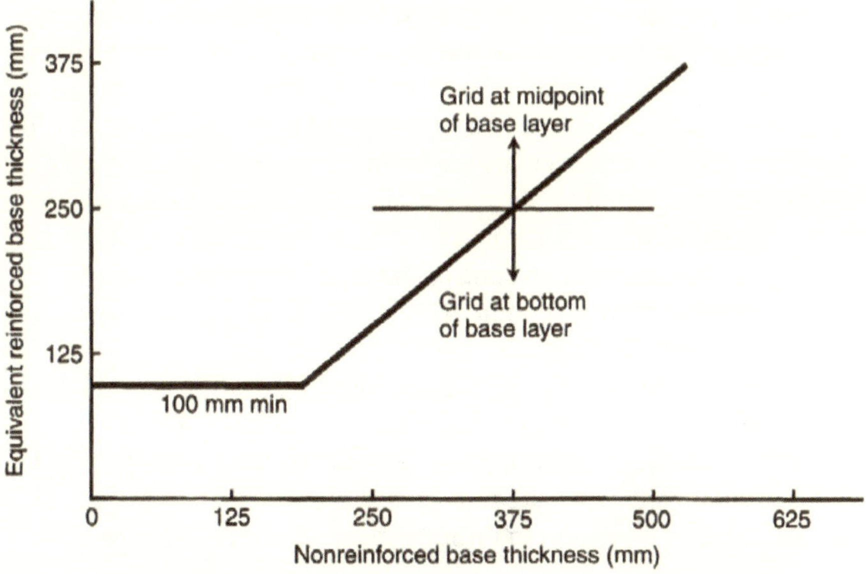

Figure 3.9 Geogrid-reinforced base course for paved highway section using HDPE geogrids, after Carroll, et al. [26].

the geogrid can be placed either in the middle or at the bottom of the base course. It is important to recognize that this curve is based on experimental data for the specific geogrid used. An equivalency between geogrids is difficult to suggest. Longitudinal and transverse rib strength, modulus in both directions, and junction strength are all included in the reinforcement mechanisms, but to what degree needs product-specific investigation.

Newer design methods are also available. They include the "linear M-E method" and "MEPDG method." See Reck (28) for a detailed description of each method.

3.2.2 Paved Roads—Pavements

There is also ongoing research on the placement of geogrids directly within the pavement itself (bitumen or concrete). This section pertains to both new construction and rehabilitation of existing pavements (i.e., the retardation of reflective cracks). Brown et al. [29] have reported that at high deformations of the pavement surface, geogrids clearly minimize rutting. However, at low deformations, the improvement is nominal. Keeping the geogrid tight during its placement (and possibly even prestressing it) appears to be logical, but it is clearly difficult to achieve. Equipment and techniques are described by Kennepohl and Kamel [30]. The material and type of geogrid is very important, since asphalt will not easily bond to the surfaces of polyethylene, polypropylene, or polyester rod (strap), but can easily do so for yarn-type geogrids, particularly if bitumen coated. The influence of geogrid shrinkage during placement of hot asphalt may be a problem for molecular stress relaxation and loss of strength or modulus for highly oriented geogrids. In spite of the above comments, success has been reported in the prevention of reflective cracking using geogrids as crack arresters [29].

The use of geogrids to retard and minimize reflective cracking within old pavements from propagating through newly placed asphalt overlays is a topic of great interest. Results of laboratory testing by Molenaar and Nods [31] suggest the use of a power law to calculate the rate of crack propagation through the new overlay thickness:

$$\frac{dc}{dN} = AK^n \tag{3.9}$$

where

$\frac{dc}{dN}$ = crack propagation rate per number of load cycles,
K = stress intensity factor, and
A, n = experimentally obtained constants.

Example 3.5 illustrates how equation 3.9 can be used in the prediction of overlay lifetime without, and then with, different types of geogrids and a geotextile.

Example 3.5:

A 100 mm asphalt overlay is to be placed on top of a severely cracked pavement having a cement treated base. The DTN for the pavement is 100,000 load repetitions (cycles) per year. The combined overlay, existing asphalt layer and base profile, yields a design stress intensity factor (K) of 10 N/mm$^{1.5}$ and constants A of 1.0×10^{-8} and n of 4.3. **(a)** Calculate the average rate of crack growth of the new asphalt overlay. At a full-propagation failure assumption, what is the lifetime (in terms of number of cycles and years) of the new asphalt overlay without reinforcement? **(b)** Redo the problem using the inclusion of various geosynthetic reinforcement materials with A values as follows:

nonwoven geotextile: $A_{GT} = 0.50\ (A_{\text{non-reinf.}})$—author estimate
polypropylene geogrid: $A_{PP} = 0.35\ (A_{\text{non-reinf.}})$—author estimate
polyester geogrid: $A_{PET} = 0.33\ (A_{\text{non-reinf.}})$—Ref. [28]
fiber glass geogrid: $A_{FG} = 0.25\ (A_{\text{non-reinf.}})$—author estimate

Solution: (a) Using the power law of equation 3.9, the crack propagation rate is calculated, from which the number of cycles and lifetime are obtained. The crack-propagation rate is

$$\frac{dc}{dN} = AK^n$$
$$= 1 \times 10^{-8} \times (10)^{4.3}$$
$$= 0.0002\, mm/cycle$$

from which the number of load cycles (non reinforced) is

$$N = \frac{T}{(dc/dN)}$$
$$= \frac{100}{0.002}$$
$$= 500{,}000 \textit{ cycles or 5 years}$$

(b) Using the various modified *A* values for different types of geosynthetic reinforcement gives rise to the table below.

Reinforcement	Crack Growth Rate (mm/cycle)	Lifetime (cycles/years)
None	2.0×10^{-4}	500,000/5
Geotextile	1.0×10^{-4}	1,000,000/10
PP geogrid	0.7×10^{-4}	1,400,000/14
PET geogrid	6.6×10^{-5}	1,500,000/15
FG geogrid	5.0×10^{-5}	2,000,000/20

The technique is very intriguing and warrants additional research in this important transportation engineering application.

While many field trials are also ongoing, it is important to note that geogrids are being used directly in new pavements and in asphalt overlays to resist thermally induced stresses. This is an important consideration when designing an asphalt pavement or its overlay. The change in temperature between night and day and from season to season, the change in temperature can be as much as 55°C. The contraction during this temperature shift is considerable, as shown in example 3.6.

Example 3.6: ______

Calculate the contraction of a 25 m long section of asphalt pavement undergoing a decrease in temperature of 55°C, assuming that the coefficient of expansion/contraction of the asphalt pavement is 12×10^{-6} per 1°C.

Solution: The calculation is as follows:

$$\Delta L = (25)\,(55)\,(12 \times 10^{-6})$$
$$\Delta L = 0.0165 \text{ m}$$

The resulting 16.5 mm could be in the form of a single crack or many smaller cracks. This depends on the condition of the pavement, primarily its oxidation since its original placement.

Bidirectional and tridirectional geogrids are being placed on existing pavements, generally with an adhesive attached or placed separately, and then covered with a bituminous overlay. Clearly, the tensile strength of the geogrid is mobilized by such thermally induced contraction stresses and probably in a very localized region(s) where the cracks initiate. Thus, the necessity for high-tensile strength is apparent. All types of polymeric geogrids are being used in this application, as are geogrids made from fiberglass. Fiberglass has some excellent tensile strength characteristics in this regard (e.g., high strength, low elongation, high modulus, and low creep). In addition, there are some ongoing attempts at using geogrids to reinforce portland-cement concrete pavements in a manner similar to that described here with asphaltic pavements.

3.2.3 Unpaved Roads

To those already involved with geosynthetics the use of geogrids in the base course of unpaved roads on soft soils is completely intuitive. Whatever the specific reinforcement mechanism (tension membrane effect for soft subgrade, lateral confinement against spreading, torsional confinement against distortion, etc.) the geogrid tends to allow for a reduced base course thickness over the nonreinforced condition. Figure 3.10 illustrates reduction percentages for various types of geogrids and geotextiles as a function of soil subgrade CBR (recall table 2.10). Here it is seen that geogrids are very effective with reductions of up to 50% of the unreinforced counterpart. Of course, one needs to initially calculate the unreinforced thickness and there are indeed methods to do so. Likewise, there should be analytic methods for calculating the reduction rather than heuristic charts such as given in figure 3.10.

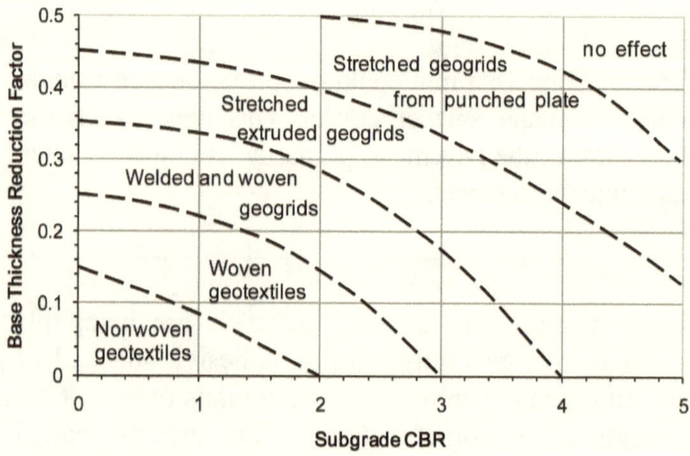

Figure 3.10 Reduction of roadway base course thickness using various geosynthetics. (After van Gurp and van Leest [32])

That said, an analytic method, and the one that will be used here, is that of Giroud, et. al. [33]. The method follows along lines similar to those described in section 2.6.1. The nonreinforced situation is handled first, and then new concepts are developed for the reinforced case. Here the mechanisms of reinforcement are increased soil strength, enhanced load spreading, and membrane support via controlled rutting. The difference in required thickness of stone base is then compared to the cost of the installed geogrid. If the latter is less expensive (as it usually is for soft soil subgrades), it is recommended for use.

For the nonreinforced case, a US Army Corps of Engineers unpaved road formula has been adapted [33] that includes the number of vehicle passages. For the geogrid-reinforced case, new concepts are developed that include the above-mentioned beneficial mechanisms attributed to inclusion of the geogrid. The effects are as follows:

1. *An increase in soil subgrade strength* from the nonreinforced case to the reinforced case as indicated by a comparison of the following equations:

$$p_e = \pi c_{uN} + \gamma h_o$$
$$p_{\lim} = (\pi + 1) c_{uN} + \gamma h$$

where

p_e = bearing capacity pressure based on the elastic limit (nonreinforced case),
$p_{\lim}$ = bearing capacity pressure based on the plastic limit (reinforced case),
c_{uN} = undrained soil strength at the *N*th vehicle passage,
γ = unit weight of aggregate,
h_o = aggregate thickness without reinforcement, and
h = aggregate thickness with reinforcement.

2. *An improved load distribution* to the soil subgrade due to load spreading, which is quantified on the basis of pyramidal geometric shape. Figure 3.11 shows the angle α_0 for the nonreinforced case versus a similar construction for the reinforced case where the new and larger angle is defined as α. The ratio of reinforced to nonreinforced situations is expressed as a ratio of $\tan \alpha / \tan \alpha_0$, which is greater than 1.0.
3. *A tensioned membrane effect*, which is a function of the tensile modulus and elongation of the geogrid and the deformed surface of the subgrade soil (i.e., the rut depth).

By taking the combined effect of the first two above-mentioned geogrid reinforcement mechanisms and comparing it to the nonreinforced case, Giroud et al. [33] have developed the design chart shown in figure 3.12. The membrane effect has been conservatively neglected. On the right side of the graph, for a standard axle load of 80 kN

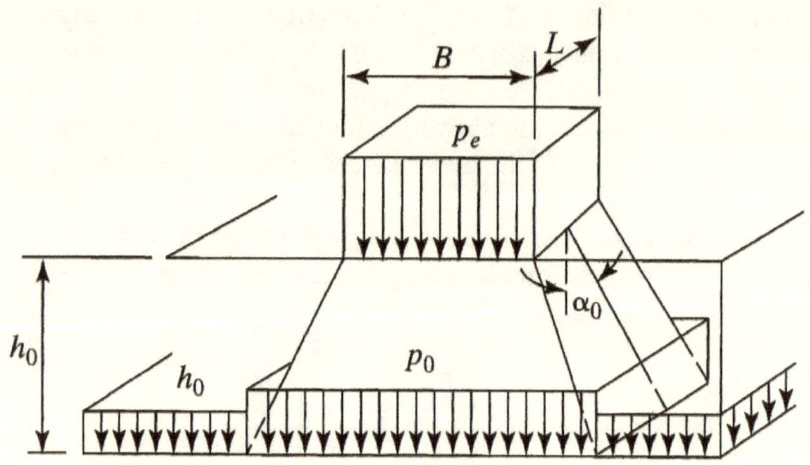

Figure 3.11 Concept of pyramidal load distribution. (After Giroud et al. [32])

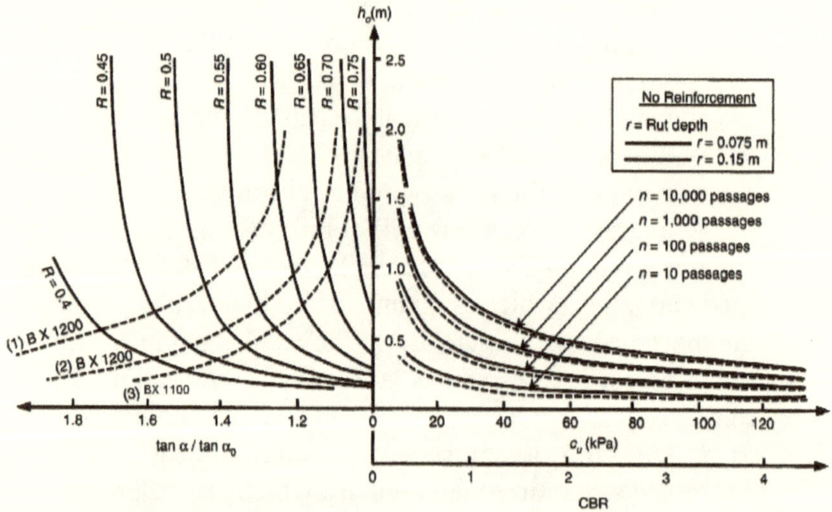

Figure 3.12 Design chart for geogrid-reinforced (left side) and nonreinforced (right side) unpaved roads. (After Giroud, et al. [32])

and any number of vehicle passes from 10 to 10,000, a thickness of nonreinforced stone base (h_0) can be obtained on estimation of the soil subgrade strength. The rut depth turns out to be relatively insignificant. This value is then extended to the left side of the figure, where it is intersects with either;

- Curve 1, for BX 1200 geogrids, which assumes a large number of vehicle passes (N > 1000) where there is a significant likelihood of aggregate contamination without the geogrid.
- Curve 2, also for BX 1200 geogrids, which assumes a low number of vehicle passes and low likelihood of aggregate contamination.
- Curve 3, for BX 1100 geogrids, which assumes a low number of vehicles passes and low likelihood of aggregate contamination.

This results in an R value that is used in the following equations to determine the aggregate thickness using geogrid reinforcement, h. The difference between h_0 (nonreinforced) and h (reinforced) is the amount of aggregate saved, Δh.

$$h = Rh_0 \text{ for } r < 150 \text{ mm and no channelized traffic pattern}$$
$$h = 0.9\,Rh_0 \text{ for } r \geq 150 \text{ mm with a channelized traffic pattern}$$

Example 3.7:

A soil subgrade has a *CBR* strength of 1.0, and is to carry 1000 standard-axle vehicle passes with a maximum rut depth of 75 mm. What is the required aggregate depth without a geogrid, the aggregate depth with BX 1200 geogrids with a low likelihood of aggregate contamination, and the difference in aggregate thickness between the two cases?

Solution: Using figure 3.12, the nonreinforced case gives

$$h_0 = 0.60 \text{ m}$$

For the geogrid-reinforced case, Curve 2 gives $R = 0.50$:

$$\begin{aligned} h &= Rh_0 \\ &= (0.50)(0.60) \\ &= 0.30 \text{ m} \end{aligned}$$

The aggregate saved is

$$\begin{aligned}\Delta h &= 0.60 - 0.30\\ &= 0.30 \text{ m}\end{aligned}$$

3.2.4 Slopes and Embankments

The use of geogrids to reinforce steep soil slopes or embankments (defined as making an angle with the horizontal of less than 70°) directly parallels the techniques and designs that were developed using geotextiles (section 2.7.2). It might be noted that one of the largest geogrid reinforced slope is an airport runway extension in Charleston, West Virginia. It is a 1(H)-to-1(V) slope and is 74 m high [34]. The use of limit equilibrium methods via a circular arc failure plane, thereby intercepting the various layers of reinforcement, was illustrated in figures 2.38 and 2.39. This allowed for the formulation of a factor of safety expression as follows:

$$FS = \frac{M_R + \sum_{i=1}^{n} T_i y_i}{M_D} \tag{3.10}$$

where

M_R = moments resisting failure due to the soil's shear strength,
M_D = moments causing failure due to gravity, seepage, seismic, dead, and live loads,
T_i = allowable reinforcement strength, providing a force(s) resisting failure,
y_i = appropriate moment arm(s), and
n = number of separate reinforcement layers.

Forsyth and Bieber [35] used this approach to design the reconstruction of a failed slope in California. They selected a desired factor of safety along with a given type of geogrid and calculated the number of layers of reinforcement to realize this value. As the example 3.8 illustrates, they used a very low allowable strength for the geogrids used (i.e., 6.67 kN/m that is only 8.4% of the ultimate value). In the context

of reduction factors as per section 3.1.5, this is equivalent to ΠRF = 11.8, which is extremely high.

Example 3.8:

For a failed soil slope of known centroid and radius resulting in a resisting moment of 2010 kN/m and a driving moment of 2570 kN/m, determine **(a)** the factor of safety without reinforcement, and **(b)** the number of layers of a specific geogrid with an ultimate strength of 78.7 kN/m and combined reduction factors of 11.8. The average centroid of the reinforcement is 14.3 m and the required factor of safety is to be 1.4.

Solution: (a) The factor of safety for the nonreinforced case is the following:

$$FS = \frac{M_R}{M_D}$$

$$= \frac{2010}{2570}$$

$FS = 0.78$; which indicates failure.

(b) The geogrid-reinforced case is as follows:

$$T_{allow} = T_{ult} / (\Pi RF)$$

$$= 78.7/11.8$$

$$= 6.67 kN/m$$

$$FS = \frac{M_R + (n)(T_{allow})(Y_{ave})}{M_D}$$

$$1.4 = \frac{2010 + (n)(6.67)(14.3)}{2570}$$

$$n = 16.6, \text{ use 17 layers}$$

When reconstructing such failed slopes, or when building steep soil slopes and embankments, the main reinforcement layers are usually interspersed with secondary reinforcement layers. These layers aid in compaction at the face of the slope and also tend to reduce surface erosion. Figure 3.13 shows a case of reconstructing a major highway ope and gaining additional space at both the toe and top of the slope.

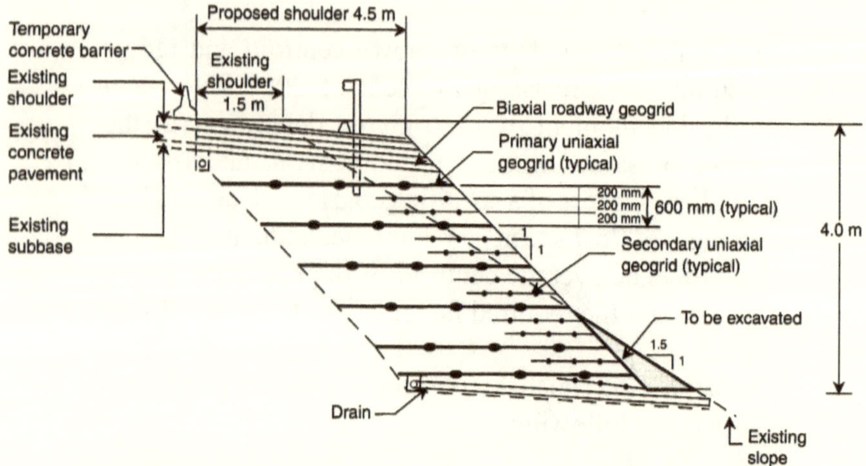

Figure 3.13 Shoulder widening of Pennsylvania Turnpike using geogrid reinforced steep soil slopes. (After Berg, et al. [36])

Example 3.8 did not go into the details of determining the location of the reinforcement layers or their lengths. For these details slope stability methods are nicely adapted to computer modeling and have resulted in a number of excellent design charts. Schmertmann et al. [37] have developed a summary of the various investigators and some of the assumptions that were made in the various studies. All used limit equilibrium methods to determine the reinforcement (geogrid or geotextile) spacing. However, only Jewell et al. [38], Ruegger [39], and Schmertmann et al. [37] have given charts for the required reinforced lengths. The former two methods use constant-length reinforcement placed parallel to the slope face, whereas Schmertmann et al. use gradually decreasing lengths as the layers proceed higher in the slope. For high embankments, where the potential failure surface is curved (e.g., a logarithmic spiral), this is both accurate and more practical. However, for low and medium walls, the more conservative approach of Jewell [38] is favored by the author and will be used here.

For reinforced slopes placed on adequately strong, level foundations, limit equilibrium can be used in the form of a two-part wedge surface as shown in figure 3.14. The design chart includes varying soil properties, slope angles, and geometric considerations (see figure 3.15). Charts are also available for pore water in the backfill soil. Example 3.9 illustrates the use of the chart under the assumption of zero pore water pressure.

Example 3.9:

We plan to construct a soil embankment at a 70° slope angle with the horizontal and 10 m height, to be reinforced with unidirectional geogrids having an ultimate strength of 180 kN/m and combined reduction factors of 4.12. The factor of safety is to be 1.4. The soil is granular, with $\gamma = 18$ kN/m^3, $\varphi = 30°$, and has no pore water pressure (i.e., $r_u = 0$). Determine the number, spacing, and length of the individual geogrid layers.

Solution: By observation, this slope at 70° to the vertical without reinforcement is in a failure state (that is, $FS < 1.0$) and is in need of some type of reinforcement. The design procedure is given in steps.

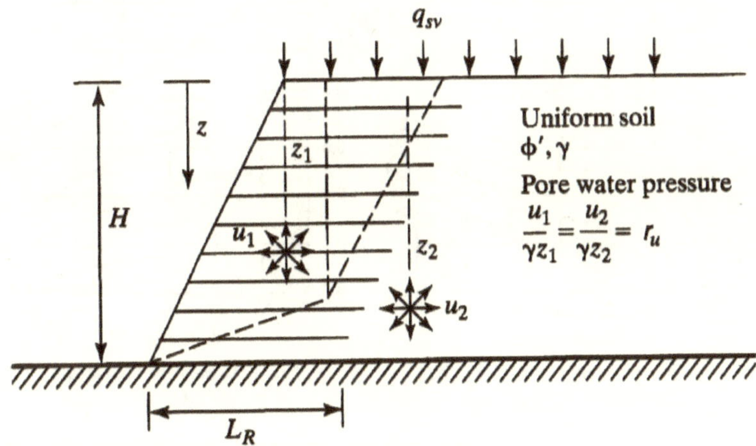

Figure 3.14 Definitions for analysis of steep reinforced soil slopes. (After Jewell [38])

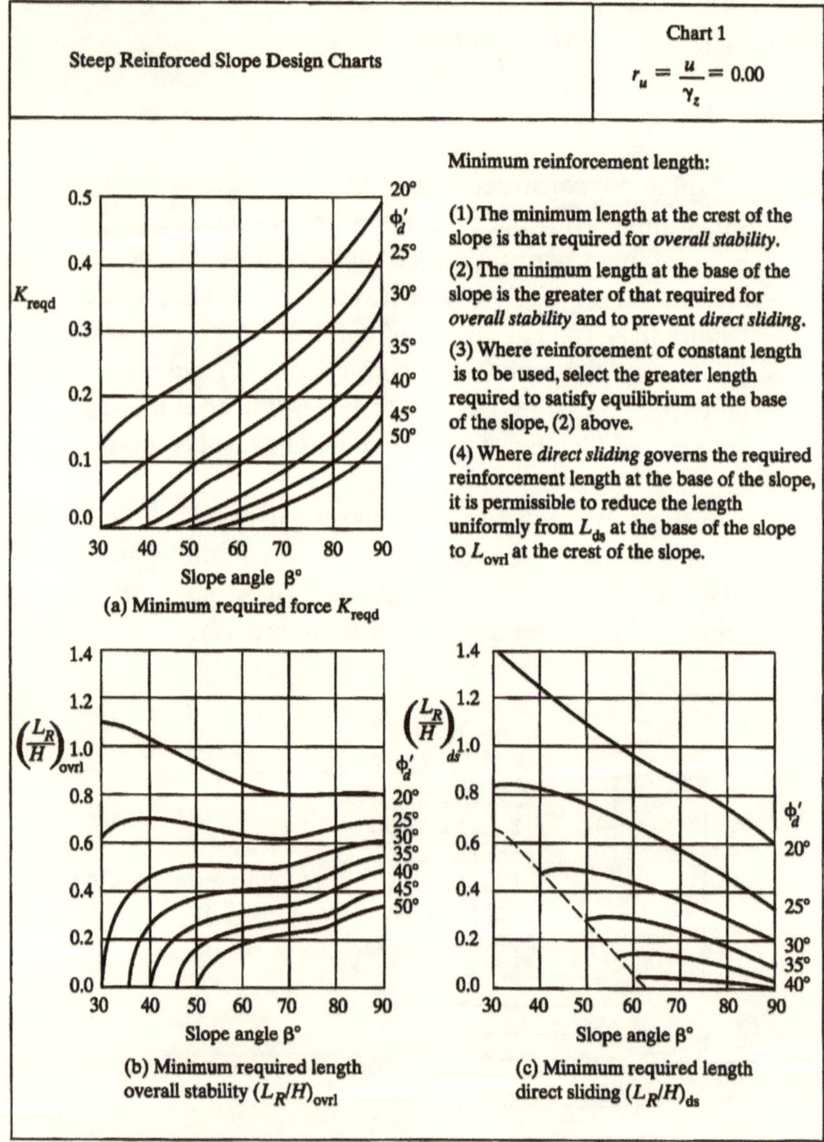

Figure 3.15 Steep reinforced soil slope design charts for zero pore water pressure. (After Jewell [38]).

(a) Calculate the allowable strength on the basis of the given reduction factors and then calculate the design strength, which includes the factor of safety

$$T_{ult} = 180\,kN/m$$

$$T_{allow} = \frac{180}{4.12} = 43.7\,kN/m$$

$$T_{des} = \frac{43.7}{1.4} = 31.2\,kN/m$$

(b) Determine the necessary values from figure 3.15 for $r_u = 0$, $\beta = 70°$ and $\varphi = 30°$. This results in the following

$$K_{reqd} = 0.19$$
$$(L_R/H)_{ovrl} = 0.51$$
$$(L_R/H)_{ds} = 0.38$$

(c) Calculate the spacing, S_v, at the base of the slope where the stresses are greatest

$$S_v = \frac{T_{design}}{K_{reqd}\gamma z_{max}} = \frac{31.2}{(0.19)(18)(10)}$$

$$S_v = 0.91m$$

If evenly spaced, the required number of geogrid layers will be

$$n = \frac{H}{S_v} = \frac{10}{0.91}$$

$$n = 11 \text{ layers}$$

(d) Select the reinforcement length:

- If $(L_R/H)_{ovrl} > (L_R/H)_{ds}$, use constant length $= (L_R/H)_{ovrl}$.
- If not, use constant length $= (L_R/H)_{ds}$ or taper the lengths from $(L_R/H)_{ds}$ at the base to $(L_R/H)_{ovrl}$ at the crest.

Since $0.51 > 0.38$ use $L_R/H = 0.51$

$\therefore L_R = 5.1$ m throughout

(e) The length at the base can be checked by conventional methods using the entire mechanically stabilized earth mass, or according to the equations set forth in [38].

(f) Check among the different geogrid behaviors in the anchorage zone behind the hypothetical shear plane. Such differences must be considered from experimental results as described in section 3.1.2. If there is concern about the use of one geogrid product versus another, the designer always has the option of lengthening the geogrids over that required by figure 3.15.

(g) Sketch the final reinforced slope and provide for miscellaneous details, as shown below; that is, use short (secondary) geogrids between the primary reinforcement and adjacent to the slope for compaction aid and against surface erosion (recall figure 3.13).

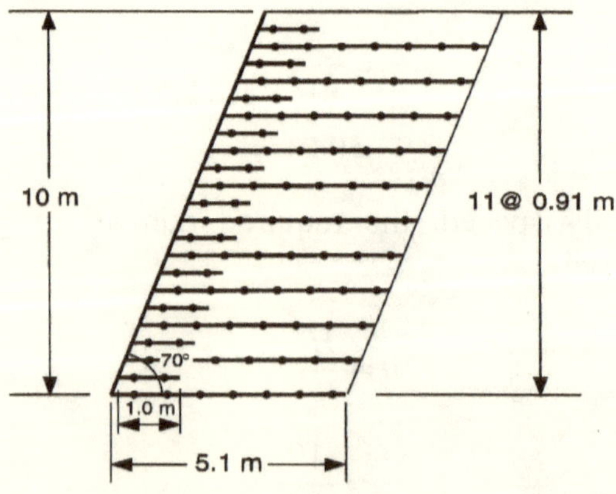

3.2.5 Reinforced Walls

A tremendous number of geogrid reinforced walls (defined as making an angle with the horizontal of 70°, or more) have been constructed in the past 15 years. The current estimate of such walls is 40,000 in the United States and more than double that worldwide. Most have been designed directly for private owners and developers, but some have involved the public sector through direct design or by the process of *value engineering*. The latter is when the low-bid general contractor offers the public agency an option for some particular segment of the project—for example, a geogrid-reinforced wall in place of a conventional reinforced-concrete wall or steel-reinforced wall. If the option is acceptable to the agency, the financial saving between the cost of the two different kinds of walls is shared equally between the agency and the contractor. It is a very effective method for the introduction of new products and concepts like geogrid (and geotextile) reinforced walls. More recently, however, some state highway departments have been designing permanent geogrid-reinforced wing-walls and bridge abutments (Abu-Hejleh et al. [40]). This section addresscs geogrid-reinforced wall facings types, costs, design, soil backfill considerations, and field performance.

The various types of permanent geogrid-reinforced wall facings are as follows. The geogrid reinforced soil mass—properly called a mechanically stabilized earth (MSE) mass—is the same in all cases.

- *Wrap-around facings:* The same as those illustrated in section 2.7.1 with geotextile walls. Note that in order to provide protection against ultraviolet light and vandalism, thus a bitumen or concrete coating is usually applied.
- *Timber facings:* Railroad ties or other large treated timbers on which the geogrid is attached by batten strips and/or held by friction when placed between the timbers.
- *Articulated precast concrete panels:* Discrete precast concrete panels with inserts for attaching the geogrid. Many aesthetically pleasing facing designs are possible.
- *Full height precast panels:* Concrete panels temporarily supported until backfilling is complete. These types of walls, however, are challenging due to vertical stresses developed

on the geogrid connections to the wall after removal of the panel support [41].

- *Cast-in-place concrete panels:* Panels that are often attached to wraparound walls that are allowed to settle and after a few months are covered with a cast-in-place facing panel. These walls are currently favored in Japan [42] where the ends of the geogrid reinforcement are sometimes embedded in gabions, which then have a concrete facing panel poured against them.
- *Gabion facings:* Polymer or steel-wire baskets filled with stone, in which the geogrid held between the baskets and fixed with rings and/or friction.
- *Welded wire-mesh facings:* Similar to gabion facings, the mesh is L-shaped with the geogrids held either mechanically or by friction to the base of the L. The geogrids can contain a seeded erosion control material (see figure 3.16) or have large aggregate placed behind it, i.e., nonvegetated.
- *Masonry block faced walls* (also called *segmental retaining walls* [SRWs]): A variety of different block types in which a geogrid is embedded between the blocks and held by pins, keyways and/or friction. Figure 3.17 presents two examples of such walls. There is a tremendous variety of facing blocks and configurations. This type of wall is seeing the greatest growth due to its pleasing aesthetics, computer-aided design, and low cost. Figure 3.18 presents survey results of retaining wall costs (Koerner et al. [43]) in which geosynthetic reinforced MSE walls are the least costly over all height ranges.

The design of the above-described walls must not be considered trivial, for many of them are critical structures that are meant to be permanent, possessing service lifetimes in excess of 100 years. Design centers around external stability of the entire mechanically stabilized earth mass (sliding, overturning, and bearing capacity), and internal stability within the reinforced mass (geogrid spacing, anchorage length and connection strength). Figure 3.19 illustrates these concepts. Each of these stability issues must be treated individually and before coming together for the final design. Example 3.10 illustrates this using a geogrid-reinforced wall with articulated precast concrete facing panels.

Figure 3.16 Steel wire mesh facing of geogrids reinforced earth wall. Note backup using bidirectional polypropylene geogrids.

Figure 3.17 Geogrid reinforced modular concrete block walls (also called segmental retaining walls).

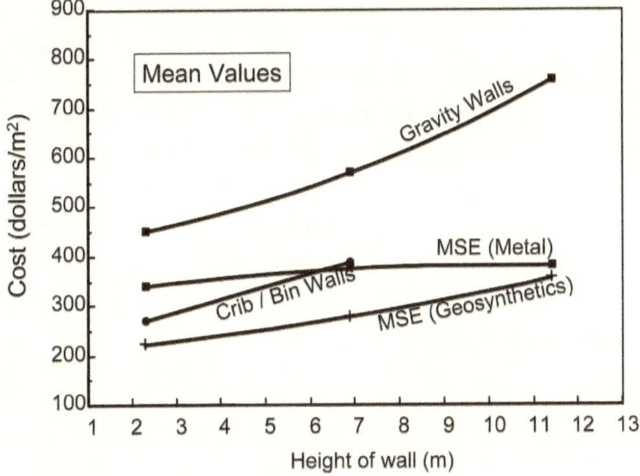

Figure 3.18 Survey results of retaining wall costs in U.S. (After Koerner, et al. [43]).

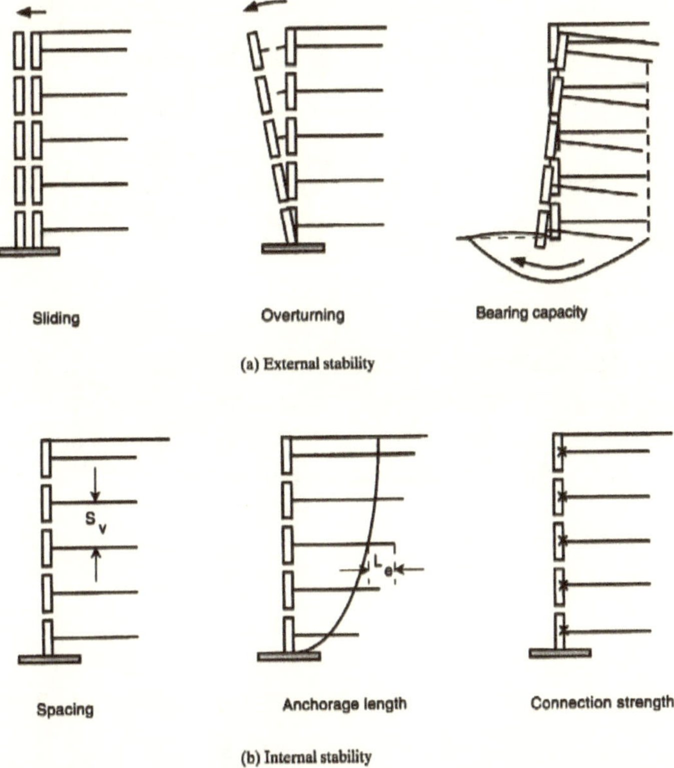

Figure 3.19 Elements of geogrid (or geotextile) reinforced wall design.

Example 3.10:

Design a 7 m high geogrid-reinforced wall where the reinforcement spacing must be at 1.0 m spacings, since the wall facing is of the articulated precast concrete type of this same dimension. The coverage ratio is 0.8 (i.e., geogrids do not cover the entire ground surface at each lift; they are slightly separated). The length-to-height ratio of the reinforced soil wall should not be less than 0.7 (i.e., $L \geq 4.9$ m). Additional details of the problem, including soil and geogrid data are given in the diagram below.

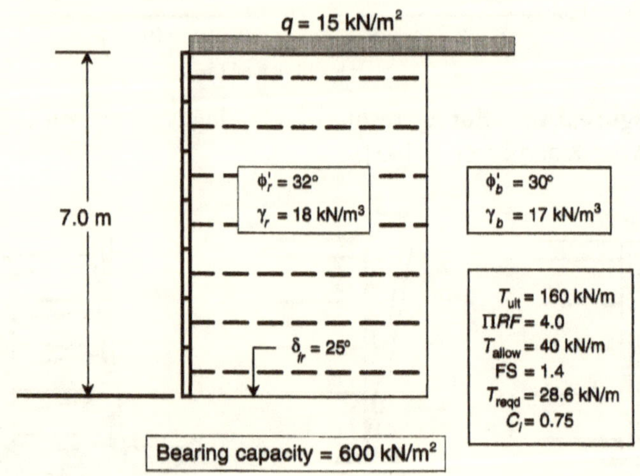

Solution:

(a) Calculate external stability, as shown in the diagram that follows, assuming $L = 4.9$ m. The coefficient of active earth pressure of the backfill soil behind reinforced zone is as follows:

$$K_{a_b} = \tan^2(45 - \phi_b/2) = \tan^2(45 - 30/2) = 0.33$$

Thus,

$$P_1 = 0.5 \times \gamma_b \times H^2 \times K_{a_b} = 0.5 \times 17 \times (7)^2 \times 0.33 = 137\,kN/m$$

$$P_2 = qK_{a_b} \times H = 15 \times 0.33 \times 7 = 34.7\,kN/m$$

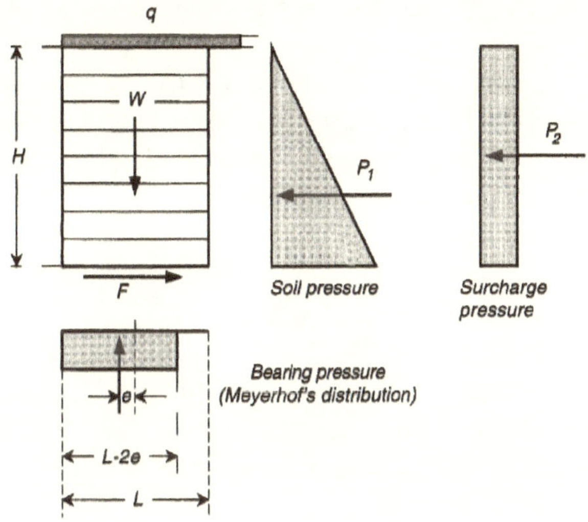

The total force is $P = 137 + 34.7 = 172$ *kN/m*

1. We now calculate the *sliding stability*

$$F = \text{Resisting force} = W \times \mu = \gamma_r \times H \times L \times \tan\delta$$

(neglecting effect of surcharge)

$$= 18 \times 7 \times 4.9 \times \tan 25° = 288 \ kN/m$$

FS_s = factor of safety against sliding

$= F/P = 288/172 = 1.67 > 1.5$; which is acceptable

2. *Overturning stability* is rarely an issue, since this type of mechanically stabilized wall is not subject to overturning because it cannot mobilize bending due to its inherent flexibility. The following calculation illustrates the conservative aspect of this mechanism.

M_s = stabilizing moment $= W \times L/2 = (18 \times 7 \times 4.9) \times (4.9/2)$

$= 1513$ kN/m

M_{ov} = overturning moment $= P_1 \times 7/3 + P_2 \times 7/2$

$= (137 \times 7/3) + (34.7 \times 7/2) = 441$ kN/m

FS_{ov} = factor of safety against overturning

$= 1513/441 = 3.43 > 2.0$; which is acceptable

3. Lastly, calculate the stresses on the *foundation soil*

e = eccentricity = $M_{ov}/(W + q \times L)$
= $441/(18 \times 7 \times 4.9 + 15 \times 4.9) = 0.64$ m
Now, this eccentricity cannot be outside of the central one-third of the footing. That is
$e < L/6 = 4.9/6$
$0.64 < 0.82$; thus, there is no tension beneath the footing
Acting length (Meyerhof's distribution)
= $L - 2 \times e = 4.9 - 2 \times 0.64 = 3.62$ m
Bearing pressure = $[(18 \times 7) + 15] \times (4.9/3.62) = 191$ kPa
FS_b = factor of safety against bearing capacity failure
= $600/191 = 3.14 > 2.0$; which is acceptable

(b) Calculate the internal stability as shown in the diagram below.

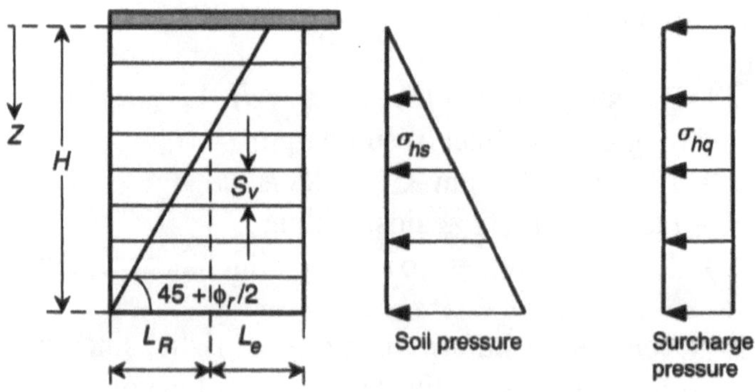

$\sigma_h = \sigma_{hs} + \sigma_{hq}$
$= \gamma z K_{ar} + q K_{ar}$
$K_{ar} = \tan^2 (45 - \varphi_r/2) = 0.31$
$\sigma_h = (18 \times z \times 0.31) + (15 \times 0.31)$
$= 5.58 z + 4.65$

1. For geogrid vertical spacing
$T_{des} = s_v \sigma_h / C_r$ (where C_r = coverage ratio)
$28.6 = s_v (5.58 z + 4.65)/0.8$

$$s_v = \frac{22.9}{5.58z + 4.65}$$

Maximum depth for $s_v = 1$ m

$$1.0 = \frac{22.9}{5.58z + 4.65} \Rightarrow z = 3.27m$$

Maximum depth for $s_v = 0.5$ m

$$0.5 = \frac{22.9}{5.58z + 4.65} \Rightarrow z = 7.37m$$

The layout pattern can now be developed based on the above-calculated maximum spacing values and the type and dimensions of the facing panels. Thus, the reinforcement for the upper three facing panels is at 1.0 m spacing and the remaining four facing panels are at 0.5 m spacing. In the diagram below, the left panel gives these details. It is based on the symmetry of geogrid connections, which means that no tension eccentricity is allowed on any facing panel. Hence, small lengths of geogrid are required for the top facing panels of the wall where the spacing interval is 1.0 m. Also, note that for cross sections located one facing panel adjacent to the illustrated design cross section, the top and bottom facing panels will be half-height, but the reinforcement spacing must be maintained as calculated above (see the right-hand panel in the diagram below.)

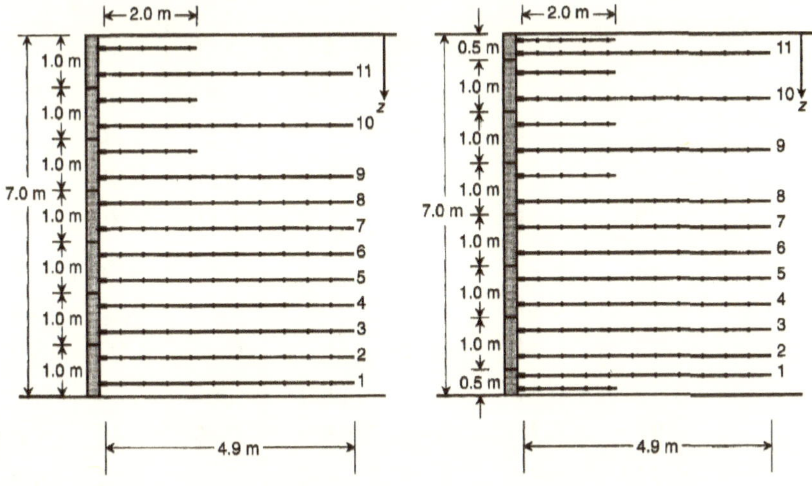

2. For the *total length* we consider the embedment plus the nonacting Rankine length.
   For the embedment length we have:

$$s_v \times \sigma_h \times FS_{pullout} = 2 \times L_e \times C_i \times \sigma_v \tan\phi' \times C_r$$

$$s_v(5.58z + 4.65)1.5 = 2L_e(0.75)(18z)(\tan 32)(0.8)$$

$$L_e = \frac{s_v(5.58z + 4.65)1.5}{(2)(0.75)(18z)(\tan 32)(0.8)}$$

$$L_e = \frac{(0.62z + 0.516)s_v}{z}$$

For the nonacting Rankine length

$$L_R = (H - z)\tan\left(45 - \frac{\phi}{2}\right)$$

$$= (7 - z)\tan\left(45 - \frac{32}{2}\right)$$

$$L_R = 3.88 - 0.554z$$

The above relationships lead to the following table

Layer Number	Depth z (m)	Spacing s_v (m)	L_e (m)	L_{emin} (m)	L_R (m)	L_{calc} (m)	L_{reqd} (m)
11	0.75	0.75	0.98	1.0	3.46	4.46	5.0
10	1.75	1.00	0.92	1.0	2.91	3.91	5.0
9	2.75	1.00	0.81	1.0	2.36	3.36	5.0
8	3.25	0.50	0.39	1.0	2.08	3.08	5.0
7	3.75	0.50	0.38	1.0	1.80	2.80	5.0
6	4.25	0.50	0.37	1.0	1.52	2.52	5.0
5	4.75	0.50	0.36	1.0	1.25	2.25	5.0
4	5.25	0.50	0.36	1.0	0.97	1.97	5.0
3	5.75	0.50	0.36	1.0	0.69	1.69	5.0
2	6.25	0.50	0.35	1.0	0.42	1.42	5.0
1	6.75	0.50	0.35	1.0	0.14	1.14	5.0

3. For *connection strength*, the tensile stress at the connection of the reinforcement to the facing panels should theoretically be very small. This is indeed the case if the panels, reinforcement, and soil backfill stay in horizontal alignment in the same manner as the wall is constructed. Unfortunately, settlement of the backfill (and perhaps foundation soil) usually occurs, thereby deforming the reinforcement and imposing stress on the wall panel connection. The amount of this stress is dependent on the backfill soil type, density, moisture content, compactive effort, and foundation conditions can be very large (see Soong and Koerner [41]).

   Due to this uncertainty, a relatively conservative approach is to use the required strength of the reinforcement as the required connection strength. For the example;

$$
\begin{aligned}
T_{conn} &\geq 1.0\, T_{des} \\
&\geq 1.0\,(28.6) \\
&\geq 28.6 \text{ kN/m}
\end{aligned}
$$

   See section 3.1.2 for the experiment necessary to obtain the allowable connection strength to the wall facing.

4. For drainage behind the wall, it is assumed that hydrostatic pressures are not developed. This assumes in turn that cohesionless sands (or gravels) are used as soil backfill in the reinforced zone. A recommended gradation is given in table 3.3. While coarser particle sizes could have been given (i.e., a gravel soil), the reduction factor for installation damage might be high and gravel is generally more expensive than sand. Comments later in this section will discuss the concern over using fine-grained (silts and/or clays) soils along with the necessity of then using back and base drains to conduct water away from the reinforced soil mass.

TABLE 3.3 RECOMMENDED BACKFILL SOIL GRADATION FOR GEOGRID OR GEOTEXTILE REINFORCEMENT APPLICATIONS (WALLS AND SLOPES) TO AVOID HYDROSTATIC PRESSURE BUILDUP AND EXCESSIVE INSTALLATION DAMAGE

Sieve Size (No.)	Particle Size (mm)	Percent Passing
4	4.76	100
10	2.0	90-100
40	0.42	0-60
100	0.15	0-5
200	0.075	0

It should be noted that most wall designers do not work from the basic principles as just illustrated. There are several widely distributed computer codes used for such wall designs [44]. The program developed by Leshchinsky [45] is most popular because of its technical accuracy, flexibility in input configurations, data generation, and graphical output. Using this particular code, several parametric variations of important design variables produced the data of table 3.4. In each of the specific analyses two types of failure surfaces are considered. One is a piecewise linear or compound failure surface per Spencer analysis that is internal to the reinforced soil zone. The other is a circular arc failure surface per modified Bishop analysis that is generally external to the reinforced zone. The resulting *FS* values are felt to be very instructive insofar as what to do, or what not to do, in order to avoid failure from occurring. In this regard, there have been wall problems with respect to both serviceability (excessive deformation) and actual failures (collapse). While the number of each is relatively small, lessons can be learned by assembling these case histories and analyzing them as a group to see if trends exist. The common situations can then be identified and avoided in the future.

Koerner and Soong [46] report 26 case histories of failures with geosynthetic reinforced walls; most of which are geogrid-reinforced masonry block walls. Twelve are of the serviceability type, wherein excessive deformation occurred. Fourteen are collapses, wherein a portion of the wall actually fell. This is not meant to suggest that more walls collapse than excessively deform; the opposite is more likely. The numbers only appear to indicate that people tend to investigate and publish collapses (vis-à-vis serviceability problems) due to the obvious finality of the situation.

TABLE 3.4 PARAMETRIC VARIATIONS OF SELECTED DESIGN VARIABLES FOR MSE WALLS WITH GEOSYNTHETIC REINFORCEMENT AS PERTAINS TO THE RESULTING FS-VALUES

(a) - Effect of Backfill Soil Shear Strength on FS-Values

Friction Angle (deg)	Internal (Spencer) Stability	External (Bishop) Stability
40	1.50	2.21
35	1.34	2.08
30	1.19	1.91
25	1.07	1.76
20	0.95	1.60

(b) - Effect of Reinforcement Layer Spacing on FS-Values

Spacing (m)	Internal (Spencer) Stability	External (Bishop) Stability
0.50	1.50	2.21
0.75	1.27	2.07
1.00	1.15	2.03
1.25	1.07	1.97
1.50	0.99	1.96
1.75	0.93	1.96

(c) - Effect of Reinforcement Length-to-Wall Height Ratio on FS-Values

"L/H" Ratio	Internal (Spencer) Stability	External (Bishop) Stability
0.7	1.50	2.21
0.6	1.41	2.16
0.5	1.31	2.07
0.4	1.21	2.01
0.3	1.09	1.88
0.2	0.98	1.86

(d) - Effect of Front Soil Exit Angle at Toe of Wall

Exit Angle at Toe (deg)	Internal (Spencer) Stability	External (Bishop) Stability
0	1.50	2.21
10	1.50	1.91
20	1.50	1.63
30	1.50	1.34
40	1.50	1.08

(e) - Effect of Elevated Water Surface Within Reinforced Soil Zone

Internal Water (% H)	Internal (Spencer) Stability	External (Bishop) Stability
0	1.50	2.21
20	1.48	1.71
40	1.39	1.47
60	1.29	1.30
80	1.10	1.10

where H = wall height

(f) - Effect of Water Filled Tension Cracks Behind the Reinforced Soil Zone

Depth (% H)	Internal (Spencer) Stability	External (Bishop) Stability
0	1.50	2.21
10	1.50	2.21
20	1.48	2.21
30	1.42	2.21
40	1.29	2.08
50	0.85	1.33

In the serviceability case histories, large-scale excessive deformation was either at the top, bottom, or throughout the wall. Within the group of seven design-related case histories, five had fine-grained backfill soils in the reinforced zone and two had granular soils throughout. The other five case histories experienced individual block, or localized, distortion. All these appear to have been caused by contractors' activities during construction. The message in both situations appears to be clear: (1) fine-grained silts and clays should be questioned insofar as backfill soils are concerned, and (2) contractors' operations must be monitored and inspected to ensure adequate construction quality control.

In the category of collapsed wall case histories, hydrostatic pressure arising from lack of drainage from fine-grained soil backfills in the reinforced zone was the overriding reason for the failures. This occurred in 10 of the 14 case histories and all are considered as design-related causes. It substantiates the concern over silt and clay backfills seen in the serviceability case histories. In three other failure case histories, contractor deficiencies were observed, again supporting the findings of the serviceability case histories insofar as lack of inspection and installation quality control are concerned.

Interestingly, in only one of the 26 case histories investigated is the problem *not* fine-grained soil backfill or inadequate construction/

inspection procedures. In this case, the wall experienced a global rotational failure behind and beneath the entire wall structure carrying it and a much larger body of soil downslope in a large-scale failure. Once again, it was a fine-grained soil problem albeit not in the reinforced soil zone per se.

If fines (silts and/or clays) are allowed for the reinforced zone backfill soil, any possible water in front, behind and beneath the reinforced zone must be carefully collected, transmitted, and discharged. Proper drainage control is absolutely critical in this regard. Furthermore, the top of the zone should be waterproofed—for example, by a geomembrane or a geosynthetic clay liner—to prevent water from entering the backfill zone from the surface. Surface water drainage as well as drainage from the retained earth zone is obviously of concern with respect to potential buildup of pore water pressures behind or within the reinforced soil zone. (See Koerner and Soong [47] for wall drainage system designs in this regard.)

In closing this section on geogrid reinforced walls, the current tendency to create live (or evergreen) walls with open facing should be mentioned. As we saw earlier in figure 3.16, the sequence is a welded wire mesh (alternatively a gabion), backed by a bidirectional geogrid and then by a seeded geosynthetic erosion control material. The reinforcing geogrids (always unidirectional types) are either attached to the steel mesh facing, or are frictionally connected by sufficient overlap length. Such walls avoid masonry block durability concerns and offer a considerably less expensive wall system. Of course, the durability of the steel wire and biodirectional geogrid backup must be considered and this is an ongoing research topic when considering one-hundred-year-permanent wall lifetimes. Durability is even more relevant when the vegetation is inhibited by using large gravel as the facing backfill material.

3.2.6 Foundation and Basal Reinforcement

Geogrids have been used to increase bearing capacity of poor foundation soils in different ways: as a continuous layer, as multiple closely spaced continuous layers with granular soil between layers, and as mattresses consisting of three-dimensional interconnected cells. The technical database for the single-layer continuous sheets has been reported by Jarrett [48] and by Milligan and Love [49]; in both cases large-scale laboratory tests are used. Figure 3.20 presents some of Milligan and Love's work graphed in the conventional nondimensionalized q/c_u versus ρ/B manner and also as $q/\sqrt{c_u}$ versus ρ/B where q is the bearing and ρ is the settlement. The latter graph is not conventional but does sort out the data nicely. Clearly shown in both instances is the marked improvement in load-carrying

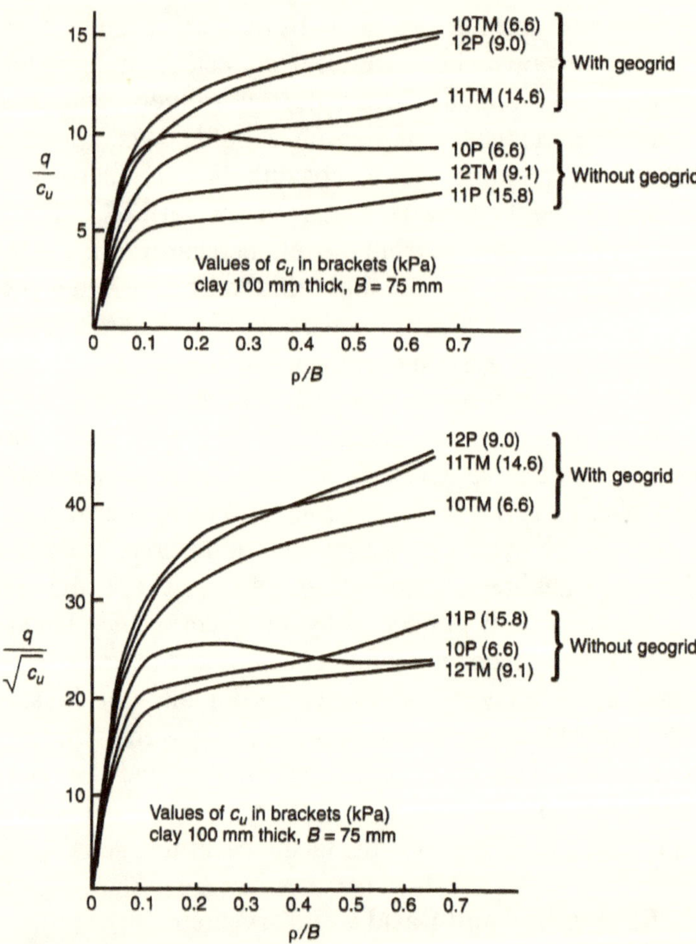

Figure 3.20 Load versus deflection curves of large laboratory tests with and without geogrid reinforcement. (After Milligan and Love [49])

capacity using geogrids at high deformation and only a nominal beneficial effect at low deformation. Beyond these observations, a precise design formulation is not currently available.

Instead of focusing on a global increase in bearing capacity, it is quite likely that single or multiple layers of geogrid (or geotextile) will aid in minimizing or eliminating differential settlement. Here localized settlements due to abruptly settling or subsiding weak zones can be spanned by the layer of reinforcement. This is known as *foundation improvement* (rather than bearing capacity via base reinforcement). Notable in this regard is a technique called *piggybacking*—the construction of new landfills above exiting landfills. The approach is to use arching theory in the calculation of the vertical stress arising from localized subsidence (i.e., differential settlement) and to provide suitably strong reinforcement.

It should be recognized that arching in natural soils overlying a locally yielding foundation is well established. In the 1930s, both Terzaghi in Austria (calculating stresses on deep tunnels) and Marston in the United States (calculating stresses on buried pipelines) developed the analytic theory. Their work resulted in the following simplified formula for vertical stress on the surface of the particular underground structure (tunnel or pipe respectively):

$$\sigma_z = 2\gamma_{ave} R\left[1 - e^{-0.5H/R}\right] + qe^{-0.5H/R} \tag{3.11}$$

where

σ_z = vertical stress on the structure or reinforcement layer,
γ_{ave} = average unit weight of material above the settlement area,
R = radius of differential settlement zone,
H = total height above the settlement area, and
q = surcharge pressure placed at the ground surface.

Note that for large values of H (typically $H \geq 6R$) the formula reduces to the following value of constant vertical stress.

$$\sigma_z = 2\,\gamma_{ave}\,R \tag{3.12}$$

Having a method to calculate the vertical stress, we can now use the value to calculate the stress in the reinforcement layer for a new landfill placed over an existing one. Note that the reinforcement can be either a geogrid or a geotextile. For support over a differential settlement area, the value of T_{reqd} is calculated as follows:

$$T_{reqd} = \sigma_z R \Omega \tag{3.13}$$

where

$$\Omega = 0.25 [(2y)/B + B/(2y)], \text{ where} \tag{3.14}$$

B = width of settlement void, and
y = depth of settlement void.

Giroud et al. [50] have combined the above equations to develop a design chart that can be used to avoid direct calculation (see figure 3.21). Note that the chart can be used for either circular voids or long extended voids.

Once the value of T_{reqd} is determined, it must be compared to T_{allow} using equation 3.6, which includes the site-specific reduction factors. Example 3.11 illustrates the technique.

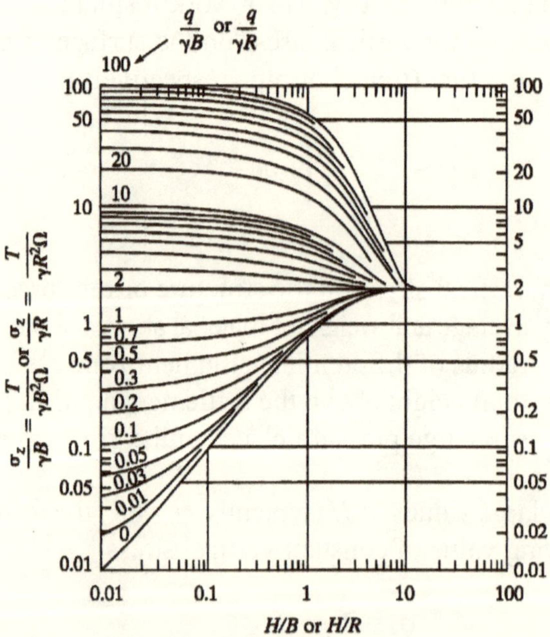

Figure 3.21 Curves of geosynthetic stress and tension that can be used for R (radius of circular void) or B (width of long voids). (After Giroud, et al. [50])

Example 3.11:

Using the Terzaghi/Marston formulation for calculating vertical stress above localized subsidence, in this case differential settlement in an old landfill of radius 1 m **(a)** calculate the required wide-width strength of a reinforcement layer if a new 30 m high landfill is to be placed on the existing one—that is, if the new landfill is to be *piggybacked* on the existing landfill. The compacted unit weight of the waste is 12 kN/m^3. **(b)** Check your calculated value against figure 3.21. **(c)** Calculate the factor of safety for a geogrid with ultimate wide-width tensile strength of 125 kN/m. In the calculations use cumulative reduction factors of 5.0.

Solution:

(a) The formula for vertical stresses in arching situations under a deep fill (such as in this example) reduces to equation 3.12. Therefore the vertical stress is calculated as

$$\begin{aligned}\sigma_z &= 2\gamma_{ave} R \\ &= 2(12)(1.0) \\ &= 24 kPa\end{aligned}$$

To transfer this vertical stress into a horizontal force, we use equation 3.13

$$T_{reqd} = \sigma_z \, R \, \Omega$$

where Ω = strain criterion [recall section 2.5.2]

$$\begin{aligned}\Omega &= 0.97 \text{ at } 5\% \text{ strain} \\ &= 0.73 \text{ at } 10\% \text{ strain}\end{aligned}$$

Assuming $\Omega = 0.73$

$$\begin{aligned}T_{reqd} &= 24 \times 1.0 \times 0.73 \\ &= 17.5 \text{ kN/m}\end{aligned}$$

(b) Check this against Fig. 3.21.

$$\frac{H}{R} = \frac{30}{1} = 30$$

$$\therefore \frac{T}{\gamma R^2 \Omega} = 2.0$$

$$T_{reqd} = 2.0(12)(1)^2(0.73)$$

$$= 17.5\,\text{kN/m; which checks.}$$

(c) The factor of safety on a geogrid with 125 kN/m ultimate strength (at 10% strain) is as follows.

$$T_{allow} = T_{ult} / \Pi RF$$

$$= \frac{125}{5}$$

$$= 25\,\text{kN/m}$$

and

$$FS = T_{allow} / T_{reqd}$$

$$= 25/17.5$$

$$FS = 1.43 \text{ which is acceptable}$$

For localized depressions that do not have large values of material covering them, as in sinkholes beneath roadways, parking lots or even buildings, equation 3.11 must be used in its entirety. Figure 3.22 presents such a study that is framed around a parking lot over a karst-prone area with H = 1.5 m, γ_{ave} = 18 kN/m^3, q = 15 kN/m^2, and R = 1.0 to 5.0 m using various geosynthetic T_{allow} values. Here is seen that the factor-of-safety decreases significantly as the sinkhole radius increases. This can be offset by higher allowable strength geosynthetics but at an obvious increase in cost. It should also be added that seams eventually become problematic when using these higher strength geosynthetics.

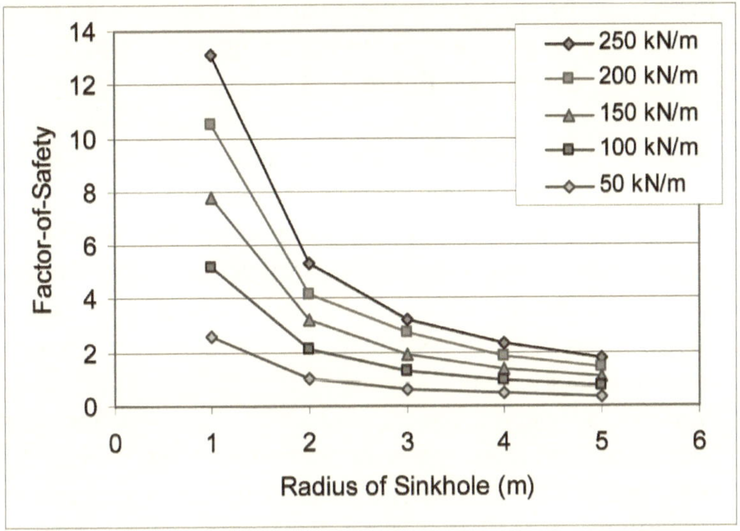

Figure 3.22 Resulting FS-values for various strength geosynthetics over sinkholes using a parking lot as an example.

In a somcwhat different context, but still focused on foundation and basal soil improvement, Edgar [51] reports on a three-dimensional *geogrid mattress* where 1.0 m wide unitized HDPE geogrids are placed vertically and interconnected to one another. Gravel is placed within the geogrid mattress as it is constructed over soft fine-grained foundation soils. In this case a 15 m high embankment was successfully constructed above the mattress. It was felt that the nonreinforced slip plane was forced to pass vertically through the mattress and therefore deeper into the stiffer layers of the underlying subsoils. This improved the foundation stability to the point where the mode of failure was probably changed from a circular arc to a less-critical plastic failure of the soft clay. The design was considered to be a successful and economic one. Another example of a 1 m high unitized geogrid mattress was constructed to support a 30 m high landfill over extremely soft mine tailings in Hausham, Germany [52]. The mattress was filled with gravel and the liner system constructed above it. The foundation soil was so soft that a nonwoven geotextile and a bidirectional geogrid had to be initially placed to provide a stable working area for the construction of the three-dimensional mattress. Such relatively thick mattresses can also be constructed by using closely spaced layers of bidirectional geogrids separated by granular soil.

In the design of such three-dimensional geogrid mattresses it is felt that a number of phenomena are occurring, all of which improve foundation soil stability (see figure 3.23).

- *Global slope stability:* This is improved by forcing the potential failure plane through the mattress and deeper into the foundation soil. It is also possible that the foundation soil may improve in strength characteristics at greater depths.
- *Bearing capacity:* This is improved in a similar manner to the point where it becomes a nonissue for mattresses greater than approximately 30 m in width.
- *Lateral extrusion (or squeeze-out):* This is undoubtedly decreased because stress concentrations have been largely eliminated via a uniform pressure distribution applied through the relatively stiff geogrid mattress.

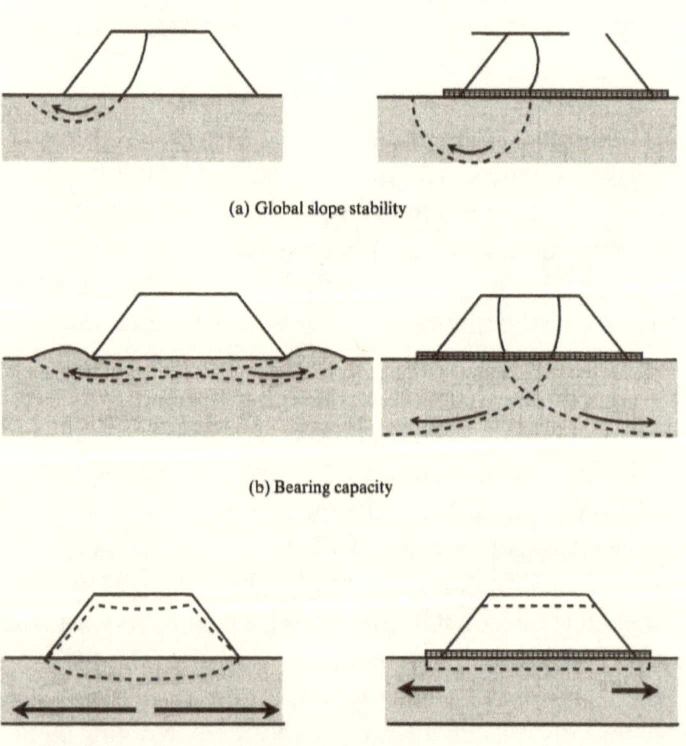

Figure 3.23 Potential improvement of embankments on soft foundation soils via three- dimensional geogrid mattresses.

In the absence of global instability, this last item is particularly important. Squeeze-out of the foundation soil is the likely service-limiting mechanism giving rise to excessive deformations. Robertson and Gilchrist [53] and Jenner et al. [54] have used slip line fields to predict the principle stresses in the soft foundation soils. Both studies give actual case histories and the monitoring feedback as to the validity of the design assumptions.

The latest application area in the context of foundation and basal soil reinforcement is the use of geogrids to span deep foundations placed through compressible soils [55,56]. The geogrids span from pile cap-to-pile cap, reducing localized settlement in the supported embankment system. From a design standpoint, the situation is exactly the same as described in section 2.7.4 using geotextiles. High-strength geotextiles and geogrids are competitive in this particular application. The technique is considered very appropriate when stone columns are used as the ground modification technique. Sometimes the stone columns are actually contained in a geogrid enclosure, which appears to be a growing application [57].

3.2.7 Veneer Cover Soils

Whenever a lined slope (geomembrane, GCL, or compacted clay) is covered with soil, a stability calculation should be made to assess the potential for sliding failure of the soil on the barrier layer. Four situations come to mind: landfill liners with leachate collection sand or gravel above them until such time that the solid waste acts as a passive resistance restraint; surface impoundment liners where the cover soil is placed over the geomembrane to shield it from ultraviolet light, heat degradation, and equipment damage; landfill covers that have topsoil and protection soil placed over the geomembrane; and general slopes and embankments containing geotextiles or erosion control materials being covered with a layer of soil. In all cases the soil layer is relatively thin (0.3 to 1.0 m), hence the sliding stability of such a veneer of cover soil is the issue.

Due to the typically low shear strength of the covering soil to the underlying geosynthetic material, numerous stability problems have arisen. The driving forces creating the instability are gravitational forces, equipment loads, surcharge loads, seepage forces and/or seismic forces. Each must be carefully considered in the context of the site-specific conditions.

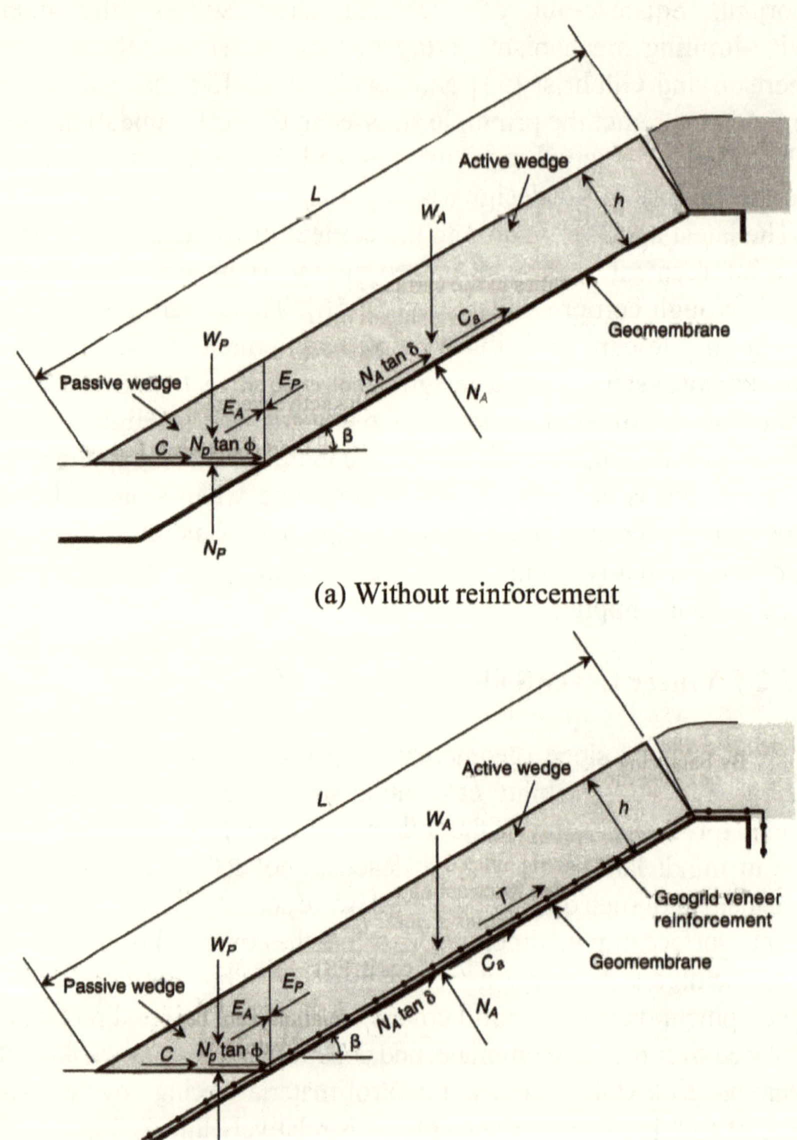

Figure 3.24 Limit equilibrium forces involved in a finite length slope analysis for a uniformly thick cover soil. (After Koerner and Soong [58].

Koerner and Soong [58] have analyzed the general situation through use of limit equilibrium and a finite slope model, as shown in figure 3.24. Consider a cover soil placed directly on a geomembrane (or other geosynthetic layer) at a slope angle β. Two discrete zones can be visualized, as shown in figure 3.24a. There is a small passive wedge near the toe of the slope resisting a long thin active wedge extending the length of the slope. It is assumed that the cover soil is of uniform thickness and constant unit weight. At the top of the slope or at an intermediate berm, we anticipate that a tension crack in the cover soil will occur, thereby breaking continuity with the remaining cover soil at the crest.

Resisting the tendency for the cover soil to slide is the interface friction and/or adhesion of the cover soil to the specific type of underlying geomembrane. The shear strength values of δ and c_a must be obtained from a laboratory direct-shear test as described earlier. Note that the passive wedge is assumed to move on the underlying cover soil so that the shear strength parameters φ and c, which come from soil-to-soil friction tests, will also be required.

By taking free bodies of the passive and active wedges with the appropriate forces being applied, the formulation for the factor of safety results. The resulting equation is not an explicit solution for the FS, and must be solved using the quadratic equation. The complete development of the equation is given in [58]. Other approaches are found in Giroud and Beech [59], Koerner and Hwu [60], and Thiel and Steward [61]. The expression for determining the factor of safety, considering the active wedge, can be derived as follows:

$$W_A = \gamma h^2 \left(\frac{L}{h} - \frac{1}{sin\beta} - \frac{tan\beta}{2} \right) \tag{3.15}$$

$$N_A = W_A \, cos\beta \tag{3.16}$$

$$C_a = c_a \left(L - \frac{h}{sin\beta} \right) \tag{3.17}$$

By balancing the forces in the vertical direction, the following formulation results:

$$E_A \, sin\beta = W_A - N_A \, cos\beta - \frac{N_A \, tan\delta + C_a}{FS} sin\beta$$

Hence the interwedge force acting on the active wedge is:

$$E_A = \frac{(FS)(W_A - N_A\,cos\beta) - (N_A\,tan\delta + C_a)sin\beta}{sin\beta(FS)}$$

The passive wedge can be considered in a similar manner:

$$W_p = \frac{\gamma h^2}{sin 2\beta} \tag{3.18}$$

$$N_P = W_P + E_P\,sin\beta \tag{3.19}$$

$$C = \frac{(c)(h)}{sin\beta} \tag{3.20}$$

By balancing the forces in the horizontal direction, the following formulation results:

$$E_P\,cos\beta = \frac{C + N_P\,tan\phi}{FS}$$

Hence the interwedge force acting on the passive wedge is

$$E_P = \frac{C + W_P\,tan\phi}{cos\beta(FS) - sin\beta\,tan\phi}$$

By setting $E_A = E_P$, the following equation can be arranged in the form of $ax^2 + bx + c = 0$, which in our case, using *FS* values, is

$$a(FS)^2 + b(FS) + c = 0 \tag{3.21}$$

where

$$a = (W_A - N_A\,cos\beta)cos\beta, \tag{3.22}$$

$$b = -[(W_A - N_A\,cos\beta)sin\beta tan\varphi + (N_A tan\delta + C_a)sin\beta cos\beta + sin\beta(C + W_p tan\varphi)] \text{ and} \tag{3.23}$$

$$c = (N_A tan\delta + C_a)sin^2\beta tan\varphi \tag{3.24}$$

The resulting *FS* value is then obtained from the following equation:

$$\mathrm{FS} = \frac{-\mathrm{b} + \sqrt{\mathrm{b}^2 - 4\mathrm{ac}}}{2\mathrm{a}} \tag{3.25}$$

where (in figure 3.24a and the above analysis)

W_A = total weight of the active wedge,
W_P = total weight of the passive wedge,
N_A = effective force normal to the failure plane of the active wedge,
N_P = effective force normal to the failure plane of the passive wedge,
γ = unit weight of the cover soil,
h = thickness of the cover soil,
L = length of slope measured along the geomembrane,
β = soil slope angle beneath the geomembrane,
φ = friction angle of the cover soil,
δ = interface friction angle between cover soil and geomembrane,
C_a = adhesive force between cover soil of the active wedge and the geomembrane,
c_a = adhesion between cover soil of the active wedge and the geomembrane,
C = cohesive force along the failure plane of the passive wedge,
c = cohesion of the cover soil,
E_A = interwedge force acting on the active wedge from the passive wedge,
E_P = interwedge force acting on the passive wedge from the active wedge, and
FS = factor of safety against cover soil sliding on the geomembrane.

When the calculated *FS* value falls below 1.0, a stability failure of the cover soil sliding on the geomembrane is to be anticipated. Thus, a value of greater than 1.0 must be targeted as being the minimum factor of safety. How much greater than 1.0 the *FS* value should be is a design and/or regulatory issue. Example 3.12 illustrates the procedure.

Example 3.12:

Given a cover soil slope of $\beta = 18.4°$ (i.e. 3H-to-1V), $L = 30$ m, $H = 900$ mm, $\gamma = 18$ kN/m^3, $c = 0$, $\varphi = 30°$, $c_a = 0$, $\delta = 18°$, determine the resulting factor of safety.

Solution:

$$W_A = \gamma h^2\left(\frac{L}{h} - \frac{1}{sin\beta} - \frac{tan\beta}{2}\right)$$

$$= (18.0)(0.90)^2\left(\frac{30}{0.90} - \frac{1}{sin18.4} - \frac{tan18.4}{2}\right)$$

$$= 14.58(33.3 - 3.17 - 0.17)$$

$$= 437kN/m$$

$$N_A = W_A\,cos\beta$$

$$= 437cos18.4$$

$$= 415kN/m$$

$$W_P = \frac{\gamma h^2}{sin\,2\beta}$$

$$= \frac{(18.0)(0.90)^2}{sin36.8}$$

$$= 24.3kN/m$$

$$a = (W_A - N_A\cos\beta)\cos\beta$$

$$= (437 - 415\cos18.4)\cos18.4$$

$$= 41.0 \text{ kN/m}$$

$$b = -[(W_A - N_A\cos\beta)\sin\beta\tan\phi + (N_A\tan\delta + C_a)\sin\beta\cos\beta + \sin\beta(C + W_P\tan\phi)]$$

$$= -[(437 - 415\cos18.4)\sin18.4\tan30 + (415\tan18 + 0)\sin18.4\cos18.4 + \sin18.4(0 + 24.3\tan30)]$$

$$= -[7.84 + 40.4 + 4.43]$$

$$= -52.7 \text{ kN/m}$$

$$
\begin{aligned}
c &= (N_A \tan\delta + C_a)\sin^2\beta \tan\phi \\
&= (415 \tan 18 + 0)\sin^2 18.4 \tan 30 \\
&= 7.8 \text{ kN/m}
\end{aligned}
$$

$$
\begin{aligned}
FS &= \frac{-b + \sqrt{b^2 - 4ac}}{2a} \\
&= \frac{52.7 + \sqrt{(52.7)^2 - 4(41.0)(7.8)}}{2(41.0)}
\end{aligned}
$$

FS = 1.11, which is too low for a final cover and an appropriate design option is to consider the use of geogrid veneer reinforcement.

Figure 3.24b illustrates a growing application of geogrid reinforcement under the generic classification of *veneer reinforcement.* The geogrid embedded in its own anchor trench at the top of the slope is placed directly on the geomembrane. Soil backfilling (with lightweight construction equipment) then proceeds from the toe to the crest of the slope. As backfill is placed, the geogrid reinforcement is tensioned and, depending on the strength of the reinforcement, some or all the gravitational stress of the soil is resisted. In the analysis that follows, the soil is assumed to be in contact with the geomembrane (acting through the apertures of the geogrid), the reinforcement is functioning at its allowable value (hence, reduction factors must be applied to the ultimate value), and the active wedge has included in it an additional vector—namely the allowable geogrid tension, T. For the active wedge, we balance the forces in the vertical direction and the following formulation results.

$$
E_A \sin\beta = W_A - N_A \cos\beta - \left(\frac{N_A \tan\delta + C_a}{FS} + T \right) \sin\beta
$$

Hence the interwedge force acting on the active wedge is

$$
E_A = \frac{(FS)(W_A - N_A \cos\beta - T \sin\beta) - (N_A \tan\delta + C_a)\sin\beta}{\sin\beta (FS)}
$$

Again, by setting $E_A = E_P$ (recall E_P from the previous analysis), the resulting formulation can be arranged in the form of equation 3.21 where

$$a = (W_A - N_A \cos\beta - T \sin\beta)\cos\beta, \tag{3.26}$$

$$b = -[(W_A - N_A \cos\beta - T \sin\beta)\sin\beta \tan\phi + (N_A \tan\delta + C_a)\sin\beta\cos\beta + \sin\beta(C + W_P \tan\phi)], \text{ and} \tag{3.27}$$

$$c = (N_A \tan\delta + C_a)\sin^2\beta \tan\phi. \tag{3.28}$$

Again, the resulting FS value can be obtained using equation 3.22. Example 3.13 illustrates the use of the above analysis.

Example 3.13:

Let us continue example 3.12, now using a geogrid with T_{ult} = 150 kN/m and cumulative reduction factors amounting to 4.5. What is the resulting factor of safety for this case of veneer reinforcement?

Solution: The W_A, N_A, and W_P values stay the same as in example 3.12. The allowable tensile strength of the geogrid reinforcement is

$$\begin{aligned} T &= T_{ult} / \Pi RF \\ &= 150/4.5 \\ &= 33.3 kN/m \end{aligned}$$

Using equations 3.26, 3.27, and 3.28, together with equation 3.25 gives the resulting FS value.

$$\begin{aligned} a &= (W_A - N_A \cos\beta - T \sin\beta)\cos\beta \\ &= (437 - 415\cos 18.4 - 33.3\sin 18.4)\cos 18.4 \\ &= 31.4 \text{ kN/m} \end{aligned}$$

$$\begin{aligned} b = -[&(W_A - N_A \cos\beta - T \sin\beta)\sin\beta \tan\phi \\ &+ (N_A \tan\delta + C_a)\sin\beta\cos\beta \\ &\quad + \sin\beta(C + W_P \tan\phi)] \\ = -[&(437 - 415\cos 18.4 - 33.3\sin 18.4)\sin 18.4 \tan 30 \end{aligned}$$

$$+(415\tan 18+0)\sin 18.4\cos 18.4$$
$$+\sin 18.4(0+24.3\tan 30)]$$
$$=-50.8 \text{ kN/m}$$

$$c = (N_A \tan\delta + C_a)\sin^2\beta\tan\phi$$
$$= (415\tan 18+0)\sin^2 18.4\tan 30$$
$$= 7.8 \text{ kN/m}$$

$$FS = \frac{-b+\sqrt{b^2-4ac}}{2a}$$
$$= \frac{50.8+\sqrt{(-50.8)^2-4(31.4)(7.8)}}{2(31.4)}$$

$FS =$ 1.45; which is acceptable

This solution for veneer reinforcement agrees well with other methods in the literature and with a finite element solution (Wilson-Fahmy and Koerner [62]).

A significant issue, however, is the input variables for the analysis. This is particularly the case for the interface friction value and for the reduction factors on the geosynthetic reinforcement. Also, if a high-strength geotextile is being used, the δ value will be for the geotextile to the geomembrane, since geotextiles do not allow for strike-through of the backfill soil. Concerning an acceptable value of the resulting factor of safety, the site-specific situation must be considered. For leachate collection soils in landfills, relatively low values of *FS* may be acceptable since the solid waste will provide a buttressing effect as it is placed in the landfill. Conversely, for final cover soils in the closure of landfills (and other permanent applications), quite high values of *FS* should be considered since the time frames for service life can be extremely long.

Lastly, in areas of anticipated earthquake activity, the slope stability analysis of a final cover soil over an engineered landfill, abandoned dump, remediated site or other permanent applications must consider seismic forces. In the United States, the Environmental Protection Agency (EPA) regulations require such an analysis for sites that have a probability of ≥ 10% of experiencing a 0.10 g peak horizontal acceleration within 250 years.

The seismic analysis of cover soils of the type under consideration is a two-part process: the calculation of a *FS* value using a pseudo-static analysis via the addition of a horizontal seismic force acting at the centroid of the cover cross section, and a mandatory permanent deformation analysis if the *FS* value in the above calculation is less than 1.0. The calculated deformation is then assessed in light of the potential damage to the cover soil section and is either accepted, or it is not and the slope will require an appropriate redesign. The redesign is then analyzed until the situation becomes acceptable.

The first part of the analysis is called a *pseudo-static approach*, which follows the previous examples except for the addition of a horizontal force at the centroid of the cover soil in proportion to the anticipated seismic activity. It is first necessary to obtain an average seismic coefficient (C_s) from a seismic zone map and subsequent procedure (e.g., Bray et al. [63]). The value of C_s is nondimensional and is a ratio of the bedrock acceleration to gravitational acceleration.

The additional seismic force is C_sW_A on the active wedge and C_sW_p on the passive wedge. By approaching the problem exactly as before and including the C_s values one obtains the following (see Koerner and Soong [58] for details):

$$a(FS)^2 + b(FS) + c = 0 \tag{3.21}$$

where

$$a = (C_SW_A + N_A \sin\beta) + C_SW_P \cos\beta \tag{3.27}$$

$$\begin{aligned} b = -[&(C_SW_A + N_A \sin\beta)\sin\beta\tan\phi \\ &+ (N_A \tan\delta + C_a)\cos^2\beta \\ &+ (C + W_P \tan\phi)\cos\beta] \end{aligned} \tag{3.28}$$

$$c = (N_A \tan\delta + C_a)\cos\beta\sin\beta\tan\phi \tag{3.29}$$

The resulting *FS* value is then obtained from equation 3.25:

$$FS = \frac{-b + \sqrt{b^2 - 4ac}}{2a} \tag{3.25}$$

If the *FS* value from such a calculation is greater than 1.0, the analysis is complete. The assumption being that cover soil stability can withstand the short-term excitation of an earthquake and still not slide. However, if the value is less than 1.0, a second part of the analysis is required.

The second part of the analysis is directed toward calculating the estimated deformation of the lowest shear strength interface in the cross section under consideration. The deformation is then assessed in light of the potential damage that may be imposed on the system.

To begin the permanent deformation analysis, a yield acceleration, "C_{sy}," is obtained from a pseudo-static analysis under an assumed *FS* = 1.0. We compare this value with the time history response assumed for the actual site location and cross section. If the earthquake time history response never exceeds the value of C_{sy}, there is no anticipated permanent deformation. However, whenever any part of the time history exceeds the value of C_{sy}, permanent deformation is expected. By double integration of the acceleration time history curve, to velocity and then to displacement, the cumulative value of deformation can be obtained (see Matasovic et al. [64]). This value is considered to be permanent deformation and is then assessed based on the site-specific implications of damage to the final system.

3.2.8 Other Geogrid Applications

Since many geogrid (and geotextile) manufacturers have technical design groups augmenting their sales and marketing groups, new geogrid applications are being developed regularly. The following lists and describes some of these activities.

- The use of geogrids within expansive or frost heaving soils to resist and/or mitigate volumetric increases [65, 66].
- Geogrid encapsulation of stone columns for improved deep foundations [67].
- Use of geogrids and deep mixing soil stabilization to mitigate surface settlement [68].
- Electrophoresis dewatering of mine tailings using geogrids as substrate layers [69].
- Geogrid reinforced backfill soils over culverts and pipelines to reduce vertical stresses on the structure [70].

- Geogrid reinforced walls and columns to construct underground stormwater detection systems [71].
- Geogrid fish pens and cages for aquaculture farming in near and deep ocean systems [72].
- Geogrid supported reactive core mats for remediation, sediment capping, gas vapor mitigation, and water treatment [73].

3.3 DESIGN CRITIQUE

The design sections just presented use geogrids in their primary function, which is as reinforcement. This primary function comes about because of a number of features regarding geogrids.

- *Economy*, as in the reduction of base course thickness in unpaved roads
- *Practicality*, as in geogrid reinforced slopes, embankments and walls
- *Necessity*, as in veneer reinforcement of cover soils on geomembranes where traditional methods of construction are not adequate

The design methods in each of the above instances are direct adaptations of traditional geotechnical engineering methods—only now the designs include a reinforcement material, namely, geogrids. The liberties taken by making this change seem reasonable and justifiable on the basis of field performance and monitoring. In fact, the feedback seems to indicate that the design methods just presented are conservative. However, before we conclude that more liberal designs are in order, the nature of the actual situation must be considered. For example, embankment slopes, reinforced walls, and landfill reinforcement applications are generally permanent structures requiring long service lifetimes, which simply demand a conservative approach.

More uncertain than the design methods are the allowable properties of the geogrid and the proliferation of a wide variety of different geogrid types. Considerations of allowable strength lead directly to the subsequent factor of safety (section 3.1.5). Bonaparte and Berg [74] subdivide applications into *permanent* versus *temporary*, and *critical* versus *noncritical.* It is a very perceptive approach, wherein the permanent and critical systems require the greatest amount of concern and caution. This, of course, is up to the designer on a site-specific basis. Regarding the variety of geogrid types, concern is warranted when a design is made and then an "or equal" specification is written. The introduction to this chapter has shown that geogrids vary considerably insofar as their physical, mechanical and endurance properties. Hence a geogrid specification must be written around a set of performance characteristics. This, too, is a problem, since most of our experience has been in writing specifications for geotextiles, not for geogrids. Although the current situation could be easily dismissed as merely growing pains, this does little good for a designer who is unsure of a product or for a manufacturers who are *very* sure of his/her product. Clearly, there is a need for generic specifications of all types of bidirectional and unidirectional geogrids.

Lastly, the designer (as well as the owner and regulator) should consider sustainability calculations whenever there exist alternative systems. This is certainly the case for walls, slopes and embankments. The design of a mechanically stabilized earth (MSE) wall or slope using geogrid reinforcement must be weighed in light of its carbon footprint versus conventional wall and slope designs. The WRAP Report [75] illustrates this situation by providing a calculation procedure for quantifying the tons of CO_2 generated from using geogrid reinforcement systems versus traditional walls and slopes for six case histories in the United Kingdom. Jones and Dixon [76] present the details of these case histories insofar as CO_2 savings are concerned (see table 3.5). Thus, it can be seen that the average CO_2 savings for the geosynthetic alternatives is an impressive 72% savings over traditional alternatives.

3.4 CONSTRUCTION METHODS

As with geotextiles, geogrids come to the job site in rolls. However, they are often narrower than geotextiles. Typically geogrid roll widths are from 1 to 3 m wide for unitized geogrids and 3 to 5 m wide for coated yarn geogrids with the bonded strap (rod) geogrid being intermediate in width. Their deployment is straightforward unless some sort of tensioning or prestressing is desired. Because of their large aperture size, sewing is not possible to join the sides or ends together and some type of mechanical system is generally employed. Unitized unidirectional geogrids can be bent and the bent end inserted into the

TABLE 3.5 SUMMARY OF CO2 SAVINGS FROM WRAP REPORT [75] AND JONES AND DIXON [76]

Description	CO_2 saving			
	Waste[1]	Fill[2]	Structure[3]	Total
1. Environmental bund Original design: Imported stone and gabion system. Geosynthetic design: Reinforced soil using on site soils.	100%	67%	96%	87%
2. Road embankment Original design: Imported stone to reduce footprint. Geosynthetic design: Reinforced soil using on site soils.	58%	36%	Increase	31%
3. Retaining wall Original design: Reinforced concrete wall. Geosynthetic design: Crib wall.	73%	73%	70%	70%
4. Retaining wall Original design: Reinforced concrete wall. Geosynthetic design: Modular block wall.	100%	100%	81%	85%
5. Retaining wall Original design: Sheet pile wall. Geosynthetic design: Steel strip reinforced soil.	-	-	84%	84%
6. Retaining wall Original design: Hollow concrete block drainage. Geosynthetic design: Geocomposite drainage.	-	Increase	82%	73%

Notes:
1. Waste material derived from the site and disposed off-site.
2. Imported material used as engineered fill on site.
3. Structural elements such as geosynthetics and conventional materials such as concrete/steel.

opening of an adjacent sheet. By placing a rod or bar down the slot that is formed, excellent load transfer is obtained. The rod is either 12 mm diameter or a tapered bar bodkin. A number of joining techniques are under development for the flexible textilelike geogrids. Of course, adequate overlap can also be used to mobilize the tensile strength of the opposing materials via shear stresses. Wire cutters will suffice to cut or trim geogrids, but a circular saw is quicker and more efficient. The flexible coated yarn geogrids can generally be cut with a sharp knife.

Unitized geogrids have been used to anchor wall-facing panels made from concrete in a manner similar to the metal strips of reinforced earth. The attachment of geogrids to facing panels involves the casting of small geogrid sections or metal hooks into the concrete panels during their fabrication. The reinforcement geogrids are mechanically connected, directly or by means of a steel dowel running lengthwise behind the hooks and attached to the ends of the geogrids. Alternatively, geogrids can be used between layers of wall sections, gabions, concrete cells, or concrete blocks to anchor the walls and reduce the earth pressure on the wall itself. Connection is by polymer or fiberglass dowels for unitized geogrids or by friction (sometimes between lock-and-key sections) for coated yarn and strap (rod) geogrids.

During the installation of geogrids for reinforcement purposes, the initial slack in the product must be removed before backfilling. This is sometimes a difficult task particularly with very flexible woven or knit geogrids. Usually, laborers will use a crowbar, pick, or steel rod to pull the geogrid taut while it is being backfilled. The amount of tension is an estimate. It should be realized that too much tension might not be advisable, especially when the geogrid is attached to wall-facing elements that are only temporarily supported. Clearly, tensioning is done on a trial-and-error basis. Its mobilization must be discussed by the parties involved before construction begins.

The above details are almost always geogrid product-specific. Thus, the manufacturer's website or technical representative must be consulted to be assured that the construction details are adequate for the proposed material system—that is, with the particular geogrid and wall-facing type. All geogrid manufacturers, to the author's knowledge, have well-trained geotechnical engineers on their full-time staff to advise consultants and owners on the details and nuances of their products.

REFERENCES

1. Kinney, T. C., "Determining the Secant Aperture Stability Modulus of a Geogrid," Shannon and Wilson Inc., Internal Report, January 7, 2000, 6 pgs.
2. Wilson-Fahmy, R., Koerner, R. M., and Fleck, J. A., "Unconfined and Confined Wide Width Testing of Geosynthetics," in *Geosynthetic Soil Reinforcement Testing* Procedures, ASTM STP 1190, edited by S. J. Cheng, ASTM, 1993, pp. 49-63.
3. Sarsby, R. W., "The Influence of Aperture Size/Particle Size on the Efficiency of Grid Reinforcement," *Proc. 2nd Canadian Symp. Geotextiles and Geomembranes*, Edmonton, Canada: The Geotechnical Society of Edmonton, 1985, pp. 7-12.

4. Ingold, T. S., "Laboratory Pull-Out Testing of Grid Reinforcement in Sand," *Geotechnical Testing J.*, Vol. 6, No. 3, 1983, pp. 212-217.
5. Koerner, R. M., Wayne, M. H., and Carroll, R. G., Jr., "Analytic Behavior of Geogrid Anchorage," *Proc. Geosynthetics '89*, IFAI, 1989, pp. 525-536.
6. Wilson-Fahmy, R., and Koerner, R. M., "Finite Element Modeling of Soil-Geogrid Interaction in a Pullout Loading Conditions," *Geotextiles and Geomembranes*, Vol. 12, No. 5, 1993, pp. 479-501.
7. Bathurst, R. J., and Simac, M. R., "Geosynthetic Reinforced Segmental Retaining Wall Structures in North America," *Proc. 5th IGS Conf.*, Special Volume, Singapore: Southeast Asia Chapter, IGS, 1994, pp. 29-54.
8. Koerner, G. R., Koerner, R. M. and Elias, V., "Geosynthetic Installation Damage Under Two Different Backfill Conditions," *Geosynthetic Soil Reinforcement Testing Procedures*, ASTM STP 1190, ed. S. J. Cheng, ASTM, 1993, pp. 163-183.
9. Hsieh, C. W. and Wu, J. H., "Installation Survivability of Flexible Geogrids in Various Subgrade Materials," Transportation Research Record No. 1772, Transportation Research Board, Washington, DC, 2001, pp. 190-196.
10. Sprague, C. J. and Allen, S. R., "Testing Installation Damage of Geosynthetics," GFR, Vol. 21, No. 6, pp. 24-27.
11. McGown, A., Andrawes, K. Z. and Kabir, M. H., "Load-Extension Testing of Geotextiles Confined in Soil," *Proc. 2nd Intl. Conf. on Geotextiles*, IFAI, 1982, pp. 93-96.
12. Ingold, T. S., Montanelli, F. and Rimoldi, P., "Extrapolation Techniques for Long Term Strengths of Polymeric Geogrids," *Proc. 5th Intl. Conf. on Geosynthetics*, Singapore: Southeast Asia Chapter, IGS, 1994, pp. 1117-1120.
13. Miyata, K., "Walls Reinforced with Fiber Reinforced Plastic Geogrids in Japan," *Geosynthetics International*, Vol. 3, No. 1, 1996, pp. 1-11.
14. Hsuan, Y. G., Koerner, R. M. and Koerner, G. R., "Field Measurements of Oxygen Temperature and Moisture Behind Segmental Retaining Walls," Proc. 7th Intl. Conf. on

Geosynthetics, September 22-27, 2002, A. A. Balkema Publ., pp. 1431-1434.
15. Farrag, K. and Shirazi, H., "Development of an Accelerated Creep Testing Procedure for Geosynthetics—Part 1: Testing," Geotechnical Testing Journal, GTJODJ, Vol. 20, No. 4, December, 1997, pp. 414-422.
16. Farrag, K., "Development of an Accelerated Creep Testing Procedure for Geosynthetics, Part 2: Analysis," Geotechnical Testing Journal, GTJODJ, Vol. 21, No. 1, March, 1998, pp. 39-44.
17. Thornton, J. S., Allen, S. R., Thomas, R. W. and Sandri, D., "The Stepped Isothermal Method for Time-Temperature Superposition and Its Application to Creep Data on Polyester Yarn," *Sixth International Conference on Geosynthetics*, Vol. 2, 1998, IFAI, pp. 699-706.
18. Greenwood, J. H. and Voskamp, W., "Predicting the Long-Term Strength of a Geogrid Using the Stepped Isothermal Method," Proc. 2nd European Geosynthetics Conference (EuroGeo 2), Bologna, Italy, 2002, pp. 329-332.
19. Yeo, S.-S. and Hsuan, Y. G., "Evaluation of Stepped Isothermal Method Using Two Types of Geogrids," Proc. EuroGeo, Scotland, 2008, Paper No. 285.
20. Liew, W. and Brown, D., "Predicting UV Durability of Exposed Geogrids With QUV Weathering Data," Proc. GRI-23 Conference, San Antonio, Texas, 2010, pp. 37-40.
21. Wrigley, N. E., "Durability and Long-Term Performance of Tensar Polymer Grids for Soil Reinforcement," *Materials Science and Technology*, Vol. 3, No. 4, 1987, pp. 161-170.
22. Koerner, R. M., Lord, A. E. Jr. and Halse, Y., "Allowable Geosynthetic Strength and Flow Considerations," *Proc. ASCE/Penn DOT Conf.*, Harrisburg, PA: Central Pennsylvania Section, ASCE, 1988, pp. 1-19.
23. Haas, R., "Structural Behavior of Tensar Reinforced Pavements and Some Field Applications," *Proc. Symp. Polymer Grid Reinforcement in Civil Eng.*, ICE, 1984, pp. 166-170.
24. Abd El Halim, A. O., "Geogrid Reinforcement of Asphalt Pavements," Ph.D. thesis, University of Waterloo, Ontario, Canada, 1983.

25. Abd El Halim, A. O., Haas, R. and Chang, W. A., "Geogrid Reinforcement of Asphalt Pavements and Verification of Elastic Layer Theory," Research Board Record No. 949, TRB, 1983, pp. 55-65.
26. Carroll, R. G. Jr., Walls, J. G. and Haas, R., "Granular Base Reinforcement of Flexible Pavements Using Geogrids," *Proc. Geosynthetics '87*, IFAI, 1987, pp. 46-57.
27. "Guide for Design of Pavement Structures," Washington, DC: AASHTO 1986.
28. Reck, N. C., "Mechanistic Empirical Design of Geogrid Reinforced Paved Flexible Pavements," *Jubilee Symposium on Polymer Grid Reinforcement*, Institute of Civil Engineering, London, England, 2009, pp. 72-78.
29. Brown, S. F., Brodrick, B. V. and Hughes, D. A. B., "Tensar Reinforcement of Asphalt: Laboratory Studies," *Proc. Symp. Polymer Grid Reinforcement in Civil Eng.*, ICE, 1984, pp. 158-165.
30. Kennepohl, G. J. A. and Kamel, N. I., "Construction of Tensar Reinforced Asphalt Pavements," *Proc. Symp. Polymer Grid Reinforcement in Civil Eng.*, ICE, 1984, pp. 171-175.
31. Molenaar, A. A. A. and Nods, M., "Design Method for Plain and Geogrid Reinforced Overlays on Cracked Pavements," *Proc. 3rd Intl. RILEM Conferrence*, L. Francken, E. Beuving and A. A. A. Molenaar, E & FN Spon., London, 1996, pp. 311-320.
32. van Gurp, C.A.P.M. and van Leest, A. J., "Thin Asphalt Pavements on Soft Soil," 9th Int. Conference on Asphalt Pavements, ISAP, Copenhagen, 2002, pp. 1-18.
33. Giroud, J.-P., Ah-Line, C. and Bonaparte, R., "Design of Unpaved Roads and Trafficked Areas with Geogrids," *Proc. Symp. Polymer Grid Reinforcement in Civil Eng.*, ICE, 1984, pp. 116-127.
34. Lostumbo, J. M., "Yeager Airport Runway Extension: Tallest Known 1H:1V Slope in US," *Proc. GeoFlorida 2010*, GSP199, ASCE, 2010, pp. 2502-2510.
35. Forsyth, R. A. and Bieber, D. A., "La Honda Repair with Geogrid Reinforcement," Proc. *Symp. Polymer Grid Reinforcement in Civil Engr.*, ICE, 1984, pp. 54-57.
36. Berg, R. R., Anderson, R. P., Rose, R. J. and Chouery-Curtis, V. E., "Reinforced Soil Highway Slopes, *Proc. TRB 69th Annual Meeting*, TRB Washington, DC: TRB, 1990, 46 pgs.

37. Schmertmann, G. R., Chouery-Curtis, V. E., Johnson, R. D. and Bonaparte, R., "Design Charts for Geogrid-Reinforced Soil Slopes," *Proc. Geosynthetics '87*, IFAI, 1987, pp. 108-120.
38. Jewell, R. A., "Application of Revised Design Charts for Steep Reinforced Slopes," *J. Geotextiles and Geomembranes*, Vol. 10, No. 3, 1991, pp. 203-233.
39. Reugger, R., "Geotextile Reinforced Soil Structures on Which Vegetation can be Established," *Proc. 3rd Intl. Conf. Geotextiles*, Vienna: Austrian Society of Engineers, 1986, pp. 453-458.
40. Abu-Hejleh, N., Wang, T. and Zornberg, J. G., "Performance of Geosynthetic-Reinforced Walls Supporting Bridge and Approaching Roadway Structures," Advances in Transportation and Geoenvironmental Systems Using Geosynthetics, J. G. Zornberg and B. R. Christoper, Eds., GSP No. 103, ASCE, 2000, pp. 218-243.
41. Soong, T.-Y., and Koerner, R. M., "On the Required Connection Strength of Geosynthetically Reinforced Walls," J. Geotextiles and Geomembranes, Vol. 15, No. 4-6, 1997, pp. 377-394.
42. Tatsuoka, F., Murata, O. and Tateyama, M., "Permanent Geosynthetic Reinforced Soil Retaining Walls Used for Railway Embankments in Japan," *Proc. Geosynthetic Reinforced Soil Retaining Walls*, ed. J. T. H. Wu, A. A. Balkema Publ., 1992, pp. 101-130.
43. Koerner, J., Soong, T.-Y. and Koerner, R. M., "Earth Retaining Wall Costs in the USA," GRI Report No. 20, GSI, Folsom, PA, 1998, 35 pgs.
44. Fonyo, B. and Sacchetti, A., "Design Software Comparison of Reinforced Steep Slopes," *Proc. 9th ICG Conf.*, Brazil, 2010, pp. 1819-1822.
45. Leshchinsky, D., "Issues and Nonissues in Block Walls as Implied Through Computer Aided Design," *Proc. GRI-12 Conference*, GII Publications, Folsom, PA, 1998, pp. 66-74.
46. Koerner, R. M. and Soong, T.-Y., "Geosynthetic Reinforced Segmental Retaining Walls," 17th PennDOT/ASCE Conf. on Geotechnical Engineering, Hershey, PA, 1999, 36 pgs.; Also Proc. GRI-14 Conference, GII Publ., Folsom, PA,

December 2000, pp. 268-297; and Journal of Geotextiles and Geomembranes, Vol. 19, No. 6, August, 2001, pp. 359-386.

47. Koerner, R. M. and Soong, T-Y., "Drainage System Design Behind Segmental Walls," 18th PennDOT/ASCE Conf. on Geotechnical Engineering, Hershey, PA, 2000, 38 pgs.; also Proc. GRI-14 Confererence, GII Publ., Folsom, PA, December, 2000, pp. 323-351; also EuroGeo III, Munich, Germany, February, 2004, pp. 355-360.
48. Jarrett, P. M., "Evaluation of Geogrids for Construction of Roadways over Muskeg," *Proc. Symp. Polymer Grid Reinforcement in Civil Eng.*, ICE, 1984, pp. 149-153.
49. Milligan, G. W. E., and Love, J. P., "Model Testing of Geogrids Under an Aggregate Layer in Soft Ground," *Proc. Symp. Polymer Grid Reinforcement in Civil Eng.*, ICE, 1984, pp. 128-138.
50. Giroud, J.-P., Bonaparte, R., Beech, J. F., and Gross, B. A., "Design of Soil Layer-Geosynthetic Systems Overlying Voids," *Geotextiles and Geomembranes*, Vol. 9, No. 1, 1990, pp. 11-50.
51. Edgar, S., "The Use of High Tensile Polymer Grid Mattress on the Mussleburgh and Portobello Bypass," *Proc. Symp. Polymer Grid Reinforcement in Civil Eng.*, ICE, 1984, pp. 103-111.
52. Rueff, H., Stoffers, U. and Leicher, F., "Deponie auf schwierigsten Untergrund," Ernst & Sohn, Bautechnik 69, Heft 5, 1992.
53. Robertson, J. and Gilchrist, A. J. T., "Design and Construction of a Reinforced Embankment Across Soft Lakebed Deposits," *Proc. Symp. Polymer Grid Reinforcement in Civil Eng.*, ICE, 1984.
54. Jenner, C. G., Bush, D. I., and Bassett, R. H. "The Use of Slip Line Fields to Assess the Improvement in Bearing Capacity of Soft Ground Given by a Cellular Foundation Mattress Installed at the Base of an Embankment," *Proc. Theory and Practice of Earth Reinforcement*, A.A. Balkema, 1988, pp. 209-214.
55. Han, J. and Gabr, M. A., "Numerical Analysis of Geosynthetic Reinforced and Pile Supported Earth Platforms over Soft Soils," *Jour. Geotechnical and Geoenvironmental Engineering*, Vol. 128, No. 1, 2002, pp. 44-53.

56. Han, J. and Akins, K., "Use of Geogrid Reinforced and Pile Supported Earth Structures," Proc. Intl. Deep Foundations Congress, ASCE, Feb. 14-16, 2002, Orlando, FL, pp. 668-679.
57. Paul, A. and Ponomarjow, A., "The Bearing Behavior of Geogrid Reinforced Crushed Stone Columns in Comparison to Nonreinforced Concrete Pile Foundations," *Proc. EuroGeo3*, Munich, Germany, 2004, pp. 285-288.
58. Koerner, R. M., and Soong, T.-Y., "Analysis and Design of Veneer Cover Soils," *Proc. 6th IGS Conf.*, IFAI, 1998, pp. 1-26.
59. Giroud, J.-P. and Beech, J. F., "Stability of Soil Layers on Geosynthetic Lining System," *Proc. Geosynthetics '89*, IFAI, 1989, pp. 35-46.
60. Koerner, R. M. and Hwu, B.-L., "Stability and Tension Considerations Regarding Cover Soils in Geomembrane Lined Slopes," *J. of Geotextiles and Geomembranes*, Vol. 10, No. 4, 1991, pp. 335-355.
61. Thiel, R. S. and Stewart, M. G., "Geosynthetic Landfill Cover Design Methodology and Construction Experience in the Pacific Northwest," *Proc. Geosynthetics '93*, IFAI, 1993, pp. 1131-1144.
62. Wilson-Fahmy, R. F. and Koerner, R. M., "Finite Element Analysis of Cover Soil on Geomembrane Lined Slopes," *Proc. Geosynthetics '93*, IFAI, pp. 1425-1437.
63. Bray, J. D., Rathje, E. M., Augello, A. J. and Merry, S. M., "Simplified Seismic Design Procedure for Geosynthetic-Lined, Solid-Waste Landfills," Geosynthetics International, Vol. 5, Nos. 1-2, 1998 pp. 203-235.
64. Matasovic, N., Kavazanjian, E. Jr. and Yan, L., "Newmark Deformation Analysis with Degrading Yield Acceleration," *Proc. Geosynthetics '97*, IFAI, 1997, pp. 989-1000.
65. Al-Omari, R. R. and Hamodi, F. J., "Swelling Resistant Geogrid - A New Approach for the Treatment of Expansive Soils," Jour. of Geotextiles and Geomembranes, Vol. 10, No. 4, Elsevier Publ. Co., 1991 pp. 295-318.
66. Alzamora, D. E., Haramy, K. and Anderson, S. A., "Use of Geosynthetics to Mitigate Frost Heave on Trail Ridge Road,"

Proc. of Geosynthetics 2009, Salt Lake City, Utah, Feb. 25-27, 2009, IFAI Publisher (on CD), pp. 578-585.

67. Bhyravajjula, R. P., Mukherjee, R. V., Chauhan, A. D. and Lakshminarayanan, M. V., "Efficacy of Geogrid in Improving the Load Carrying Capacity of Stone Columns," *Proc. 9th Int. Geosynthetics Conf.*, Guaruja, Brazil, 2010, pp 1929-1934.
68. Ogisako, E., "Estimation of Settlement of Ground Using Geogrid and Deep Mixing Soil Stabilization," Proceedings 7ICG - Nice, France, A. A. Balkema Publ., 2002, pp. 413-318.
69. Pavlakis, J., Fourie, A. B. and Jones, C. J. F. P., "Electrophoretic Dewatering of Mine Tailing Using Geosynthetics," Proceedings 7ICG - Nice, France, A. A. Balkema Publ., 2002, pp. 1581-1584.
70. Yamazaki, S., Mohri, Y., Matsushima, K., Hori, T. and Fujita, N., "Field Test of a Pipeline Buried at Shallow Depth with a Geogrid," Proc. 8th Intl. Conf. on Geosynthetics, Yokohama, Japan, Millpress Publ., Rotterdam, 2006, (on CD only).
71. Sheridan, T. G., "Geosynthetic Materials Play a Role in New Underground Stormwater Detention System," Geosynthetics Magazine, Vol. 28, No. 3, 2010, pp. 35-41.
72. Weggel, J. R., "Wave and Current Loading on Geosynthetic Fish Pens and Cages," Proc. GRI-21 Conference on Geosynthetics in Agriculture and Aquaculture, 2008, (Papers on CD).
73. Meric, D. et al., "Testing the Efficiency of a Reactive Core Mat to Remediate Subaqueous Contaminated Sediments," *Proc. GeoFrontiers 2011*, Dallas, Texas, 2011, pp. 895-904.
74. Bonaparte, R. and Berg, R., "Long-Term Allowable Tension for Geosynthetic Reinforcement," *Proc. Geosynthetic '87*, IFAI, 1987, pp. 181-192.
75. Waste and Resources Action Programme (WRAP) Report, "Sustainable Geosystems in Civil Engineering Applications," Project MRF116, Capita Symonds, Banbury, U. K., 2009, $\simeq$ 300 pgs.
76. Jones, R. and Dixon, N., "European Perspectives on Sustainable Development Using Geosynthetics," *Proc. GRI-24 Conference*, GII Publ., Folsom, PA, 2011, pp. 1-7

PROBLEMS

3.1 List and describe the basic similarities and differences between geogrids and geotextiles.

3.2 List and describe the basic similarities and differences between geogrids and geonets.

3.3 Flexible geogrids are coated with latex, bitumen, or polyvinyl chloride. What is the reason for such coatings?

3.4 Derive an anchorage resistance formula on the basis of figure 3.4 to include all three components of pullout resistance. (Hint: See [5])

3.5 From the anchorage analysis in section 3.1.2:

a. Using the formula derived in problem 3.4 for a geogrid in granular soil of $\varphi = 35$ deg. at 19.5 kN/ m^3 and 2.0 m depth, what is the ultimate pullout resistance per meter width for a length of 600 mm? The physical properties of the geogrid are as follows: 6.0 mm wide ribs at 50 mm centers in both directions, the rib thickness is 3.0 mm, and the friction angle mobilized by the rib's surface to the soil is 30°.

b. What is the relative proportion of the three components of anchorage strength calculated in part (a)?

c. If the wide-strip tensile strength of the above geogrid was 45 kN/m, could the anchorage force calculated in part (a) be mobilized? If not, what portion of it could be?

3.6 Repeat the calculations of problem 3.5(a) using a rib width of 8.0 mm, then 10.0 mm, and then 12.0 mm. Plot the anchorage strength versus rib width.

3.7 What junction strength is required for the results of problems 3.5(a) and 3.6, if all the available anchorage strength is mobilized? (Hint: The transverse rib shear and transverse rib bearing must both be transmitted through the junction.)

3.8 Discuss the usefulness (or uselessness) of performing an in-isolation junction strength test as described in figure 3.2. If you feel it has no relevancy, what alternate test do you suggest to evaluate the significance of the transverse ribs of geogrids?

3.9 In light of the geogrid junction strength using a clamp as shown in figure 3.2:

a. What effect would normal stress on the junction have on the test results?

b. What is the role of long-term stress on the junctions?

c. How can long-term junction strength be evaluated?

3.10 Regarding unitized polyolefin geogrids:

a. Phenomenologically describe what happens to the molecular structure of polyethylene and polypropylene at or slightly above room temperature, when they are uniformly stretched in a continuous direction.

b. What influence does cold working have on the mechanical properties of the final product?

3.11 What is the effect of very high temperature on the following mechanical properties of geogrids?

a. Modulus

b. Tensile strength

c. Elongation at failure

d. Creep behavior

3.12 Explain the two different behaviors shown in figure 3.8 in that the HDPE geogrid is governed by creep strain and the PET geogrid is governed by creep rupture. Use the glass transition temperatures of table 1.3 in your answer.

3.13 If the ultimate tensile strength of a geogrid evaluated in wide-strip tensile strength test is 60 kN/m, what allowable strength should be used in design for the following situations?

a. temporary unpaved access road for construction vehicle use for approximately one year

b. stone base reinforcement in a permanent paved road for 30 year lifetime

c. differential settlement beneath a wall footing (i.e., improved bearing capacity) for 50 to 100 year lifetime

3.14 In using geogrids for reinforcement of paved roads, a possible mechanism involving increased bearing capacity is often mentioned. On a conceptual basis, how does this work?

3.15 How does the geogrid effectiveness factor (*GEF*) of section 3.2.1 relate to the fabric effectiveness factor (*FEF*) of section 2.10.2?

3.16 For the power law of equation (3.9), used to calculate the crack propagation through pavement overlays, what type of laboratory experiment would you devise to arrive at the K, A, and n values required in the analysis?

3.17 Regarding geogrid reinforced unpaved roads:

a. What is the required aggregate thickness for an unpaved road carrying ten thousand 80 kN vehicle passes with rut depths of 0.15 m on a soil subgrade whose undrained shear strength is 20 kN/m^2 (use figure 3.12)?

b. Using BX 1200 geogrid reinforcement, channelized traffic, and high likelihood of aggregate contamination, how much aggregate can be saved?

c. What increase in load-spreading angle does this represent (assuming that the nonreinforced case is $\alpha_0 = 26.6°$)?

3.18 Repeat problem 3.17 assuming that the soil subgrade has $CBR = 2.0$.

3.19 Repeat problem 3.17 using UX 1200 geogrids.

3.20 Using the approach indicated by figures 3.14 and 3.15, determine the number, spacing, and length of the geogrids needed to stabilize the embankment below using a factor of safety of 1.3. Use combined reduction factors of 4.3 on the ultimate geogrid strength of the candidate geogrid of 180 kN/m to arrive at an allowable strength.

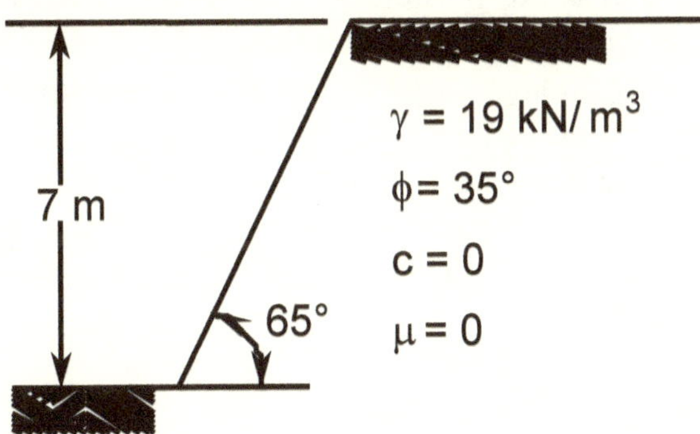

3.21 Determine the factor of safety against sliding and overturning (i.e., the external stability) for the wall below, carrying a surcharge load of 7.2 kPa.

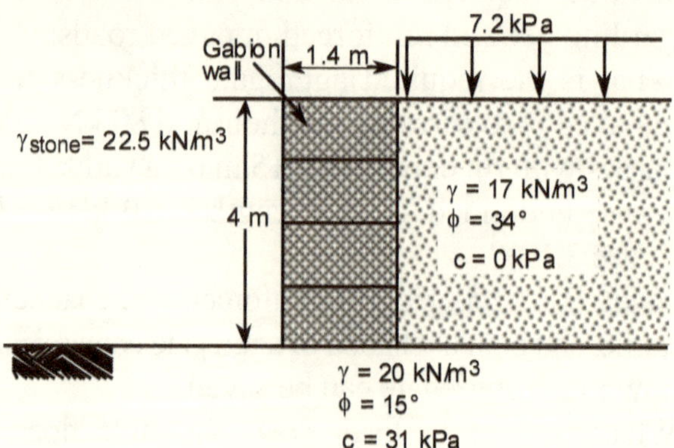

3.22 List advantages and disadvantages between the wire mesh/ geogrid faced walls of figure 3.16 and the modular block walls of figure 3.17.

3.23 When using geogrids for reinforcement of masonry block facing retaining walls, what are the various ways of attaching the geogrids to such block facings?

3.24 In considering the factor-of-safety decreases in *each* of the six parametric variations given in table 3.4, explain the major impact of avoiding low *FS* values.

3.25 Figure 3.20 shows that at high deformations the load-carrying capacity of shallow footings can be increased considerably using geogrids. How can this feature be used without having the footing undergo large settlement? (Draw some sketches of how this could be accomplished.)

3.26 Using the design guide of figure 3.21, solve for T_{reqd} as in example 3.11 (section 3.2.6) for the following parametric variations) i.e., hold all values constant except the target parameter).

a. Vary H from 1 to 30 m

b. Vary R from 1 to 10 m

c. Vary q from 7 to 100 kPa at H/B = 0.1

3.27 For example 3.12, concerning the nonreinforced cover soil (section 3.2.7), recalculate the *FS* for δ values of 13, 15, 21 and 24° and plot the response along with that for 18° given in the example.

3.28 Recalculate problem 3.26 varying the cover soil thickness, using 300, 600 and 1200 mm to compare with that given in example 3.12 for 900 mm (use $\delta = 18°$).

3.29 For example 3.13, for geogrid-reinforced cover soil (section 3.2.7), recalculate the factor of safety for T_{ult} values results using 100, 200 and 250 kN/m and plot the results along with that for 150 kN/m given in the example.

3.30 Given a 30 m long slope with uniform thickness cover soil of 300 mm at a unit weight of 18 kN/m^3. The soil has a friction angle of 30 deg. and zero cohesion, i.e., it is sand. The cover soil is on a geomembrane as shown in figure 3.22a. Direct shear testing has resulted in an interface friction angle of 22 deg. with zero adhesion. The slope angle is 3(H)-to-1(V), i.e., 18.4 deg. A design earthquake appropriately transferred to the site's cover soil results in an average seismic coefficient of 0.10. Using equations 3.21, 3.27-3.29 and 3.25 calculate the *FS* value and comment accordingly.

3.31 What are the various ways of transferring load from one sheet of geogrid to the next for

a. Homogeneous unidirectional geogrids?
b. Coated yarn geogrids?
c. Polymer rod (strap) geogrids?

Chapter 4

Designing with Geonets

4.0 INTRODUCTION

A geonet can be defined as follows:

> Geonet: A geosynthetic material consisting of integrally connected parallel sets of ribs overlying similar sets at various angles for in-plane drainage of liquids or gases. Geonets are often laminated with geotextiles on one or both surfaces and are then referred to as drainage geocomposites. They are competitive with other drainage geocomposites to be described in chapter 8.

As originally described in section 1.5, geonets are formed by a continuous extrusion process into a netlike configuration of parallel sets of homogeneously interconnected ribs. Figure 1.17 shows how the extruded core is opened up into the final netlike configuration. Figure 4.1 gives a closer view of the details of three categories of geonets. The following are illustrated:

- *Biplanar geonets:* These are the original and most common types and consist of two sets of intersecting ribs at different angles and spacings. The ribs themselves are of different sizes and shapes for different styles (figure 4.1a).
- *Triplanar geonets:* These have parallel central ribs with smaller sets of ribs above and beneath mainly for geometric stability (figure 4.1b).
- *Other geonets:* These newer geonet structures have either box shaped channels or protruding columns from an underlying support network (figure 4.1c).

Each of the above categories have variations within themselves (mainly thickness) and new product development by various manufacturers is quite active (see Austin [1]).

All geonets that are currently available are made from polyethylene resin. The density varies from 0.94 to 0.96 mg/l, with the higher values forming the more rigid products. In this regard, the resin is true high-density polyethylene (HDPE) unlike the density used in HDPE geomembranes that is really medium density. The resin is formulated with 2.0 to 2.5% carbon black (usually in a concentrated form mixed with a polyethylene carrier resin), and 0.25 to 0.75% additives that serve as processing aids and anti-oxidants.

(a) Biplanar geonets

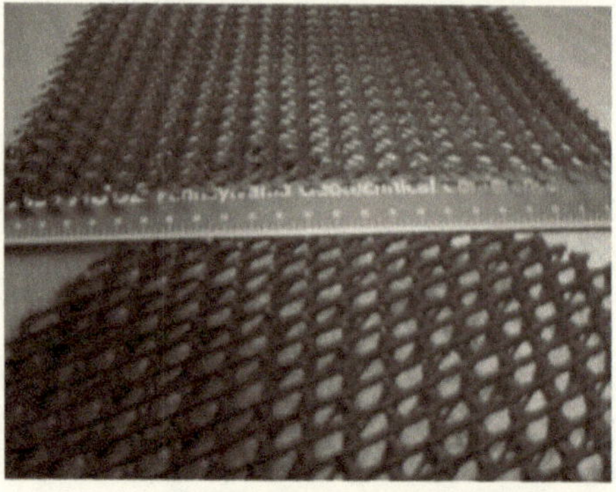

(b) Triplanar geonets

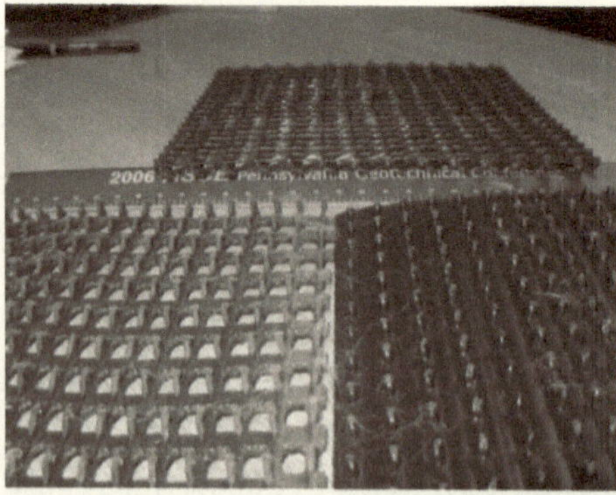

(c) Other geonets (boxshaped channels and protruding columns)

Figure 4.1 Various categories of geonets.

4.1 GEONET PROPERTIES AND TEST METHODS

Since the primary function of a geonet is to convey liquid within the plane of its structure, the in-plane hydraulic flow rate, or transmissivity, is of paramount importance (see Williams et al. [2]). However, other features, which may influence this value over the service lifetime of the geonet, are also of importance. Thus, a number of physical, mechanical, endurance, and environmental properties will also be presented in this section.

4.1.1 Physical Properties

The density or specific gravity of the polymer is an important property and it can be evaluated either by ASTM D1505 or D792. The former is preferred if an accuracy of at least 0.005 mg/l is required. Another physical property needed to characterize a geonet is its thickness, which can be determined using ASTM D5199 or ISO 9863. While there is no listing for geonets as such, it is recommended that geonet thickness be measured under a normal pressure of 20 kPa. (This is the same pressure that is recommended to measure geomembrane thickness.) Note that geonets are not nearly as sensitive to thickness variation under normal pressure as are geotextiles; they are similar to geomembranes in this regard.

Mass per unit area can be determined using ASTM D5261 or ISO 9864. For a 5 mm thick, solid-rib extruded biplanar geonet, the mass per unit area is usually in the range of 800 to 1600 g/m^2. It is not a design property per se, but it is important from a manufacturer's point of view.

Other physical properties such as rib dimensions, planar angles made by the intersecting ribs, cross-planar angles made at the juncture locations, aperture size and shape, and so on, can be measured directly and are straightforward to obtain.

4.1.2 Mechanical Properties

A number of mechanical properties of geonets are important to consider.

Tensile Strength and Elongation. The wide-width tensile strength curves of figure 4.2 results from testing of a 5.0 mm thick biplanar geonet in the machine and cross-machine directions. A 200 mm wide × 100 mm long test specimen was used in these tests.

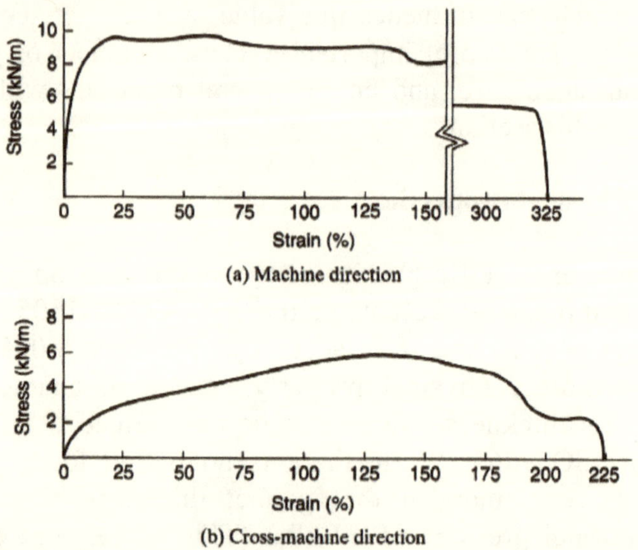

Figure 4.2 Tensile strength behavior of 5.0 mm thick solid rib biplanar extruded geonet.

The strain rate was 10 mm/mm. The average of a series of such unconfined wide-width tensile tests gave the following information:

- Machine direction: peak strength = 9.9 kN/m, strain at peak = 23%, strain at failure = 290%
- Cross-machine direction: peak strength = 5.6 kN/m, strain at peak = 170%, strain at failure = 240%

Note the differences in behavior, suggesting that there is a preferential direction in strength between the machine and cross-machine directions. This is even more the case for triplanar and boxlike geonets. If greater tensile strength of a geonet is desired, consideration should be given to other types of geonets oriented in the proper direction.

Compressive Strength and Deformation. Of greater importance than the above-described in-plane tensile strength tests on geonets is the cross-plane compressive strength. This is because of the influence that compressive deformation or collapse has on the ability of the geonet to conduct liquid within its planar structure. There are a number of approaches to measuring a geonet's compressive strength. Using 150 mm square test specimens normally loaded under a constant strain rate load of 0.05 mm/min, the curves of figure 4.3 are produced. Here it is shown that both the solid-rib extruded biplanar and the (quite rare) foamed-rib extruded biplanar geonets are initially stiff but

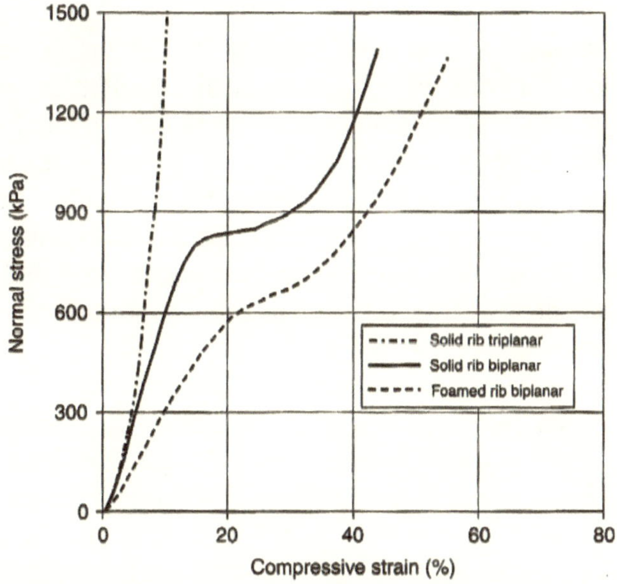

Figure 4.3 Compressive test data of various types of geonets.

begin to deform at normal stress between 600 and 850 kPa. This occurs because the parallel sets of ribs making up the respective geonets are not exactly perpendicular to one another at the junctions. Thus, at high-compressive stress, there is a *lay-down* or *roll-over* tendency, which gives rise to the characteristic behavior shown. Note that the geonet can still convey liquid beyond this point, but to a somewhat diminished degree. Newer types of round-ribbed biplanar geonets do not have this tendency; Yeo and Hsuan [3]. The extruded triplanar geonet (with a still higher flow-rate capacity) shows no inflection point since the central and thicker set of ribs is perpendicular to the normal stress. Thus, its response to normal stress increases with compressive strain in direct proportion to the density of the resin.

Box-type and columnar type geonets each have unique compressive strengths and must be evaluated accordingly. It should also be noted that inclined compressive stresses have been evaluated [4].

Note that the long-term creep strength is not indicated by these short-term tests. High normal stress, indentation, and creep are discussed in Corcoran et al. [5]. Since creep testing is very time-consuming, temperature-accelerated methods should be considered, among which is the stepped isothermal method (SIM) that has been applied to geonet compression testing (see Thornton et al. [6]).

Shear Strength. A geonet's capacity for sustaining shear stress along a plane within its own thickness without collapsing may be of concern in certain situations. This arises when there are high opposing shear stresses acting on the top and bottom of the geonet, tending to put the material into a state of pure shear. This is not felt to be much of a problem for the currently available biplanar and triplanar geonets. Considerably more concern should be focused on the interface friction behavior with respect to the materials above and beneath the geonet. For an application of a geonet against a geomembrane or other impermeable surface like concrete, the shear strength is generally very low. Even a geotextile placed on the surface of a geonet (with no bonding) has a low shear strength. Due to these low shear strengths (friction angles ranging from 5 to 15°), it is common practice to bond the geotextile(s) to the geonet at the manufacturer's facility. Bonding has been done in the past using adhesives, but most manufacturers currently use thermal attachment methods (i.e., hot wedge, infrared, etc.). The test method commonly used as a quality control test to determine the degree of attachment is the tensile peel strength of the geotextile to the geonet according to ASTM D413. The more directly relevant interface shear test method is ASTM D5321 or ISO 12957.1, and it is obviously a product-specific and site-specific test that must be performed for each set of conditions that arise (see Lydick and Zagorski [7]). Unfortunately, there is no universal relationship between the tension peel and interface shear tests. It is an interesting research area. That said, minimum uniformly bonded peel strengths often seen in specifications to avoid delamination and a subsequent interface shear failure are between 90 and 180 N/m. Higher values require greater melting of the geonet with proportionate loss of planar flow (or transmissivity) that is not desirable.

4.1.3 Hydraulic Properties

The in-plane hydraulic test to determine planar flow rate, or transmissivity, of geonets should be performed according to ASTM D4716 or ISO 12958. Both test methods use a planar transmissivity device and not the radial transmissivity device that was described and illustrated in section 2.3.4 on geotextiles. This is necessary because the flow regime in a geonet is surely turbulent (consisting of irregular flow paths and eddies) whereas in a geotextile it is probably laminar (thus, allowing for radial flow and integration over the flow path length). Some relevant points regarding the test method are as follows.

- The specimen size must be 150 by 150 mm, or larger. Kolbasuk et al. [8] have shown that sample size and aspect ratio (i.e., length-to-width ratio) can have significant effect on the resulting measured flow rates.
- As an index test, the geonet test specimen is sandwiched between rigid plates. However, the cross-section can be varied as desired to approach becoming a site-specific performance test.
- Normal stresses are applied for a 15 min duration, and flow rates are measured at different hydraulic gradients during a subsequent 15 min duration.
- Stress is then increased to the next level and the cycle is repeated.
- De-aired water at 5 ppm, or lower, of dissolved oxygen is not explicitly specified, although for critical situations it might be necessary.
- A standard laboratory temperature of 20°C is required.

Following this procedure and using a 200 mm square test specimen with a 6.3 mm thick biplanar geonet in the test device shown in figure 4.4, the flow-rate curves of figure 4.5 resulted. Here, an increasing flow rate can be observed at each higher hydraulic gradient evaluated. The flow rates of figure 4.5 are extremely high in comparison to the flow rates in soil. For example, 300 mm of sand, having a permeability coefficient of 0.1 cm/s at a hydraulic gradient of 1.0, can carry only 0.02 m^3/min-m. Thus, geonets can handle large flow rates compared to soil due to the higher velocity of flow within the significantly larger flow channels. Furthermore, the machine direction flow rate in triplanar and boxlike geonets is even higher than in biplanar geonets. From these

(a) A flow rate testing device

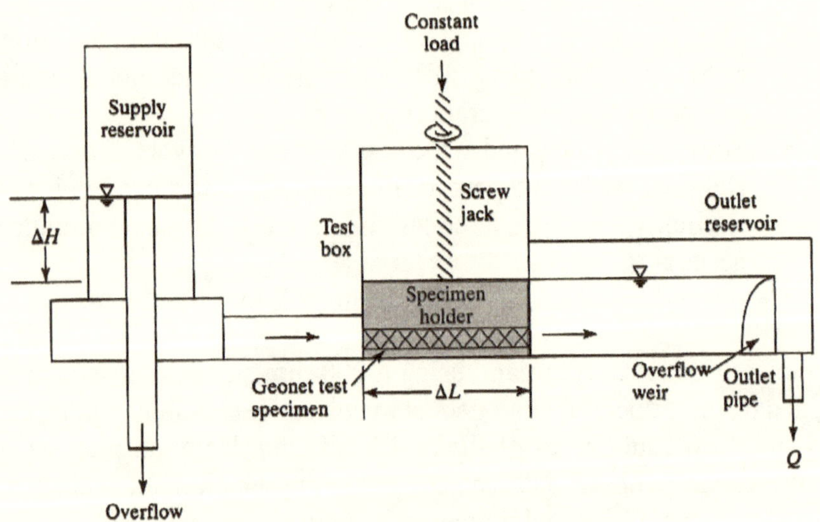

(b) Schematic diagram of flow rate testing device

Figure 4.4 Permeability device for measuring transmissivity (parallel in-plane flow) of a geonet.

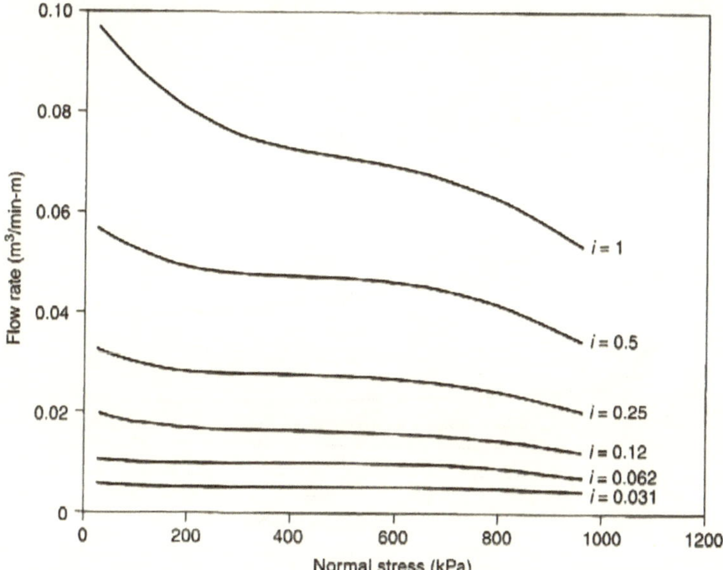

Figure 4.5 Flow rate behavior of a 6.3-mm thick biplanar geonet sandwiched between two 1.5-mm HDPE geomembranes.

data we can calculate a transmissivity value, the assumption being that the system is saturated at all times and flow is laminar. However, the flow regime in geonets is turbulent in its behavior. Thus, the discussion in this chapter will generally be using flow-rate values rather than transmissivity values.

The data shown in figure 4.5 are of the *index* test type. Site-specific situations, however, can be included in the test procedure so as to make the test results more *performance* oriented. To do so, we must have the representative conditions above and below the test specimen, and use liquids of the type (and sometimes at the temperature) to be conveyed in the actual system. Figure 4.6 illustrates the influence of one such variation. Here a cross-sectional profile consisting of a layer of clay (kaolinite clay at 15% water content), a 540 gm/m^2 nonwoven needle-punched polyester geotextile, a geonet, and a 1.5 mm HDPE geomembrane is used. Since the geonet is the same type as that producing the data in figure 4.5, the results can be compared directly, the only difference being the clay soil / geotextile separator placed over the geonet. The marked decrease in the flow rates of figure 4.6

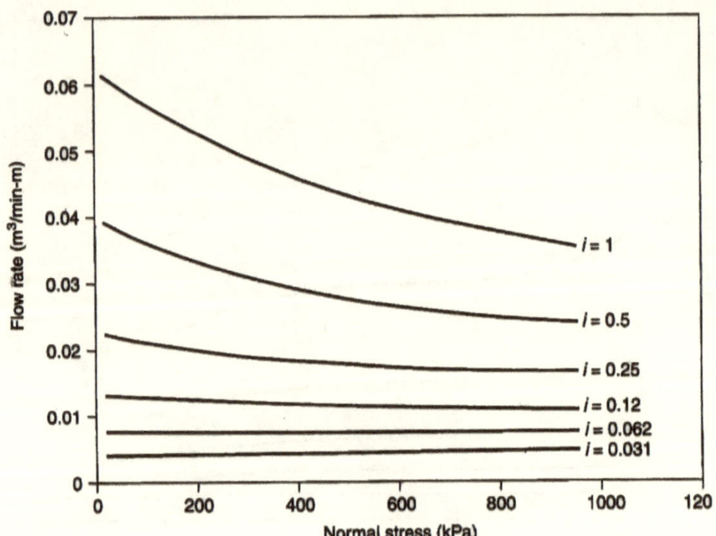

Figure 4.6 Flow rate behavior of a 6.3-mm thick biplanar geonet sandwiched between a 540 g/m2 nonwoven needle-punched geotextile with clay above and a 1.5-mm HDPE geomembrane below.

as compared to figure 4.5 comes from intrusion of the geotextile into the core space of the geonet via the pressure applied to the clay. Numerically, the comparison is seen to be as high as 40% from the index test values. Additionally, the geotextile must be capable of sustaining the applied stress, suggesting that long-term tests are required to adequately assess such situations. Sustained load deformation, or creep, will be discussed later.

It should be emphasized that *flow rates per unit width* values are not *transmissivity* values. To convert flow rate per unit width to transmissivity we use Darcy's formula, which tacitly assumes saturated conditions and laminar flow, neither of which are rigorously met with the typical flow regime in a geonet. Yet current US EPA leak detection regulations [9] state the following:

- For landfills and waste piles, the geonet's *transmissivity* (θ) must be

$$\theta \geq 3 \times 10^{-5}\ m^2/s$$

- For surface impoundments, the geonet's *transmissivity* (θ) must be

$$\theta \geq 3 \times 10^{-4}\ m^2/s$$

We convert from flow rate per unit width to transmissivity (recall section 2.3.4) as follows.

$$q = k\,i\,A \tag{4.1}$$

$$q = k\,i\,(W \times t)$$

$$q/W = i\,(k \times t)$$

$$q/W = i\,\theta \tag{4.2}$$

Thus, it is seen in equation 4.2 that the units of q/W and θ are identical, but only at $i = 1.0$ are the numeric values the same. A hydraulic gradient of 1.0 occurs when the geonet is placed vertically, as in the lining of a tank wall, not on the sloping surfaces of a typical landfill, surface impoundment, or highway slope. Example 4.1 illustrates the numeric conversion.

Example 4.1

A geonet tested in the laboratory under site-specific conditions resulting in a flow rate per unit width of 0.65×10^{-4} m^2/s at a hydraulic gradient 0.040. **(a)** What is the equivalent transmissivity in units of m^2/s and **(b)** does this value meet the EPA criteria for landfills and surface impoundments?

Solution: Using equation 4.2 and converting units we have the following:

(a) The equivalent transmissivity is

$$\theta = \left(\frac{q}{W}\right)\left(\frac{1}{i}\right)$$

$$= 0.65 \times 10^{-4}(1/0.040)$$

$$\theta = 16.2 \times 10^{-4}\ m^2/s$$

(b) Comparing the above to the EPA criteria,

$$162 \times 10^{-5} > 3.0 \times 10^{-5} \text{ m}^2/\text{s}$$

which is easily acceptable for landfills, with a resulting $FS = 54$. Furthermore,

$$16.2 \times 10^{-4} > 3.0 \times 10^{-4} \text{ m}^2/\text{s}$$

which is also acceptable for surface impoundments, with a resulting $FS = 5.4$.

4.1.4 Endurance Properties

The major endurance properties of concern when using geonets have to do with the long-term sustained deformation of the material and its ability to continue to transmit the required in-plane flow rate. Four issues are relevant: type of polyethylene resin, creep behavior of the geonet structure, intrusion of adjacent materials into the geonet's apertures, and the possible extrusion of clay through a covering geotextile. Each will be discussed.

Type of Polyethylene Resin. Depending on the type of polyethylene resin, primarily characterized by its density, the geonet will have different mechanical and endurance properties. The high-density resins (e.g., greater than 0.950 mg/l), will result in relatively high-modulus, high-strength, and high-creep resistance. Conversely, lower-density resins (e.g., less than 0.945 mg/l) will be more flexible and can deform under high-compressive stresses more easily. The lower-density resins, however, often have the advantage of having better stress crack resistance properties. (Stress crack resistance is particularly critical for HDPE geomembranes, as will be discussed in chapter 5.) The importance of stress crack resistance for a geonet is a design issue and related to the site-specific situation.

Creep Behavior. Sustained load (or creep) is the reduction in thickness of a geonet under an applied compressive stress. Here the density of the resin (as described above), type of structure, and composition of the rib junctions are all significant. Figure 4.7 presents geonet creep data at 480 kPa for 1000 hours. Note that even this is too short a time for conventional practice because the extrapolation of trends beyond one order of magnitude is questionable. Thus, data for 10,000 hr., extrapolated to 100,000 hr. (≅ 11 years), just begins to get

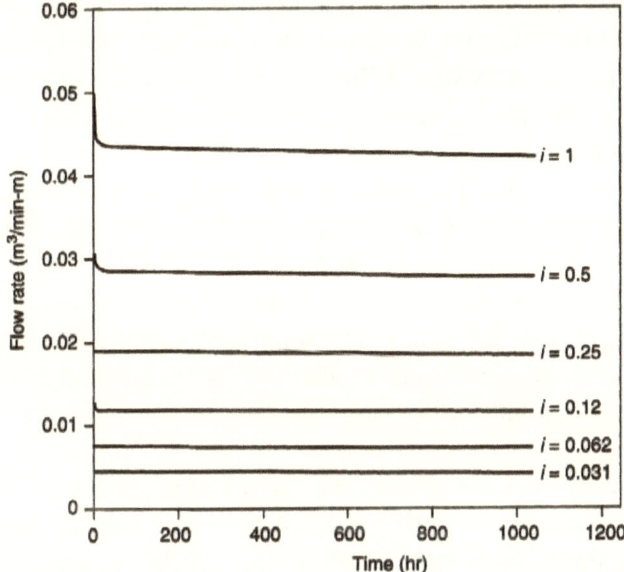

Figure 4.7 Long-term creep test results of same cross section as Figure 4.6, except under sustained load at 480 kPa.

into the frame of the life expectancy of many engineered systems using geonets. To shorten the testing time, other techniques such as time-temperature superposition (including the stepped isothermal method [6]) appear to be appropriate. See Yeo and Hsuan [4] in this regard.

Intrusion of Adjacent Materials. All geonets will necessarily be covered on their upper and lower surfaces with geotextiles, geomembranes, concrete wall surfaces, or some other material. If the geonet's surfaces are not covered, the adjacent soil will invade its apertures, rendering flow impossible. *Intrusion* refers to the deformation of flexible covering materials, primarily geotextiles, occupying some of the geonet's void space, as illustrated by the flow-rate reductions between figures 4.5 and 4.6. As seen by a comparison, intrusion into the core space is a very real phenomenon causing flow-rate decrease. See Hwu et al. [10] and Eith et al. [11] for data in this regard. It should be mentioned that thermal bonding of the geotextile to the geonet has a tendency to decrease some of the intrusion.

Now superimpose on these short-term reductions, sustained compressive stresses for extended times. Figure 4.7 presents the results of such a test series, where each test has load maintained for

1000 hours. Note that the response curves are essentially horizontal. Thus, creep intrusion (as contrasted to elastic, or initial, intrusion) is not an issue. The geotextile in this case is a nonwoven needle-punched polyester of 550 gm/m^2 mass per unit area with clay above. Clearly, the selection of the geotextile is important in sustaining the applied compressive stresses. The initial amount of intrusion, not necessarily the long-term behavior, is primarily a function of the geotextile's initial modulus. All other things being equal, those nonwoven geotextiles with high modulus will have the minimum amount of initial intrusion and, hence, higher flow rates. The long-term creep behavior depends on the polymer type, stress level, distance between geonet ribs, and so on, and is best quantified by appropriate laboratory testing.

Extrusion of Clay Materials. If a compacted clay liner or a geosynthetic clay liner with bentonite is placed adjacent to a geonet composite, there is a possibility of the clay particles *extruding* through the geotextile's voids into the geonet. This would be serious insofar as the flow-rate capability of the geonet; it must be avoided under all circumstances. Such a situation has happened in laboratory tests with woven monofilament geotextiles on the geonet and could happen with woven slit-film geotextiles as well. Conversely, nonwoven geotextiles have generally been effective in preventing soil extrusion. The minimum mass per unit area of the covering geotextile is a site-specific issue.

4.1.5 Environmental Properties

A series of environmentally related issues can have impact on the flow-rate performance of geonets. The first that comes to mind is temperature. Under high temperatures, flow rates increase over standardized laboratory test conditions. The converse is true for cold temperatures. These are usually minor effects and can be calculated on the basis of viscosity corrections (recall section 2.3.4). Perhaps more important is that creep of the polymers (geonet and adjacent geotextile or geomembranes) increases under increasing temperature. Simulated testing under these conditions is possible but costly. In lieu of such testing, a conservative design approach regarding creep is warranted.

The second environmental consideration focuses on the nature of the liquid being transmitted. If chemicals or leachate are being transmitted, a number of questions arise. One of these is the chemical resistance of the polymers being used for the geonet and the covering geotextiles and/or geomembranes to the site-specific liquid. Here the choice of high-density polyethylene for geonets is fortunate, since it is very resistant to most aggressive leachates. Again, laboratory testing can be performed using the actual or simulated leachate, but this can create concerns in a laboratory that is not equipped to handle contaminated liquids. ASTM is nicely set up with standards in this regard—i.e., laboratory immersion per D5322, geonet testing per D6388, and geotextile testing per D6389. The turbidity and viscosity effects of leachate (versus water) used in testing are other considerations, but these are often of second-order importance compared to some of the other issues being raised. Furthermore, they can be corrected by straightforward density and viscosity relationships (recall section 2.3.4).

The third environmental consideration has to do with biological growth within the geonet and/or on the geotextiles that allow liquid to enter the geonet. In most transportation-related systems, such as roads and walls, the problem does not appear to be too serious. In systems related to landfill leachates, agricultural wastes, and wastewater sludges, the issue should be addressed. At the bottom of a landfill, temperatures can be high, ample organic material (as a biological food source) is available, and bacteria and fungi could indeed thrive. Whether oxygen is available or not only dictates whether aerobic or anaerobic conditions prevail. Agricultural wastes and wastewater sludges have extremely high microorganism counts as evidence by their BOD value. Procedurally, we must use a high flow-rate factor of safety or have systems designed so that flushing is possible. This area simply begs for future inquiry. Biological growth on geotextiles has been addressed and a design procedure is available (see Koerner et al. [12]).

The fourth environmental consideration, resistance to light and weather, is not felt to be a serious concern for most situations in which geonets are used. Polyethylene is quite resistant to weather-related degradation, and carbon black is included in all the known products. Nevertheless geonets should be covered as soon as possible after placement. For geotextile covered geonets, the situation is controlled by the (more severe) ultraviolet degradation of the geotextile (recall section 2.3.6).

4.1.6 Allowable Flow Rate

As described in section 2.1.3, the very essence of the design-by-function concept is the establishment of an adequate factor of safety. For geonets, where flow rate is the primary function, this takes the following form.

$$FS = \frac{q_{allow}}{q_{reqd}} \tag{4.3}$$

where

FS = factor of safety (to handle unknown loading conditions or uncertainties in the design and testing methods),
q_{allow} = allowable flow rate as obtained from laboratory testing, and
q_{reqd} = required flow rate as obtained from design of the actual system.

Alternatively, we could work from transmissivity to obtain the equivalent relationship;

$$FS = \frac{\theta_{allow}}{\theta_{reqd}} \tag{4.4}$$

where θ is the transmissivity, under definitions as above. As mentioned previously, however, it is preferable to design with flow rate rather than transmissivity because of nonlaminar flow conditions in geonets and the intuitive nature of the term.

Concerning the allowable flow rate or transmissivity value, which comes from hydraulic testing of the type described in section 4.1.3, we must assess the realism of the test setup in contrast to the actual field system. If the test setup does not model site-specific conditions adequately, then adjustments to the laboratory value must be made. This is usually the case. Thus, the laboratory-generated value is an ultimate value that must be reduced before use in design; that is,

$$q_{allow} < q_{ult}$$

One way of doing this is to ascribe reduction factors on each of the items not adequately assessed in the laboratory test. For example,

$$q_{\text{allow}} = q_{\text{ult}}\left[\frac{1}{RF_{IN} \times RF_{CR} \times RF_{CC} \times RF_{BC}}\right] \tag{4.5a}$$

or if all the reduction factors are considered together.

$$q_{\text{allow}} = q_{\text{ult}}\left[\frac{1}{\Pi RF}\right] \tag{4.6a}$$

where

- q_{ult} = flow rate determined using ASTM D4716 or ISO 12958 for short-term tests between solid platens using water as the transported liquid under laboratory test temperatures,
- q_{allow} = allowable flow rate to be used in equation 4.3 for final design purposes,
- RF_{IN} = reduction factor for elastic deformation, or intrusion, of the adjacent geosynthetics into the geonet's core space,
- RF_{CR} = reduction factor for creep deformation of the geonet and/or adjacent geosynthetics into the geonet's core space,
- RF_{CC} = reduction factor for chemical clogging and/or precipitation of chemicals within the geonet's core space,
- RF_{BC} = reduction factor for biological clogging within the geonet's core space, and
- ΠRF = product of all reduction factors for the site-specific conditions.

The determination of "q_{ult}" has been quite controversial in the past. Thus, a modification of equations 4.5a and 4.6a has been recommended (see GRI-GC8 Test Method) that incorporates site-specific boundary conditions, some amount of creep and with it the intrusion reduction factor. The result (which is recommended by the author) is as follows:

$$q_{\text{allow}} = q_{100}\left[\frac{1}{RF_{CR} \times RF_{CC} \times RF_{BC}}\right] \tag{4.5b}$$

or if all the reduction factors are considered together

$$q_{allow} = q_{100}\left[\frac{1}{\Pi RF}\right] \tag{4.6b}$$

where

q_{allow} = allowable flow rate to be used in equation 4.3 for final design purposes

q_{100} = initial flow rate per ASTM D4716 or ISO 12958 specifically determined under simulated conditions for one-hundred-hour duration

RF_{CR} = reduction factor for creep to account for long-term behavior

RF_{CC} = reduction factor for chemical or precipitation clogging

RF_{BC} = reduction factor for biological clogging

Some guidelines as to the various reduction factors to be used in different situations are given in table 4.1. Please note that some of these values are based on relatively sparse information. Other reduction factors, such as overlapping connections, temperature effects, and liquid turbidity, could also be included. If needed, they can be included on a site-specific basis. On the other hand, if the actual laboratory test procedure has included the particular item, it would appear in the above formulation as a value of unity. Examples 4.2 and 4.3 illustrate two of the uses of geonets and serve to point out that high reduction factors are warranted in critical situations.

Example 4.2

What is the allowable geonet flow rate to be used in the design of a secondary leachate collection (or leak detection) system? Assume that laboratory testing at proper design load and proper hydraulic gradient gave a short-term between-rigid-plates value of 2.5×10^{-4} m²/s.

TABLE 4.1 RECOMMENDED REDUCTION FACTOR VALUES FOR EQS. 4.5a and 4.5b DETERMINING ALLOWABLE FLOW RATE OR TRANSMISSIVITY OF GEONETS

Application Area	Reduction Factor Values in Equation 4.5			
	RF_{IN}	RF_{CR}*	RF_{CC}	RF_{BC}
Sport fields	1.0 to 1.2	1.0 to 1.5	1.0 to 1.2	1.1 to 1.3
Capillary breaks	1.1 to 1.3	1.0 to 1.2	1.1 to 1.5	1.1 to 1.3
Roof and plaza decks	1.2 to 1.4	1.0 to 1.2	1.0 to 1.2	1.1 to 1.3
Retaining walls, seeping rock, and soil slopes	1.3 to 1.5	1.2 to 1.4	1.1 to 1.5	1.0 to 1.5
Drainage blankets	1.3 to 1.5	1.2 to 1.4	1.0 to 1.2	1.0 to 1.2
Infiltrating water drainage for landfill covers	1.3 to 1.5	1.1 to 1.4	1.0 to 1.2	1.5 to 2.0
Secondary leachate collection (landfill)	1.5 to 2.0	1.4 to 2.0	1.5 to 2.0	1.5 to 2.0
Primary leachate collection (landfills)	1.5 to 2.0	1.4 to 2.0	1.5 to 2.0	1.5 to 2.0

*These values are sensitive to the density of the resin used in the geonet's manufacture. The higher the density, the lower the reduction factor. Creep of the covering geotextile(s) is a product-specific issue. The magnitude of the applied load is also of major importance.

Solution: Average values from table 4.1 are used in equation 4.5a (however, note the large reduction).

$$q_{\text{allow}} = q_{\text{ult}}\left[\frac{1}{RF_{IN} \times RF_{CR} \times RF_{CC} \times RF_{BC}}\right] \quad (4.5)$$

$$= 2.5 \times 10^{-4}\left[\frac{1}{1.75 \times 1.7 \times 1.75 \times 1.75}\right]$$

$$= 2.5 \times 10^{-4}\left[\frac{1}{9.11}\right]$$

$$q_{\text{allow}} = 0.27 \times 10^{-4}\, m^2 / s$$

Example: 4.3

What is the allowable geonet flow rate to be used in the design of a capillary break beneath a roadway to prevent frost heave? Assume that laboratory testing was done at the proper design load and hydraulic gradient and that this testing yielded a 100 hour duration value of 2.0×10^{-4} m^2/s with simulated field conditions beneath and above the geonet test specimen.

Solution: Since better information is not known, average values from table 4.1 are used in equation 4.5b.

$$q_{allow} = q_{100}\left[\frac{1}{RF_{CR} \times RF_{CC} \times RF_{BC}}\right] \tag{4.5}$$

$$= 2.0 \times 10^{-4}\left[\frac{1}{1.1 \times 1.3 \times 1.2}\right]$$

$$= 2.0 \times 10^{-4}\left[\frac{1}{1.72}\right]$$

$$q_{\text{allow}} = 1.17 \times 10^{-4}\, m^2 / s$$

4.2 DESIGNING FOR GEONET DRAINAGE

This section will be subdivided into a discussion of required theory (which somewhat repeats previously described issues, due to its importance), drainage examples in the waste containment-related field and drainage examples in the transportation-related field.

4.2.1 Theoretical Concepts

Design-by-function requires the formulation of a factor of safety as follows:

$$FS = \frac{\text{allowable (test) value}}{\text{required (design) value}}$$

For geonets serving as a drainage medium, the targeted value is flow rate and the above concept becomes equation 4.3.

$$FS = \frac{q_{allow}}{q_{reqd}} \tag{4.3}$$

where

q_{allow} = allowable flow rate (as discussed in section 4.1), and
q_{reqd} = required flow rate (to be discussed here).

As stated previously, if we desire an alternative to the flow rate, calculations can be based on Darcy's formula (assuming saturated conditions and laminar flow) obtaining the transmissivity, θ. This important concept is repeated from equations 4.1 and 4.2:

$$q = kiA \quad (4.1)$$

$$q = ki\,(W \times t)$$

$$q = (kt)iW$$

$$kt \equiv \theta = \frac{q}{iW} \quad (4.2)$$

where

q = volumetric flow rate (m^3/s),
k = coefficient of permeability (m/s),
i = hydraulic gradient (dimensionless),
A = flow cross-sectional area (m^2),
θ = transmissivity (m^2/s),
W = width (m), and
t = thickness (m).

As seen in equation 4.2, q/W and θ carry the same units and are directly related to one another by means of the hydraulic gradient *i*. At a hydraulic gradient of 1.0, they are numerically identical. At all other values of hydraulic gradient they are not equal. Also note that the system should be saturated and flow must be laminar in order to use transmissivity. When in doubt, it is usually best to use flow rate per unit width.

4.2.2 Environmental-Related Applications

Geonets are widely used in landfill-liner systems as the primary leachate collection systems on side slopes and sometimes bottom slopes, and as leak detection systems between the primary and secondary liners (see references 13-15). They are also commonly placed above the barrier layer in final cover systems for drainage of surface water infiltrating the cover soil, Koerner and Soong [16]. All three of these applications will be illustrated by means of numeric examples. As such, geonets are placed with either a geomembrane on both sides, or a geomembrane beneath and a geotextile above. In this latter case, the geotextile has either granular soil or clay above it. Note that the inverse cross section is also possible—that is, there is a geomembrane above and a geotextile with soil below. The flow-rate results illustrated in figures 4.5 and 4.6 indicate some of these alternatives. Lastly, the geonet can have geotextiles on both sides generally for the purpose of enhanced friction against the opposing surface(s) or against textured geomembranes.

Example 4.4

> Determine the flow-rate factor of safety of a geonet used a primary leachate collection system above a geomembrane and beneath a sand covering layer in a landfill cell whose plan view shown in the following diagram. The geonet delivers its flow to the central perforated header pipe leading to the downgradient removal sump. The landfill is 35 m high with a waste unit weight of 13 kN/m^3. The geonet being considered has been tested with the results shown in figure 4.5. Site-specific reduction values will be necessary so as to obtain the allowable flow rate. The design inflow rate is 30,000 l/ha-day, which was the average leachate collected in all the New York state landfills in 2000. Note that this value is very much site-specific with hydrology, type of waste, and liquids management at the site all being major considerations.

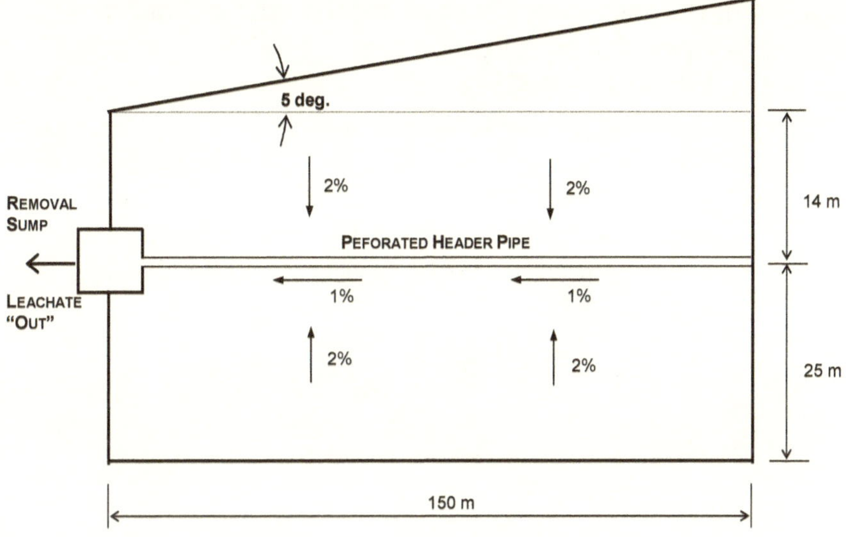

Solution:

(a) The maximum length of geonet perpendicular to the central drainage pipe is calculated:

$$L_H = 14 + (150) \tan 5°$$
$$= 14 + 13.1$$
$$L_H = 27.1 \text{ m} > 25, \text{ use } 27.1 \text{ m}$$

(b) The inflow rate is converted to units compatible with the test values:

$$r = 30{,}000 \text{ l/ha-day}$$
$$= \frac{1000}{30{,}000\ (1000)(100)(100)(100)}$$
$$r = 0.003 \text{ m}^3\text{/day-m}^2$$

(c) The required (and maximum) flow rate per unit width is calculated (see Qian et al. [17]):

$$q_{reqd} = \mathrm{r\ L_H}$$
$$= [0.003\ (27.1)]/\ [(24)(60)]$$
$$q_{reqd} = 0.0000565 \text{ m}^3\text{/min-m}$$

(d) The laboratory-obtained maximum flow rate is estimated from figure 4.5 at $\sigma_n = 35 \times 13 = 455$ kPa, and $i = 0.02$:

$$q_{max} = 0.003 \text{ m}^3\text{/min-m}$$

(e) This is now reduced to an allowable value using average values from table 4.1 except for intrusion, which is assumed to be at a maximum value in equation 4.5a.

$$q_{allow} = q_{ult}\left[\frac{1}{2.0 \times 1.7 \times 1.75 \times 1.75}\right]$$

$$= 0.003\left[\frac{1}{10.4}\right]$$

$$q_{all} = 0.000288 \text{ m}^3/\text{min}-\text{m}$$

(f) Finally, the *FS* value is obtained

$$FS = \frac{q_{allow}}{q_{reqd}}$$

$$= \frac{0.000288}{0.0000565}$$

$$FS = 5.1; \text{ which is acceptable}$$

Example 4.5

Assuming that there are no regulations governing the situation, is the geonet whose response is shown in figure 4.5 adequate for the following landfill leak detection and collection system? The geonet lies between two HDPE geomembranes (there is no pipe drainage system), and the design flow is 100 times *de minimis** leakage. (*De minimus* leakage is

---

* US EPA Regulatory note: The primary (or upper) geomembrane of a hazardous waste facility should allow no more than *de minimis* leakage of all polluting species through the liner itself. The concept of *de minimis* comes from the legal principle *de minimis non curat lex* (the law does not concern itself with trifles). *De minimis* leakage is considered to be the amount that is of no threat to human health or the environment. It is recognized that geomembranes, since they are not impermeable, will allow some transmission of waste constituents, by such means as vapor permeation or via very small imperfections. The actual level of *de minimis* leakage of a constituent is

approximately 10 l/ha-day). The minimum slope of the bottom of the 300 m long landfill is 6%, and the landfill when completed will be 50 m high with a unit weight of waste being 11 kN/m³.

Solution:

(a) The required flow rate converted to comparable units is

$$q_{reqd} = \frac{(100)(10)(0.001)}{(10{,}000)(24 \times 60)} \times 300$$

$$= 2.08 \times 10^{-5} \text{ m}^3/\text{min}-\text{m}$$

(b) The ultimate flow rate is taken from figure 4.5 (at $\sigma_n = 50 \times 11 = 550$ kPa and $i = 0.06$).

$q_{ult} = 0.01$ m³/min-m

and from table 4.1 it is reduced to obtain an allowable value using average values in equation 4.5a throughout:

$$q_{allow} = q_{ult}\left[\frac{1}{1.75 \times 1.7 \times 1.75 \times 1.75}\right]$$

$$= 0.01\left[\frac{1}{9.11}\right]$$

$$q_{allow} = 0.00110\ m^3/\text{min}-\text{m}$$

(c) Therefore, the factor of safety is

$$FS = \frac{q_{allow}}{q_{reqd}}$$

$$= \frac{110 \times 10^{-5}}{2.08 \times 10^{-5}}$$

$FS = 53$, which is more than adequate.

---

specific to the site, the constituent's toxicity, and the mobility and biodegradability of the constituent. Specific levels for individual constituents have not been set, although the US EPA believes that total *de minimis* leakage should be no more than 1 gallon per acre per day (gpad). This is approximately 10 l/ha-day. Current legislation [9], however, avoids setting a general action leakage rate (ALR) and, instead, requires each permit application to set its own site-specific ALR value.

(d) Note that for the data of figure 4.6, for the geotextile and clay over the geonet, the flow rate is approximately half of the above and the factor of safety is reduced accordingly.

Example 4.6

What is the factor of safety of a geonet placed above a geomembrane and beneath a geotextile in the final cover of a completed solid-waste facility (landfill or waste pile). As shown in the diagram below, the slope is 10% and the cover soil fill height is 1.25 m. Assume that the slope length is 120 m long. One hundred hour simulated boundary condition in-plane flow tests show that the flow rate is 1.8×10^{-4} m^2/s at a hydraulic gradient of 0.10 and at a normal pressure of 50 kN/m^2 (which includes soil weight plus equipment loads).

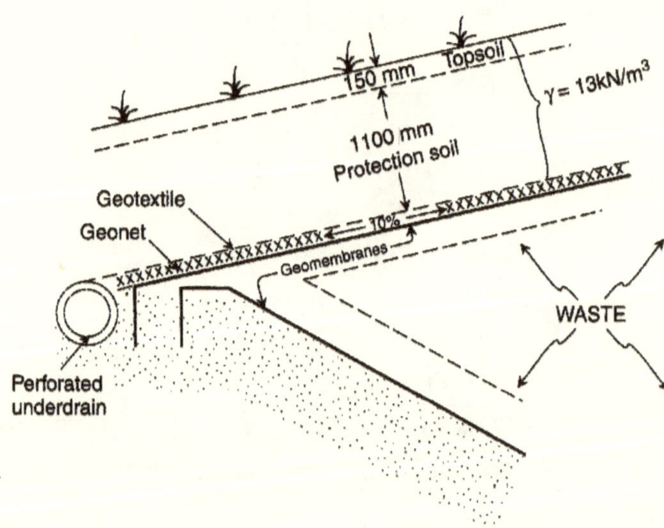

Solution:

(a) The allowable flow rate is taken from equation 4.5b and table 4.1, using average values

$$q_{allow} = 1.8 \times 10^{-4} \left[\frac{1}{1.25 \times 1.1 \times 1.75} \right]$$

$$= 1.8 \times 10^{-4} \left[\frac{1}{2.41} \right]$$

$$q_{allow} = 0.75 \times 10^{-4} \, \mathrm{m}^2/\mathrm{s}$$

(b) The required flow rate either must be approximated or must be determined by using a liquid-mass balance, including local hydrological data, soil storage, leakage, and so on. This was done in this case, following an hourly rainfall modeling procedure (see Koerner and Daniel [18]), giving a required flow rate 0.17×10^{-4} m^2/s at the end of a 120 m long section of closure. Note that the computer model entitled Hydrologic Evaluation of Landfill Performance (HELP) is widely used by regulatory personnel and consultants working in the solid-waste field (Schroeder ct al. [19]), but it can only track seepage on a daily basis and is not felt to adequately model intense storms. That said, the HELP model is recommended for determination of leachate arriving at the base of a landfill for subsequent drainage as depicted in example 4.4.

(c) The final factor of safety becomes

$$FS = q_{allow} / q_{reqd}$$

$$= \frac{0.75 \times 10^{-4}}{0.17 \times 10^{-4}}$$

$$FS = 4.4$$

This value of factor of safety, being well above 1.0, is acceptable. However, seepage failures in landfill final covers have been so troublesome that very high *FS* values are required [16]. The final decision on acceptability is site-specific.

4.2.3 Transportation-Related Applications

Geonets have been well established as an alternative drainage material to granular soils in the environmental field, and there is no reason why they should not be used in transportation-related applications as well. This section addresses two such applications, both of which are very relevant.

Example 4.7

Given an area that is to be retrofitted with a new pavement because of past problems with frost heave, the new scheme is intended to have a geonet with thermally bonded geotextiles on both sides and will be placed immediately beneath the depth of maximum frost penetration (see the sketch below). Based on the rising capillary water, the required flow rate to be conveyed is estimated to be 0.17×10^{-4} m²/s. A candidate geonet has been selected, and tests performed at a gradient of 0.05 have resulted in a flow rate of 0.83×10^{-4} m²/s. What is the factor of safety?

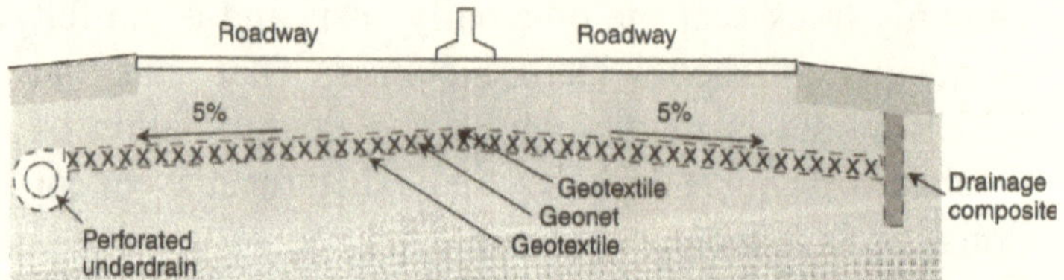

Solution:

(a) Assuming that the laboratory tests were 100-hour duration between field simulated lower and upper boundaries, the value of 0.70×10^{-4} m²/s must be reduced according to equation 4.5b and table 4.1. Using average values for the various reduction factors, we have

$$q_{allow} = 0.70 \times 10^{-4}\left[\frac{1}{1.1 \times 1.3 \times 1.2}\right]$$

$$q_{allow} = 0.70 \times 10^{-4}\left[\frac{1}{1.72}\right]$$

$$= 0.408 \times 10^{-4}\, m^2/s$$

(b) Now the actual flow-rate factor of safety can be determined.

$$FS = q_{allow} / q_{reqd}$$

$$= \frac{0.408 \times 10^{-4}}{0.17 \times 10^{-4}}$$

$$FS = 2.4; \text{ which is adequate.}$$

Example 4.8

Determine the drainage factor of safety for the geonet behind an 8 m high cantilever retaining wall as shown in the diagram below. The geonet being used has a response curve shown in figure 4.5. In the field it will be placed against the concrete on one surface, and it will have a geotextile on the surface against the backfill soil. The soil backfill is a silty sand (ML-SW) with $k = 5.0 \times 10^{-5}$ m/s. Note this is the same problem that was attempted using a geotextile in section 2.9.3 (example 2.26), where the factor of safety was found to be 0.0062.

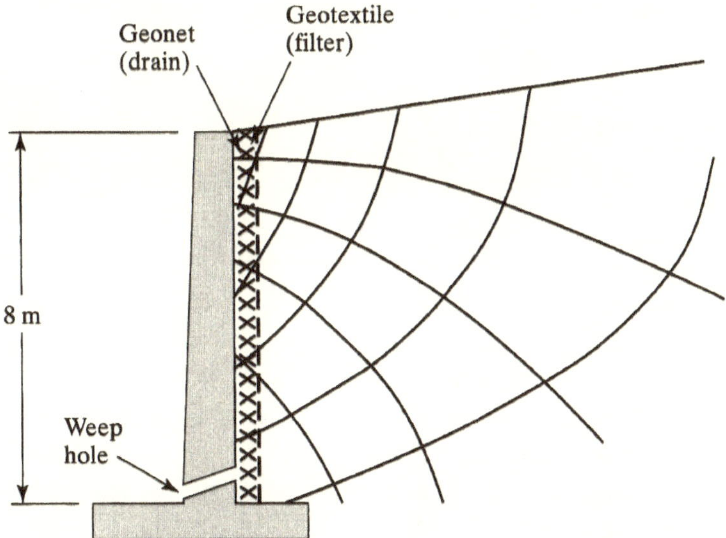

Solution:

(a) Calculate the maximum flow rate coming to the geonet. From the flow net sketched on the figure above, we have

$$q = kh\left(\frac{F}{N}\right)$$
$$= \left(5.0\times10^{-5}\right)(8)\left(\frac{5}{5}\right)$$
$$= 4.0\times10^{-4}\, m^2/s$$

(b) Determine the flow gradient within the geotextile:

$$i = \sin 90°$$
$$= 1.0$$

(c) Calculate the required transmissivity. Note that transmissivity is used so that the results can be directly compared to example 2.26 using geotextiles:

$$q = kiA$$
$$= ki$$
$$= (kt)(i\times W)$$
$$(kt) = \theta = \frac{q}{i\times W}$$
$$= \frac{4.0\times10^{-4}}{1.0\times1.0}$$
$$\theta_{reqd} = 4.0\times10^{-4}\ \text{m}^2/\text{s}$$

(d) From the laboratory data of figure 4.5, obtain the ultimate flow rate, convert it to a transmissivity value, and then reduce it to an allowable transmissivity. The normal pressure is obtained first: $\sigma_n \cong 0.5\,(8)(18) = 72$ kPa, and at $i = 1.0$ gives

$$q = 0.094\ \text{m}^2/\text{s/min} = 15.6 \times 10^{-4}\ \text{m}^2/\text{s}$$

For a vertical wall, where $i = 1.0$, the flow rate per unit width (q/W) is identical to the transmissivity θ (recall equation 4.2), so

$$\theta_{ult} = 15.6 \times 10^{-4} \text{ m}^2/\text{s}$$

Reducing the ultimate value according to equation 4.5a and table 4.1 (with arbitrary selected values),

$$\theta_{allow} = 1.56 \times 10^{-3} \left[\frac{1}{1.4 \times 1.3 \times 1.2 \times 1.2} \right]$$

$$= 1.56 \times 10^{-3} \left[\frac{1}{2.62} \right]$$

$$\theta_{allow} = 0.595 \times 10^{-3} \text{m}^2/\text{s}$$

(e) Knowing both allowable and required values of transmissivity, the factor of safety can now be calculated.

$$FS = \frac{\theta_{allow}}{\theta_{reqd}}$$

$$= \frac{0.595 \times 10^{-3}}{0.40 \times 10^{-3}}$$

$FS = 1.48$; which is marginally acceptable

Thus, the geonet characterized by the figure 4.5 data is adequate to drain the wall, whereas multiple layers of geotextiles were not adequate. However, even this factor of safety is somewhat low and a higher flow capacity drainage composite should be investigated—for example, a high flow-rate geonet or other type of drainage geocomposite. The problem will be repeated a third time in chapter 8.

4.3 DESIGN CRITIQUE

The design examples just presented focused entirely on *FS* values based on either flow rate or transmissivity. This was done to reinforce the concept that in-plane drainage is the primary and unique function of geonets. In this regard, the geonet must be properly specified. At least three items are necessary to make a proper flow-rate assessment: flow rate (which is preferred to transmissivity), the normal stress, and intrusion or extrusion.

Regarding the laboratory obtained flow rate the present trend is to perform one-hundred-hour duration transmissivity tests with the specimen between field simulated bonding conditions. This results in an initial creep adjusted and intrusion eliminated value, recall equation 4.5b.

A few words about the applied normal stress are also in order. To avoid rib lay-down and/or creep deformation, the compressive strength capability of the geonet must be higher than the design value. This value should be approximately 1.5 times (for short-service lifetimes) to 2 or more times (for long-service lifetimes in critical situations). Thus, the structural stability of the geonet must be ensured against creep deformation and/or collapse (Narejo and Allen [20]). The actual flow-rate value used for design purposes, however, can be taken from the curves at the design load at which the system will be operating. It should be noted that newer geonets do not exhibit rib lay-down and thus eliminate this concern.

Lastly, when soil is adjacent to the geonet, the type of geotextile covering it is of great significance with respect to intrusion or extrusion. While for most situations the geotextile is usually designed as a filter, it must also span the apertures of the geonet without excessively intruding or collapsing into the core space. There will always be some intrusion, and just how much is allowable depends on the site-specific situation. This can be evaluated experimentally, and values given in table 4.1 reflect a series of such experiments. A possible method of minimizing intrusion could be the use of a high-modulus woven monofilament geotextile. For environmental-related facilities, however, this might not be appropriate. If hydrated clay or bentonite is above the geotextile and is under high pressure from the weight of the landfill above it, the clay will be extruded through the open spaces in the woven geotextile directly into the geonet openings. This is completely unacceptable. Thus, a nonwoven needle-punched geotextile with a labyrinth of overlapping fibers is generally used for geonet coverings. However, some amount of intrusion must be anticipated and adequately accounted for (the rib spacings are important in this regard). A nonwoven heat-bonded geotextile might be a compromise geotextile with both high modulus (to prevent excessive intrusion) and high fiber overlapping (to prevent extrusion) [21]. The major problem in using this type of geotextile appears to be thermal bonding the geotextile to the geonet. In general, geotextiles should be thermally bonded to geonets so as to avoid a potential shear plane. The bonding, however, should not be excessive since flow in the geonet can be compromised from excessive geonet melting during fabrication. Whatever the geotextile type and depending on the actual stress level, a 200 g/m^2 geotextile should be the minimum mass per unit area.

These same considerations must be expressed when a geosynthetic clay liner (GCL) is placed over or under a geonet. The geotextile of the GCL facing the geonet must be viewed in the same light as other geotextiles in this discussion. Depending on the "standard" type of geotextile on the GCL facing the geonet, it is possible that a higher mass per unit area geotextile is necessary or an additional geotextile may be required between the GCL and the geonet.

4.4 CONSTRUCTION METHODS

Geonets are supplied in rolls from 2.0 to 6.7 m wide. They should be placed and covered in a timely manner. While UV and heat effects are not as severe in geonets as they are in geotextiles (because of thicker ribs in contrast to thin yarns and fibers), it is good practice not to leave the material exposed and subjected to accidental damage or contamination of any variety. Contamination can occur from soil, miscellaneous sediment, construction debris, ingrowing vegetation, and so on.

The rolls are usually placed with their roll directions oriented up-and-down slope, rather than along (or parallel to) them. There are two reasons for this: First, the machine direction has the greatest strength (recall figure 4.2) and flow rate; second, such orientation eliminates seams along the flow direction. If triplanar or boxlike channel geonets are being used for their high flow in the machine direction, the proper orientation is critical during placement. For very long slopes or along the base of a facility, flow must continue unimpeded from one geonet to the next. When geotextiles are laminated to the geonet, the geotextiles must be stripped back from the overlapped area such that the upgradient geonet is directly on the downgradient geonet in shingled manner. There can be no geotextile sandwiched within this overlap area [22].

The seaming or joining of geonets is difficult. Assuming stress does not have to be transferred from one roll to the next, plastic electrical ties, threaded loops, and wires have all been used with a relatively small overlaps of 50 to 100 mm. Overall, there is room for improvement in this regard (see Zagorski and Wayne [23]). Metal hog rings should never be used when geonets are used adjacent to geomembranes. There are questions as to what influence overlapping has on the geonet's flow rate. The connection of geonets to perforated drainage pipes is difficult and extremely important [22]. The geonet's outlet must be free draining at all times even in winter under freezing conditions.

Notwithstanding the above concerns, geonets are very impressive with respect to their flow-rate capability, ease of construction, savings in airspace, and overall economy in many facilities where drainage must be accommodated. Some aspects of geonets will be revisited in chapter 8 when we discuss additional geocomposite drainage geocomposites.

REFERENCES

1. Austin, R. A., "The Manufacture of Geonets and Composite Products," *Proc GRI-8 on Geosynthetic Resins, Formulations and Manufacturing*, IFAI, 1995, pp. 127-138.
2. Williams N., Giroud, J.-P., and Bonaparte, R., "Properties of Plastic Nets for Liquid and Gas Drainage Associated with Geomembranes," *Proc. Intl. Conf. Geomembranes*, IFAI, 1984, pp. 399-404.
3. Yeo, S.-S. and Hsuan, Y. G., "Effect of Geometry of Different Types of Biplanar Geonets," *Proc. 9th ICG Conference*, Brazil, 2010, pp. 1171-1174.
4. Yeo, S.-S. and Hsuan Y. G., "The Short—and Long-Term Compression Behavior of Geonets and Geocomposites Under Inclined Conditions," Geosynthetics International, Vol. 14, No. 3, 2007, pp. 154-164.

5. Corcoran, G. T., Cheng, S.-C. J. and Spear, A. D., "High Normal Stress Compression of Geosynthetic Lining Systems," *Proc. 5th IGS Conf.*, A. A. Balkema Publ., 1994, pp. 837-840.
6. Thornton, J. S., Allen, S. R., Siebken, J. R., "Long Term, Compressive Creep Behavior of High Density Polyethylene Geonet," *Proc. of the 2nd European Geosynthetics Conference and Exhibition*, October 15-18, 2000, Bologna, Italy, pp. 869-874.
7. Lydick, L. D., and Zagorski, G. A., "Interface Friction of Geonets: A Literature Survey," *J. Geotextiles and Geomembranes*, Vol. 10, Nos. 5-6, 1991, pp. 167-176.
8. Kolbasuk, G. M., Lydick, L. D., and Reed, L. S., "Effects of Test Procedures on Geonet Transmissivity Results," *J. Geotextiles and Geomembranes*, Vol. 11, Nos. 4-6, 1992, pp. 153-166.
9. EPA 40 CFR 260, 264, 265, 270 AND 271, *Federal Register*, Vol. 57, No. 19, *Rules and Regulations*, January 29, 1992, pg. 3463.
10. Hwu, B.-L., Sprague, C. J., and Koerner, R. M., "Geotextile Intrusion into Geonets," *Proc. 4th Intl. Conf. on Geotextiles, Geomembranes and Related Products*, A. A. Balkema, 1990, pp. 351-356.
11. Eith, A. W., and Koerner, R. M., "Field Evaluation of Geonet Flow Rate (Transmissivity) Under Increasing Load," *J. Geotextiles and Geomembranes*, Vol. 11, Nos. 5-6, 1992, pp. 153-166.
12. Koerner, G. R., Koerner, R. M., and Martin, J. P., "Geotextile Filters Used for Leachate Collection Systems: Testing, Design of Field Behavior," *J. Geotechnical Eng. Div., ASCE*, Vol. 120, No. 10, 1994, pp. 1792-1803.
13. Bonaparte, R., Williams, N., and Giroud, J-P, "Innovative Leachate Collection Systems for Hazardous Waste Containment Systems," *Proc. Geotechnical Fabrics Conf. '85*, IFAI, 1985, pp. 10-34.
14. Lundell, C. M. and Menoff, S. D., "The Use of Geosynthetics as Drainage Media at Solid Waste Landfills," *Proc. Geosynthetics '89*, IFAI, 1989, pp. 10-17.
15. Eith, A. W. and Koerner, R. M., "Field Evaluation of Geonet Flow Rate (Transmissivity) Under Increasing Load," *Jour.*

Geotextiles and Geomembranes, Vol. 11, Nos. 4-6, 1992, pp. 153-166.
16. Koerner, R. M. and Soong, T.-Y., "Analysis and Design of Veneer Cover Soils," *Proc. 6th International Geosynthetic Society Conference*, IFAI Publ., March 25-29, 1998, 1-26.
17. Qian, X., Koerner, R. M. and Gray, D. H., *Geotechnical Aspects of Landfill Design and Construction*, Prentice Hall Publishing Co., Upper Saddle River, NJ, 2002, 717 pgs.
18. Koerner, R. M., and Daniel, D. E., *Final Covers for Solid Waste Landfills and Abandoned Dumps*, ASCE Press, 1997, 256 pgs.
19. Schroeder, P. R., Dizier, T. S., Zappi, P. A., McEnroe, B. M., Sjostrom, J. W., and Peyton, R. L., "The Hydrologic Evaluation of Landfill Performance (HELP) Model: Engineering Documentation for Version 3," EPA/600/R-94/168b, US E.P.A., Risk Reduction Eng. Lab., Cincinnati, OH, 1994.
20. Narejo, D. and Allen, S., "Using the Stepped Isothermal Method for Geonet Creep Evaluation," *Proc. EuroGeo3*, Munich, Germany, 2004, pp. 539-544.
21. Ramsey, B. and Narejo, D., "Using Woven and Heat-Bonded Geotextiles in Geonet Geocomposites," Proc. GeoFrontiers, GSP 130-142, ASCE, 2005, (on CD).
22. Koerner, R. M. and Koerner, G. R., "Geocomposite Drainage Material Connections and Attachments," *Proc. GRI-22 Conference*, Salt Lake City, UT, GSI Publ., Folsom, PA, 2009, pp. 57-65.
23. Zagorski, G. A., and Wayne, M. H., "Geonet Seams," *J. of Geotextiles and Geomembranes*, Vol. 9, Nos. 4-6, 1990, pp. 207-220.

PROBLEMS

4.1 Geonets are used specifically for their in-plane drainage capability. Give the reasons they are not used for the following:

(a) separation
(b) reinforcement
(c) filtration
(d) containment (moisture barrier)

4.2 In their use for the drainage function, what keeps the adjacent soil from getting in their apertures and blocking flow?

4.3 If a geotextile is placed adjacent to a geonet, what function(s) does the geotextile provide? How does the combination of geotextile and geonet accommodate flow?

4.4 All the geonets described in this chapter are made of polyethylene. Could they be made from other polymers? Why do you suppose they are made from polyethylene?

4.5 It is noted in figure 4.1 that the aperture size varies from product to product. What effect does aperture size have on flow and intrusion?

4.6 In the typical extruded biplanar geonets, the vertical axes of the intersecting ribs are not quite perpendicular to one another. What implications does this have for the compressive load carrying capacity of the geonet?

4.7 For triplanar geonets, the flow rate is significantly higher than for biplanar geonets.

- **(a)** Where in a typical landfill configuration can these geonets be used?
- **(b)** Why is knowledge of the slope direction critical to know?
- **(c)** Why is the flow in the cross-machine direction not particularly important?

4.8 In considering the "other" geonets shown in figure 4.1c, what concerns might one have with boxlike channel types and with protruding column types?

4.9 The shear strength between a geotextile and a geonet can be quite low and troublesome when used in side-slope design. Describe two methods by which the geotextile can be attached to the geonet to avoid the potential problem. Include the advantages and disadvantages of each method.

4.10 The flow-rate reduction between figures 4.5 and 4.6 are up to 40% for 500 kPa normal stress. Beyond this stress level, the reductions are lower. Why are they lower at the higher stress levels?

4.11 For a geotextile covering a geonet that has a clay liner placed above it, discuss the difference between extrusion and intrusion.

4.12 Regarding the placement and use of a compacted clay liner over a geotextile bonded onto a geonet:

(a) What are the implications of using a lighter-weight geotextile for the results shown in figure 4.6?

(b) What would happen if a woven monofilament geotextile of ≥6% open area were used?

(c) What would happen if a nonwoven heat bonded geotextile were used?

4.13 Equations 4.5a and 4.5b give somewhat different approaches toward the calculation of $q_{allow.}$ Describe the differences in the two approaches.

4.14 The ultimate flow rate of a geonet being considered for the primary leachate collection system on a landfill side slope is 7.3×10^{-4} m²/s. Using equation 4.5a the maximum values in table 4.1, what is the allowable flow rate?

4.15 Using a geonet beneath an artificial surface in a tennis court requires an allowable flow rate of 1.6×10^{-4} m²/s. What would be the necessary ultimate flow rate using the average values from table 4.1?

4.16 Recalculate example 4.5 (section 4.2.2) concerning secondary leachate collection systems for design flows from 1 to 4000 times de minimus. Plot the resulting factors of safety against required flow rate.

4.17 A geonet is being considered for primary leachate collection on the 40 m long side slopes of a landfill. Using the data of figure 4.6, interpolating at a normal stress of 700 kPa and a hydraulic gradient of 0.184, what is the factor of safety for a flow rate of 25,000 1/ha-day? The cumulative reduction factors should be 8.0. (Note that this leachate flow rate is typical of primary leachate flow rates in the state of New York.)

4.18 Recalculate the *FS* values example 4.6 (section 4.2.2) concerning a geonet in a landfill cover, where the required flow rate varies from 1×10^{-3} to 1×10^{-5} m²/s. Plot the resulting factors of safety against slope angle.

4.19 What are the long-term normal stress implications for a geonet's flow-rate capability?

4.20 What are the long-term implications for a geonet's flow-rate capability in landfill liner design if the $FS < 1.0$?

www.ingramcontent.com/pod-product-compliance
Lightning Source LLC
Chambersburg PA
CBHW031812170526
45157CB00001B/31

* 9 7 8 1 4 6 2 8 8 2 8 8 5 *